POSITRON STUDIES OF SOLIDS, SURFACES, and ATOMS

POSITRON STUDIES OF SOLIDS, SURFACES, and ATOMS

A Symposium to celebrate STEPHAN BERKO's 60th Birthday

Brandeis University, December 12, 1984

edited by

Allen P. Mills, Jr.
William S. Crane
Karl F. Canter

World Scientific

Published by

World Scientific Publishing Co. Pte. Ltd.
P. O. Box 128, Farrer Road, Singapore 9128

Library of Congress Cataloging-in-Publication data is available.

POSITRON STUDIES OF SOLIDS, SURFACES, AND ATOMS

ISBN 9971-978-44-X

Printed in Singapore by Kim Hup Lee Printing Co Pte Ltd.

POSITRON STUDIES OF SOLIDS, SURFACES, and ATOMS

A Symposium to Celebrate STEPHAN BERKO's 60th Birthday
Brandeis University
Gerstenzang 123

Wednesday, December 12, 1984

9:00 AM		Welcoming Remarks -- President E. Handler Dean A. P. Carter

Session I: A. Bansil, discussion leader

9:10 AM	P. E. Mijnarends:	"Positron and Electrons in Energy Bands in Solids" *Netherlands Energy Research Foundation ECN, 1755 ZG Petten*
9:45 AM	A. T. Stewart:	"Positron Studies of Simple Metals" *Queen's University, Kingston, Ontario, Canada*
10:20 AM	R. N. West:	"Positron Studies of Metal Alloys" *University of East Anglia*
10:55 AM	Break	

Session II: R. Nieminen, discussion leader

11:10 AM	K. G. Lynn:	"Low Energy Positron Measurements of Defect Profiles" *Brookhaven National Laboratory Upton, NY 11973*
11:45 AM	P. M. Platzman:	"Surface Positrons and the Many Positron-Many Electron System" *AT&T Bell Laboratories Murray Hill, NJ 07974*

12:20 PM	Lunch: Feldberg Room, The Stein	
	Section III:	L. O. Roellig, discussion leader
2:10 PM	K. F. Canter	"Slow Positron Optics" *Brandeis University* *Waltham, MA 02254*
2:45 PM	A. P. Mills:	"Positron Ejection from Ionic Solids and the Possibility of Moderators for Slow Muons" *AT&T Bell Laboratories* *Murray Hill, NJ 07974*
3:55	Break	
	Section IV:	R. Drachman, discussion leader
4:10	B. L. Brown:	"Positrons in a Low Density Gas: Understanding the Galactic Center Annihilation Line and Forming Monoenergetic Beams of Positronium" *AT&T Bell Laboratories* *Murray Hill, NJ 07974*
4:45 PM	A. Rich:	"Polarized Positron Measurement of Electron Helicity in Chiral Molecules" *University of Michigan, Ann Arbor, MI 48109*
5:20 PM	Closing Remarks	
5:30 PM	Sherry:	Abelson 333
7:00 PM	Banquet:	The Faculty Club (reservations only)

LIST OF PARTICIPANTS

Abbott, L. F.

Bansil, Arun
Beardsley, G. M.
Berko, Stephan
Brown, B. L.

Canter, K. F.
Carr, P. H.
Chen, D. M.
Chowdhary, Anita
Chu, Steven
Crane, W. S.
Cushner, Stephan

Das, Shyam
Deutsch, Martin
Drachman, R. J.

Farmer, W. S.
Freidenreich, Brian

Goldstein, J. S.
Gross, E. P.
Gullikson, Eric

Haghgooie, Mohammad
He, Y. J.
Heller, Peter
Hitelman, Stewart
Hohenemser, Christoph
Horsky, T. N.
Huang, W. F.

Kamal, Shatdal
Krithivas, G.

Langer, R. M.
Lee, R. R.
Li, Xue-Song
Lippel, P. H.
Liu, Sen
Lynn, K. G.

Mader, I. J.
Meyer, R. B.
Mijnarends, P. E.
Mills, Jr., A. P.

Nieminen, R. M.

Odagaki, Takashi
Olafsson, Olafur J.
Oldenbourg, Rudolf

Pendleton, Hugh
Platzman, P. M.

Rich, Arthur
Roellig, L. O.
Rueckner, W. H. E.

Schnitzer, H. J.
Schoepf, D. C.
Schultz, P. J.
Schweber, S.
Sferlazzo, Piero
Sharma, S. C.
Shulman, M. A.
Sinclair, Frank
Sneddon, Leigh
Stewart, A. T.
Surko, C. M.

Tawel, Raoul
Terrell, J. H.

Vehanen, Asko
von Stetten, Eric

Wardle, J. F. C.
Weber, Marc
Weingart, J. T.
Weiss, A. H.
West, R. N.

1. J. J. Mader
2. R. M. Nieminen
3. L. O. Roellig
4. P. M. Platzman
5. P. J. Schultz
6. K. G. Lynn
7. Marc Weber
8. D. M. Chen
9. J. T. Weingart
10. Asko Vehanen
11. A. H. Weiss
12. M. A. Shulman
13. Mohammad Haghgooie
14. G. Krithivas
15. R. N. West
16. W. H. E. Rueckner
17. D. C. Schoepf
18. B. L. Brown
19. R. R. Lee
20. Arthur Rich
21. A. T. Stewart
22. Martin Deutsch
23. Stephan Berko
24. A. P. Mills, Jr.
25. P. H. Lippel
26. P. E. Mijnarends
27. S. C. Sharma
28. R. J. Drachman
29. Arun Bansil
30. H. J. Schnitzer
31. W. S. Crane
32. R. M. Langer
33. J. H. Terrell
34. P. H. Carr
35. K. F. Canter
36. Steven Chu
37. Christoph Hohenemser
38. W. F. Huang
39. R. B. Meyer
40. Stephen Cushner
41. Takashi Odagaki
42. Y. J. He
43. Eric Gullikson
44. W. S. Farmer
45. Leigh Sneddon
46. G. M. Beardsley
47. T. N. Horsky
48. Frank Sinclair
49. J. F. C. Wardle
50. Raoul Tawel
51. Piero Sferlazzo

FOREWORD

To Our Dear Colleagues:

The idea of a festive Symposium to help celebrate Stephan Berko's 60th birthday drew a unanimously enthusiastic response from his students and from the many other people who have worked directly with him. Accordingly, ten scientists agreed to meet at Brandeis University on December 12, 1984 to speak about exciting recent aspects of their research on positrons. The talks were limited to ten to allow us to have a one day meeting. The subject matter included studies of metals, alloys, defects, slow positrons, muons, astrophysics, and helicity measurements. While the choice of speakers was dictated largely by their availability and locality, the meeting was expected to be interesting to many since the presentations were to emphasize the latest results and speculations. The talks and discussions were recorded, transcribed, and edited into the present volume. Hopefully we have arrived at a happy medium between the spontaneity of the actual transcriptions and the stiffness of the usual formal articles.

Besides the lectures, there are some appendices that were included to make this book more interesting. There is also a partial list of the participants and a photograph that we offer as a partial recompense to those who were unable to attend. Conspicuously absent from this volume are numerous embarrassing although much deserved public testimonials. However, our gathering for the symposium and the publication of this collection is an obvious demonstration of our appreciation for Steve Berko and an appropriate birthday present for him.

The editors of this volume would like to thank several people for their efforts on behalf of the symposium. In particular, we are grateful to the president of Brandeis University, Evelyn Handler for introducing the meeting, Dean A. P. Carter for her remarks of appreciation about Steve Berko, Prof. Jack Goldstein for hosting the meeting and taking the photograph of the attendees, and especially to Mrs. Mimi Fricks for her devotion to ensuring that the symposium would be a success.

The Editors
January, 1986

CONTENTS

RELATED PAPERS

APPENDIX

POSITRON STUDIES OF SOLIDS, SURFACES, and ATOMS

Stephan Berko: A Perspective

Introduction

I am glad I was given the opportunity of writing the preface to the book resulting from the symposium celebrating Stephan Berko's sixtieth birthday. Like so many of his colleagues, I knew only certain segments of his scientific work. I knew of his accomplishments in the field of positron-matter interactions, but I never had the opportunity to appreciate at one time the various facets of Stephan Berko as a colleague, friend, professor, scientist, husband and parent. The purpose of this preface is to share my recently gained knowledge with the readers.

Early Life

Stephan Berko was born on December 16, 1924 in Oradea, Rumania. He spent his youth in Sighet, a small town on the northern frontier of Transylvania, now part of Rumania, but belonging to Hungary during World War II. Sighet produced two world-famous persons: Joseph Szigeti, one of the leading violinists of this century, and Elie Wiesel, the author and social commentator. Berko received his early education in Sighet, living with his Hungarian speaking family. His father and grandfather were surgeons. Later he attended the Jewish High School in Debrecen, Hungary's second largest city. It was a remarkable education, with former university professors as teachers and an Olympic gold medalist as a coach in self-defense classes.

After graduation in 1943, Berko worked as a technical assistant in a small microwave factory in Budapest, returning to Sighet in 1944, after the occupation of Hungary by German troops. In the spring of 1944 he and his entire family were transported together with the large Jewish population of Sighet to the Auschwitz concentration camp. Berko was transferred to a small camp built on top of the ruins of the bombed-out Warsaw Ghetto and finally to Dachau, Germany. He was the only one of his immediate family to survive.

After being liberated by the US Army in 1945, he joined a United Nations welfare team as a volunteer to organize multi-lingual schools for several thousand children in a Displaced Persons camp in Munich. In 1946 he helped organize a Jewish Student Union for surviving victims of World War II and enrolled at the Technical University of Munich, where he became an assistant in mathematics. Having won a fellowship from the US Hillel Foundation through exams administered in Germany, he was sent in early 1948 to the University of Virginia. He received a B.A. degree in Physics in 1950 and was elected to Sigma Xi, Phi Beta Kappa and the University of Virginia Raven Society. Three years later he received a Ph.D degree under the direction of Frank L. Hereford, who later became the President of the University of Virginia. From 1953 to 1961 he held a postdoctoral NRC Fellowship at Princeton University (1953-54), a Sloan Foundation Fellowship at the Bohr Institute, Copenhagen (1958-59), and advanced from Assistant Professor to Professor at the University of Virginia.

During his stay at the University of Virginia he developed a close friendship with Clinton J. Davisson, who joined the Physics Department after retiring from Bell Laboratories. Davisson was his mentor; he and the rest of the Physics Department, chaired by Jesse W. Beams, aided him in making the transition from a displaced person to a respected American physicist. During the summer of 1950 he worked as a research assistant to W. F. G. Swann at Swarthmore's Bartol Research Laboratories. There he met his future wife, Ellen Schmidt, a teacher of Italian and French literature. They have three daughters, Barbara Ann, Deborah, and Jacqueline, and three grandchildren.

Brandeis Years

A new phase in his life began in 1961 when Berko moved to Brandeis University to initiate an experimental program. His efforts with his close colleague, Edgar Lipworth, gave balance to the Brandeis Physics Department, which was highly theoretical at that time. He was Chairman of the Physics Department in 1965-67 and again in 1971-74. During the first period the Department applied for an NSF Center of Excellence grant, and during the second period the grant was made available; thus he received the fruits of his earlier efforts and was able to further strengthen the experimental program of the Department.

He has taken two leaves from Brandeis. The first was spent at the Solid State Physics Institute, University of Paris, Orsay, France in 1968. In 1976-77 while a Guggenheim Fellow at the Laue Langevin Institute, Grenoble, France, he investigated the possibility of using positrons from a ^{64}Cu source located in the core of the high flux reactor, but found the background too high for angular correlation of annihilating radiation (ACAR) experiments. Today, however, he is part of a team which uses a high intensity slow positron beam, constructed by Kelvin Lynn, based on the same source (^{64}Cu) but placed outside the High Flux Beam Reactor at Brookhaven; successful ACAR experiments designed to study surfaces are being performed.

In 1977 Berko was elected Fellow of the American Academy of Arts and Sciences and in 1978 he was appointed to a William R. Kenan Jr. Professorship at Brandeis.

Research

Although most of Berko's research has been about positron-matter interactions, the subject of his Ph.D thesis was measuring the deflection of cosmic ray mu-mesons and high energy electrons in magnetized iron to determine whether the deflection was due to $\vec{B}$ or $\vec{H}$ (it was $\vec{B}$). It should be noted that even this first research was in the field which was to occupy him for the next thirty years, the use of elementary particles to study solids.

After receiving his Ph.D he did research in nuclear physics. It was at Princeton that he first encountered positrons and positronium through the early work of R. H. Dicke. He developed and built fast timing equipment for the cyclotron, and gained experience that he later applied to the field of positron annihilation. At Copenhagen, where he initiated coulomb excitation experiments of heavy nuclei using a cyclotron, he overlapped with R. E. Bell, the Canadian physicist who devised many of the early innovations in measuring short life times.

After several publications on positronium chemistry, he published in 1957 his first angular correlation of annihilation radiation (ACAR) paper with J. S. Plaskett and R. E. Kelley. The next year he published a truly seminal paper with J. S. Plaskett titled "Correlation of Annihilation Radiation in Oriented Single Metal Crystals." S. DeBenedetti and A. T. Stewart had already published their research on polycrystalline solids. Berko and Plaskett pointed out the usefulness of looking for anisotropies in oriented single crystals in order to study the electronic structure and in particular the Fermi surface of metals. It should be mentioned that the first ACAR paper in 1957 reported a large anisotropy in graphite; to this day graphite has the largest anisotropy ever obtained with ACAR.

In 1960 Berko published a paper with W. Brandt and W. W. Walker on "Positronium Decay in Molecular Substances." This paper is noteworthy for two different reasons: it introduced the important concept of an excluded volume for positrons, and it induced Werner Brandt to enter the field of positron annihilation.

Other notable papers by Berko and collaborators include:

- The theory of ACAR experiments using spin polarized positrons- "Positron Annihilation in Ferromagnetic Solids," (1966).
- The test of many-body theory by measuring alkali metal lifetimes- "Positron Lifetimes in Metals," with H. L. Weisberg (1967).
- One of the first papers on positron trapping in metals (in dislocations)- "Angular Distribution of Annihilation Radiation from Plastically Deformed Aluminum," with J. Erskine (1967).
- The first mention of delocalized positronium states-"Evidence for Positronium-Like Bloch States in Quartz Single Crystals," with A. Greenberger, A. P. Mills, and A. Thompson (1970).
- Several papers on positron annihilation in disordered alloys-for references see his review "Fermi Surface in Disordered Alloys; Positron Annihilation Experiments" (1979).
- The first paper to show that positronium can be produced by a beam of low energy positrons-"Efficient Positronium Formation by Slow Positrons Incident on Solid Targets," with K. F. Canter and A. P. Mills (1974).
- The first observation of positronium Lyman-α radiation, and measurements of its first excited state. These papers were cited by the APS as among the most important papers published in 1975-"Observations of Positronium Lyman-α Radiation," with K. F. Canter and A. P. Mills (1975); and "Fine Structure Measurements in the first Excited State of Positronium," with A. P. Mills and K. F. Canter (1975).
- The first high precision two dimensional ACAR studies-"Momentum Density Measurements by Positron Annihilation in Metals and Alloys; Recent Experiments with a Multicounter Two-Dimensional Angular Correlation Apparatus," with J. Mader (1975); "Electronic Momentum Density by Two-Dimensional Angular Correlation of Annihilating Radiation in Aluminum," with J. Mader, H. Krakauer, and A. Bansil (1976); and "Momentum Density Measurements with a New Multicounter Two-Dimensional Angular Correlation of Annihilation Radiation Apparatus," with M. Haghgooie and J. J. Mader (1977).
- The development of techniques for three dimensional image reconstruction from two dimensional ACAR in order to study Fermi surfaces-"Studies of the

Fermi Surface of V_3Si by 2D ACAR Measurements," with W. S. Farmer, F. Sinclair and G. M. Beardsley (1979), and with W. S. Farmer and F. Sinclair (1980); and "Reconstructed 3D Positron-Electron Momentum Density in Cu: Fermi Surface and Many-Body Enhancement Studies," with F. Sinclair and W. S. Farmer (1982).

- The application of the 2D ACAR technique to the study of the electronic structure at point defects in metals-"Positron Annihilation Spectroscopy of the Equilibrium Vacancy Ensemble in Aluminum" and "A Monovacancy-Divacancy Model Interpretation of Positron Annihilation Measurements in Aluminum," with M. J. Fluss, B. Chakraborty, K. R. Hoffmann, P. Lippel and R. W. Siegel (1984).
- The ACAR measurement of positrons on the surface of a metal-"Positron or Positronium-Like Surface State on Al(100)?," with K. G. Lynn, A. P. Mills Jr., R. N. West, K. F. Canter and L. O. Roellig (1985).

These publications were selected from more than 75 contributions Berko has made to date. His present activity is such that we can expect many more from him.

Teaching Contributions

Many undergraduate students, PhD candidates, and visiting scholars have benefited from Berko's masterful lecturing, his uncompromising integrity, and his sense of what research is worthwhile. Berko speaks highly of his students and feels that their support and hard work have been essential to his successful research efforts. Besides teaching in the classroom at Brandeis, Berko has lectured extensively in the United States, Canada, Europe, Japan and India. An especially important part of his teaching is contained in various review papers such as the 1981 Fermi School lecture notes reproduced in this volume. Two demonstration movies made by him for college teaching, "Electron-Positron Annihilation" in 1970 and "Single Photon Polarization" in 1972, are still being shown. It is characteristic that of all his accomplishments, Berko singled out the organization of the schools in the Displaced Persons camps after World War II as having given him the most satisfaction.

L. O. Roellig

TALKS

Positrons and Electrons in Energy Bands in Solids

P. E. Mijnarends

Netherlands Energy Research Foundation ECN
1755 ZG PETTEN, The Netherlands

First of all I would like to say how very, very honored and pleased I am to be here on this very special birthday. I would like to thank the organizers of this symposium for inviting me and making it possible for me to be present here today.

When one is asked to give a talk on the subject in the context of a birthday like this one, it is unavoidable that one's thoughts go back to the fifties when life was easy and simple. As Alec Stewart[1] and others have shown, the angular correlations of metals as widely different as lithium and lead have the shape of inverted parabolas (Fig. 1). Hence, the Fermi surface in these metals had to be a sphere and the radius of the Fermi surface, the Fermi momentum, could be obtained directly from the position of the kinks in the angular correlation. Of course there were also a number of metals which had bell-shaped angular correlations, notably the transition metals, but those were not amenable to detailed interpretation and therefore they were largely left alone.

Steve (Berko) was the first one to make life more complicated for the positron physicist. In 1958 he published an article in the Physical Review (together with J. S. Plaskett) on the anisotropy of the angular correlations in aluminum and copper.[2] That paper became a classic in positron annihilation and it was one of the first papers I read when I entered the field in late 1961. When I looked at the smallness of the anisotropy in Cu (Fig. 2), I remember a kind of feeling that was later expressed by one of your singers, Peggy Lee, in a song: "Is that all there is?".[3] It then became clear to me that positron annihilation in solids meant looking for very small effects.

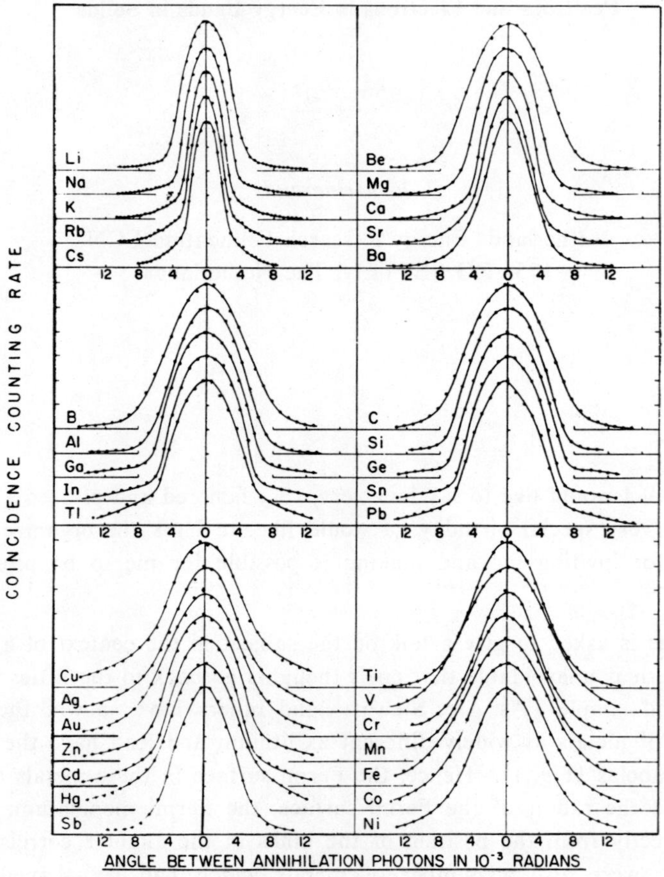

Fig. 1. Observed angular correlation of photons from positrons annihilating in various elements (normalized to equal height). From ref. 1.

A quarter century has passed since then. This period has seen an enormous progress both in experimental techniques and in the techniques of band-structure calculation. On the experimental side of the momentum density work, the most important thing that happened, I think, was the pioneering work of Steve (Berko) and his group on the measurement of angular correlations in two dimensions,[4,5] which led to the present-day 2D machines. On the theoretical side our understanding of the interaction between positrons and electrons in energy bands has deepened considerably to the point where a set of general rules is beginning to emerge which enables one to predict a momentum density, given a specific

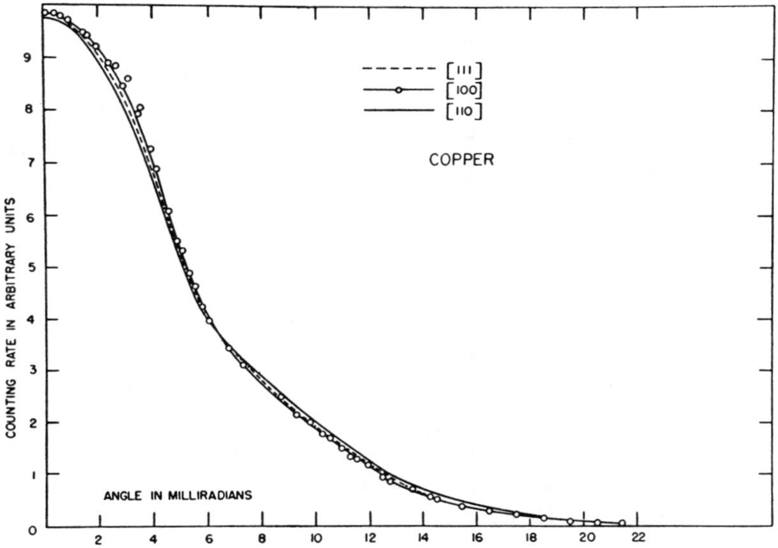

Fig. 2. Angular correlations measured in Cu oriented along the [111], [110] and [100] directions. Experimental points are included only for the [100] curve. From ref. 2.

electronic structure.[6] It would be easy to give a miniature course on this subject lasting the whole day, but in view of the limited time I shall restrict myself to showing you a few recent results which I think are of some interest.

NiMnSb

The first system that I shall talk about belongs to a class of novel materials that have recently aroused some interest in the Netherlands. They are a kind of Heusler alloys having the chemical formula XMnY and the Cl_b structure[7] which

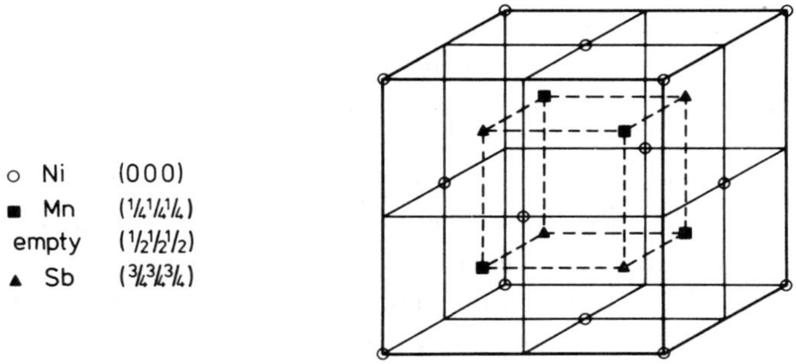

Fig. 3. Cl_b–type crystal structure.

is shown in Fig. 3. X can be some transition metal atom, e.g. Fe, Co, Ni, Cu, Pd, and Y stands for an atom from groups III to VI of the periodic system such as Ga, In, Sn, Sb, etc. These alloys are ternary magnetic intermetallic compounds. Real Heusler alloys, of course, have the chemical formula X_2MnY and the $L2_1$ structure, which is obtained from the one in Fig. 3 by putting the extra transition metal atoms at the empty lattice positions.

The interest in the Netherlands for these materials came from industry, where people were looking for materials with a large magneto-optical Kerr effect for the storage of information in thermomagnetic recording media, i.e. for erasable optical recording.[8,9] De Groot et al. at the University of Nijmegen have done a band-structure calculation for NiMnSb and similar materials[10] and they found that NiMnSb is indeed a very interesting material with unusual properties. The material is a ferromagnet, so spin degeneracy has been lifted and there are two electron spin populations, a majority and a minority one (Fig. 4). The majority-spin electrons have a band structure with a conduction band, which intersects the Fermi level. Hence, for this spin direction the material is a normal metal. There is nothing special about that! For the minority-spin electrons we see a band structure with a band gap about 1 eV wide with the Fermi level lying in the gap. This spin population has the electronic structure of a semiconductor. Again this is nothing special; there are hundreds of semiconductors. But of course, the interesting thing is that both of these characters, metallic and semiconducting, occur within one and the same material.

The question came up, how far it would be possible to verify certain aspects of the band structure with positron annihilation. We also wondered to what extent the empty lattice positions would act as vacancies, because, as we all know, the positron likes to sit in a vacancy and that might spoil the whole game. One of my co-workers, Karl Hanssen, calculated the positron state and positron wave function[11] by the KKR method and indeed, as expected, he found the positron wave function to be very heavily concentrated on the vacant sites. The high peaks in Fig. 5a are the positions of these vacant sites; the valleys correspond to the various atoms in the crystal. How important this effect is can be seen by putting Ni atoms in the 'vacancies'. In that way one obtains a kind of model 'Ni_2MnSb' (Fig. 5b). Indeed the positron wave function has now become much more evenly distributed. There is no sign anymore of the high peaks which were there originally, so these vacant sites have a tremendous effect. In the beginning this bothered us a little because we wondered, will one see much of any Fermi breaks in this material? So Hanssen went on and did a semi-relativistic[12] KKR

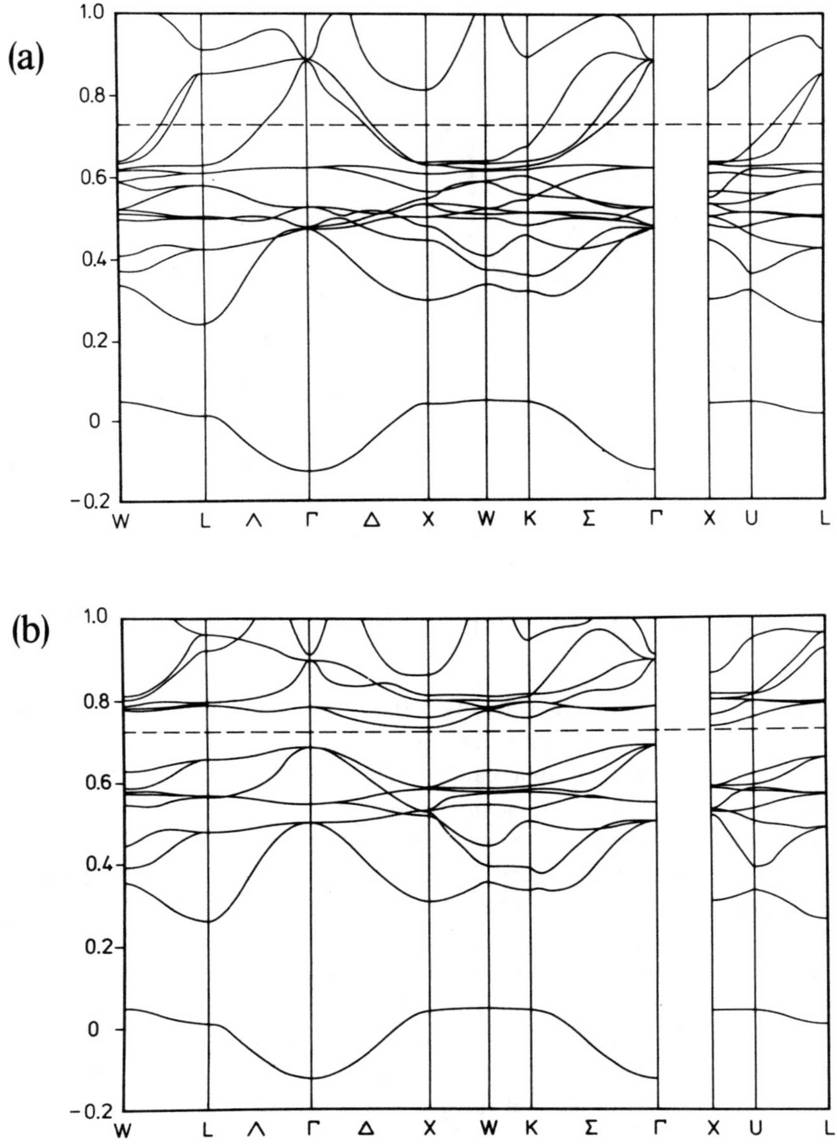

Fig. 4. Band structure of NiMnSb for the majority (a) and minority (b) spin electrons. Energies in Rydberg.

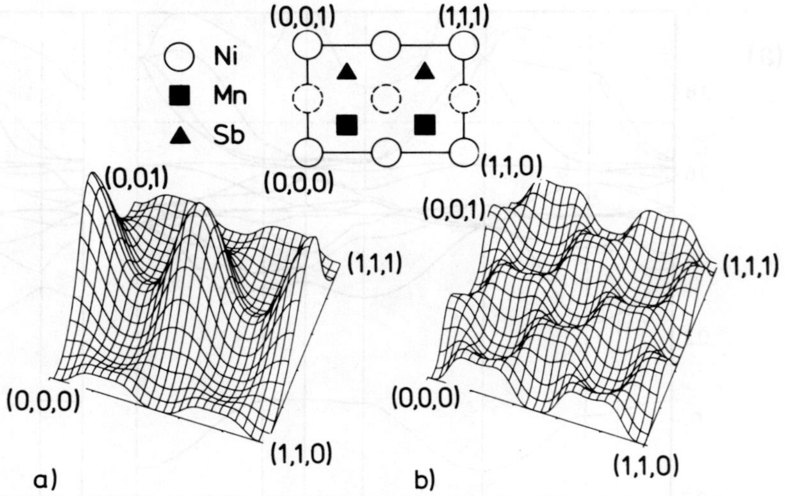

Fig. 5. a. Positron probability density $|\psi_+(\underline{r})|^2$ in a {110} plane of NiMnSb (Cl_b crystal structure). The high peaks correspond to the empty lattice positions, the minima to the positions of the atoms (see inset). b. Same for 'Ni_2MnSb' ($L2_1$ crystal structure), obtained by placing additional Ni atoms at the empty lattice positions (dashed in the inset) of NiMnSb. From ref. 11.

calculation of the momentum density[13] (both the two-photon momentum density and the electron momentum density, the latter measurable by Compton scattering[14]) for the two spin bands.[15] Here (Fig. 6) you see the two-photon momentum density for the majority-spin electrons along the [100] direction. Indeed the majority-spin momentum density shows Fermi breaks (the minority-spin momentum density does not because there are no partly filled minority-spin bands). There is a small one at 1, a large one at 2, and a small one at 3, so the question was, are the Fermi breaks at 1 and 3 so small because of the presence of the 'vacancies'? It is interesting that if one looks at the momentum density for Compton scattering (Fig. 7), which is not affected in the same way by the presence of the vacant sites, it turns out that the Fermi breaks are practically of the same size. Actually, the one at 1 is slightly smaller in the case of Compton scattering and the one at 3 is slightly larger, but the differences are only small. Why is this? It turns out that most of the momentum density at low momenta comes from the low-lying Sb-like Δ_1 band (labeled (a)) between −0.11 and +0.05 Ry, i.e., far below the Fermi level. When one goes to somewhat higher momenta the contribution (a) drops off as those of other bands come up. The solid curves (b) and (c) belong to Δ_1 bands somewhere in the middle of the

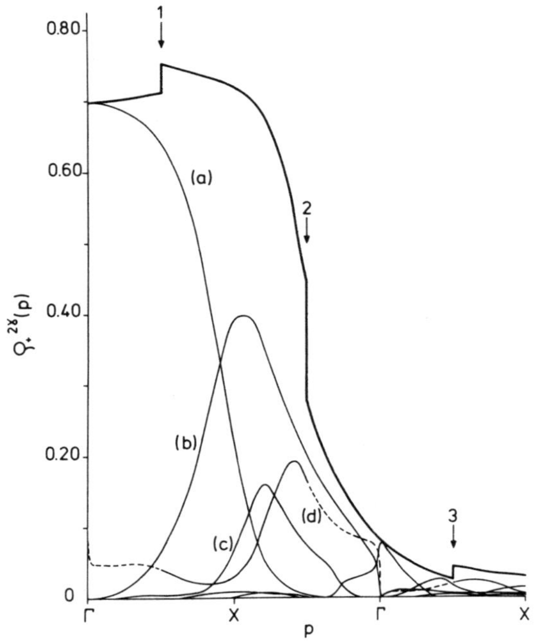

Fig. 6. Two-photon momentum density $\rho_{+}^{2\gamma}$ (p) along the [100] direction for positrons annihilating with the majority-spin electrons in NiMnSb. The thin curves represent the contributions of the various bands of Δ_1 symmetry (the other bands do not contribute), the dashed parts correspond to unoccupied states. The heavily drawn curve gives the total of all bands. The numbers 1 through 3 label the Fermi breaks. The distance ΓX corresponds to a momentum of 4.10×10^{-3} mc. From ref. 11.

band structure. Curve (d) which is partly dashed and partly solid gives the contribution of the Δ_1 band that goes through the Fermi level. The dashed part corresponds to states above the Fermi level. So it looks as if the positron affinity to the vacant sites does not have much effect on the size of the Fermi breaks.

This suggests that it should be possible to see these Fermi breaks in a carefully performed high-resolution 2D angular correlation experiment.

It is hard to grow good single crystals of these materials, but Dr. K. H. Y. Buschow was so kind as to let us have some crystals. Here (Fig. 8) are the very first, very preliminary experimental data in 2D of the spin density in this ferromagnet, i.e. the 2D angular correlation for minority-spin electrons minus that for majority-spin electrons. The spin density consists of a kind of Christmas pudding with a moat around it. There is considerable structure at the top of the

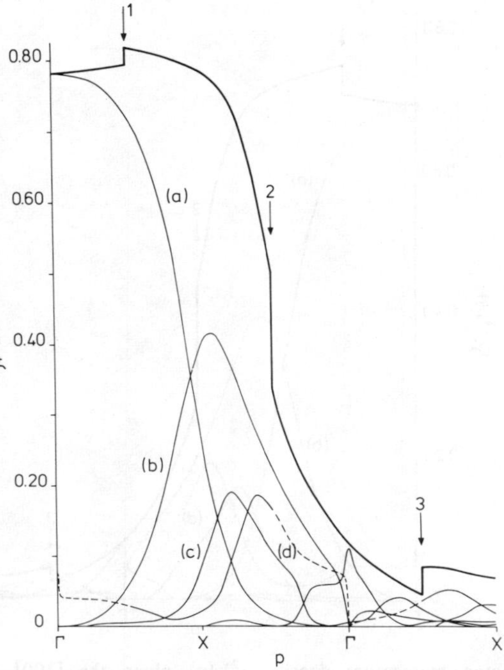

Fig. 7. Majority-spin electron momentum density ρ_+ (p) in NiMnSb. Otherwise as Fig. 6. From ref. 11.

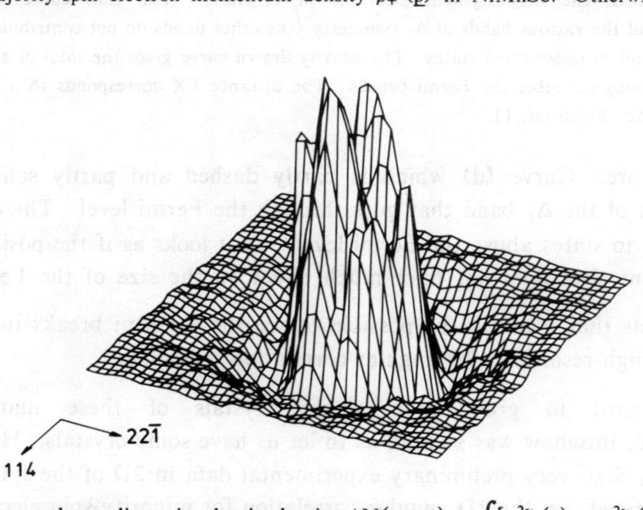

Fig. 8. Measured two-dimensional spin density $\Delta N(p_y, p_z) = \int [\rho_-^{2\gamma}(p) - \rho_+^{2\gamma}(p)] dp_x$ in the $(\bar{1}10)$ plane in NiMnSb, not corrected for the momentum sampling function. Each division is 10^{-3} mc.

pudding. Some of it will no doubt be due to the insufficient statistics of the measurement since the picture represents only 5 days of measurements per field direction (the data were taken last week). However, if one looks at the symmetry of the structure (we have not done any symmetrizing of the data and even the momentum sampling function of the machine is still included) one sees that some of this structure must be real. We look forward to collecting more data on this material and doing a proper analysis.

Disordered Alloys

Another subject in which we are interested is that of electronic structure, momentum density, and positron annihilation in substitutionally disordered alloys. This work was done in collaboration with Arun Bansil from Northeastern University. In a non-dilute disordered alloy the lack of long-range order causes the mean free path of the electrons to be finite. Hence, the wave vector k is not a good quantum number anymore, i.e. there is disorder scattering of the electrons, and the lifetimes of the electron states become finite. The energy levels become complex, with the imaginary parts representing the disorder smearing, and new energy levels may appear.[16,17]

I would now like to illustrate the effects of alloying with some calculations of the band structure and momentum density, first for pure copper and then for disordered $Cu_{1-x}Ge_x$ alloys. (For the sake of simplicity, the alloy calculations have been based on the average t-matrix approximation and in the momentum density calculations a constant positron wave function has been assumed.[18] These simplifications do not affect the essence of the results).[19] The questions to be considered are: What happens with the momentum density, what effect does the smearing of the bands have (i.e. the fact that the bands are complex), and, in cases where new bands are formed, how do these new bands affect the momentum density? But before one can answer these questions it is necessary to go back one step and look at pure copper (Fig. 9). To many of you this will be a well-known picture. It shows the band structure of pure copper along ΓX, together with the momentum density along the [100] direction. Owing to the crystal symmetry only two bands can contribute to the momentum density along that direction.[20] These bands are formed by the hybridization of the parabolic sp-like conduction band with one of the d-bands of the same (Δ_1) symmetry representation. As a result, the momentum density consists of two contributions. The density corresponding to the lowest band, labeled 2, is large at low momenta but drops off quickly as the conduction band hybridizes with the d band. The contribution of the other Δ_1 band starts out as a d-like contribution and grows as the d band hybridizes with the sp-like conduction band. The positron sees only

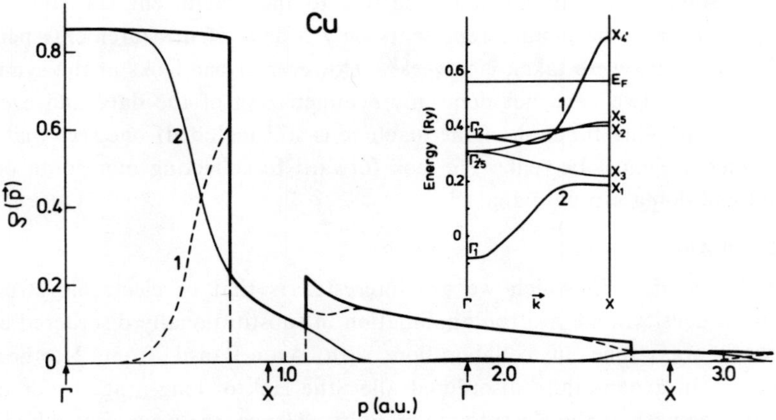

Fig. 9. Band structure (inset) and momentum density along the [100] direction in Cu. The positron wave function has been assumed constant. The labels 1 and 2 indicate the contributions of the corresponding Δ_1 bands.

the sum total of the densities in these two bands which is flat, that is, to the positron the electronic structure of copper looks very much like that of a nearly-free-electron gas with some d contribution being visible in the gap between the Fermi break in the first Brillouin zone and that in the next zone. The important lesson to be learned from this is that whatever hybridization one has at some distance below the Fermi level, one does not see any of it with positrons. Only when the Fermi level lies in the neighborhood or in the middle of this hybridization region one would expect to see effects of hybridization.

What happens if we now add a few percent germanium to copper? As Prasad and Bansil[21] have shown, a low-lying impurity band associated with Ge appears around −0.25 Ry (Fig. 10). One may now ask: "How will the appearance of this band affect the momentum density?" The somewhat disappointing answer is: "It doesn't". To see this we have split the band structure into three regions, each corresponding to one of the broadened bands of Δ_1 symmetry. The lowest one, labeled 3, corresponds to the impurity band. The next one, labeled 2, contains the lower part of the conduction band, and the highest region, labeled 1, the upper part. We see that at low momenta the impurity band gives a fair-sized contribution even for a germanium percentage as low as 3.3%, but it all goes at the expense of the contribution of the conduction band. All that seems to happen is that there is a transfer of momentum density from the conduction band to the impurity band while the sum total remains the same. Thus, one does not see direct evidence of the presence of the impurity band in the momentum density.

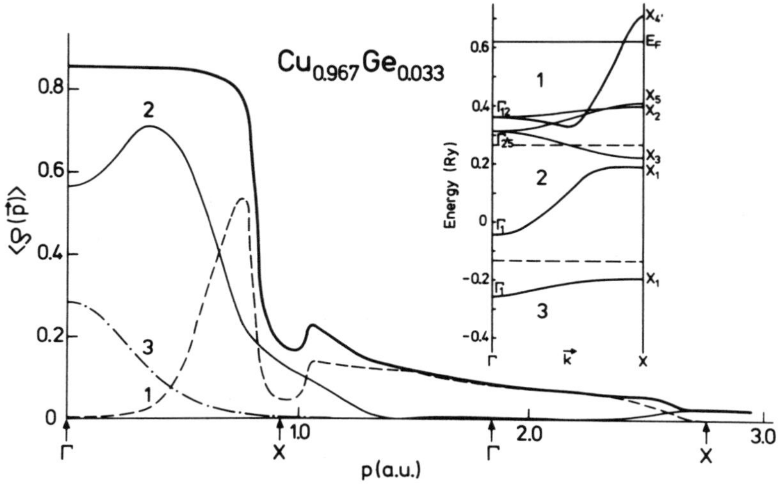

Fig. 10. Band structure (inset) and configurationally averaged momentum density (both in the average t-matrix approximation) along the [100] direction in disordered $Cu_{0.967}Ge_{0.033}$. The positron density has been assumed constant throughout the alloy. The curves labeled 1 through 3 represent the contributions to the momentum density of the correspondingly marked energy regions of the band structure. The damping of the bands (not shown) is greatest for the lowest Δ_1 band and for the states near $X_{4'}$ (cf. ref. 21).

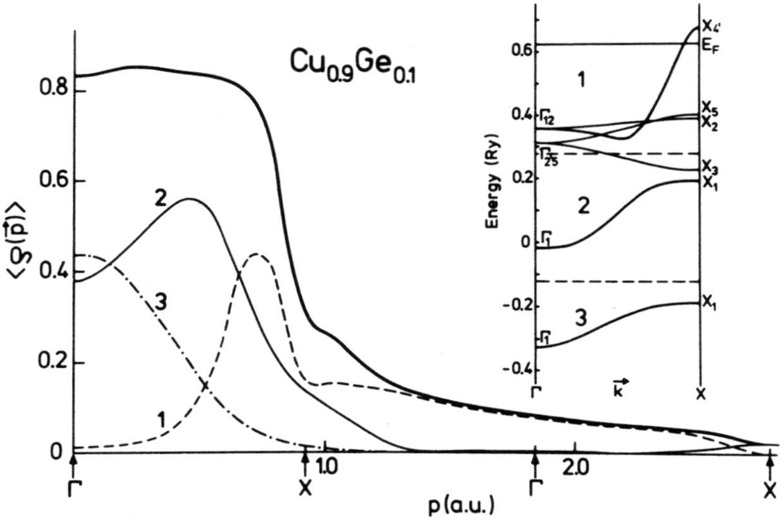

Fig. 11. Same as Fig. 10, but for $Cu_{0.9}Ge_{0.1}$. The slight dip in the total density near p=0 may be an artifact of the calculations.

What one does see, however, is two other things. In the first place, if germanium is added to copper, or rather if one replaces copper atoms by germanium atoms, one introduces extra electrons. A fraction of these will be accommodated in the impurity band, the remainder will cause the Fermi level to go up. This means that the Fermi surface expands and that the gap (at X) between this Fermi surface and the image of the Fermi surface in the next Brillouin zone becomes more narrow.[22] Secondly, the smearing at the top of the conduction band causes the Fermi break, which was originally very sharp, to be smeared out. Actually, the slope of the Fermi break is directly proportional to the lifetime of the electron states at the Fermi surface, i.e. one can use positrons to measure the lifetime of the electron states.

What happens if we increase the germanium concentration to 10% (Fig. 11)? The trend just continues, the contribution of the impurity band becomes larger at the expense of the contribution from the conduction band, so the sum total is practically flat. The Fermi surface grows further, the gap narrows down more and the smearing of the states near $X_{4'}$ becomes even greater so that the gap practically disappears. The only remnant of it is the little dip just below 1 a.u. This heavy smearing should be easily observable with positron annihilation.

This brings me to the last topic of my talk, the state of the positron in disordered alloys.

The Positron State in Disordered Alloys

The positron, like the electrons, 'sees' the substitutional disorder in the alloy. This is important because I just said that one can use, or expects to be able to use, positrons to measure the lifetime of electron states at the Fermi surface. However, if one does that, one should be aware that the positron itself may contribute to the smearing of the Fermi break. Arun Bansil and I have investigated this point. To begin with, we have looked at the smearing of the positron state in a disordered alloy of 50% copper and 50% nickel using the coherent potential approximation.[19] To that end, we have plotted the positron spectral momentum density for various momenta (Fig. 12). The zero of the energy scale is the bottom of the positron conduction band. If one would do this for a perfect metal like copper one would expect to see delta functions at the energies corresponding to the selected momenta. In this disordered alloy one again sees peaks, but the peaks have now acquired a certain width; they have been broadened due to positron disorder scattering. The smearing increases as one goes up to higher momenta and higher energies. At the bottom of the band the spectral momentum density is extremely sharp but it still has a finite width.

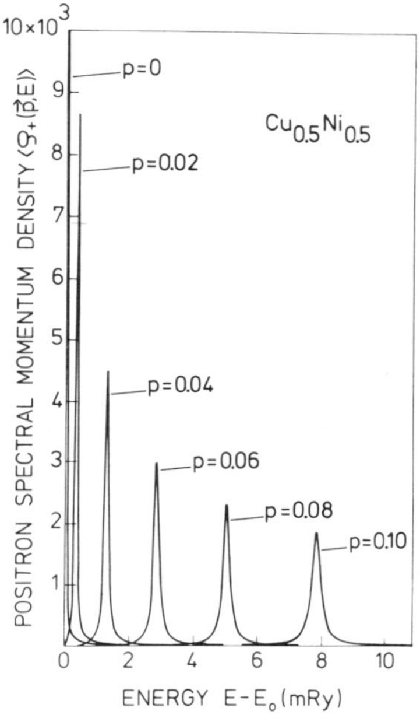

Fig. 12. Positron spectral momentum density in disordered $Cu_{0.5}Ni_{0.5}$ for momenta p (in units of $2\pi/a$) along the [100] direction, calculated in the coherent potential approximation. Energies on the horizontal scale are given with respect to the band edge at $E_0 = 0.50143$ Ry.

On the low-energy side of that peak there is a band edge, in other words, there are no positron states below the minimum of the band. At the temperature T=0 the positron will be sitting at the bottom of the band. What happens if we increase the temperature? We know from pure metals that the positron will acquire a certain amount of thermal motion which is governed by the Boltzmann factor.[23] The positron will start moving around near the bottom of the energy band and hence the positron momentum density, which in a pure metal at T=0 is a delta function, will be broadened into a Gaussian peak of finite width.[24] If one does the same thing in a disordered alloy one will first of all find this thermal broadening. In addition to that there will be a second contribution to the broadening of the positron momentum density coming from the damping of the positron conduction band. In Fig. 13a we have plotted the positron momentum distribution. The dashed curves would apply if there were only thermal broadening and no disorder scattering. If one now includes disorder scattering one obtains the solid curves. As you see, they are slightly wider than the curves from thermal motion only. The effect is very small because the damping of the

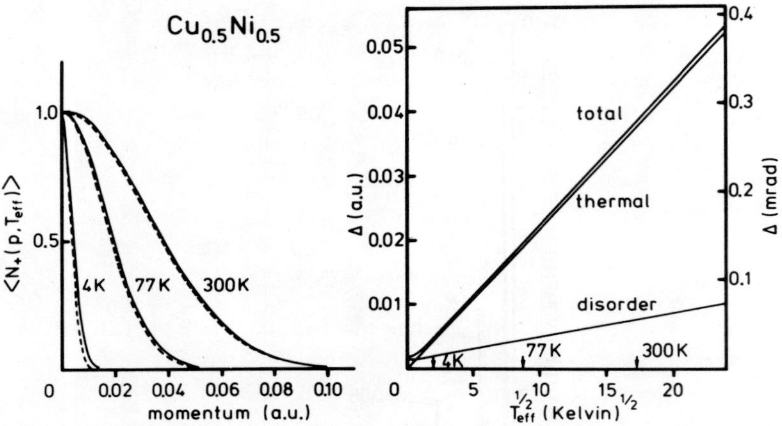

Fig. 13. a. Positron momentum density for p along [100] in disordered $Cu_{0.5}Ni_{0.5}$ at various effective temperatures $T_{eff} = m^*T$, where m^* is the effective mass of the positron. The dashed curves give the effect of thermal motion alone.

b. Disorder scattering and thermal motion contributions to the halfwidth $\Delta = 1/2$ FWHM of the positron momentum density in $Cu_{0.5}Ni_{0.5}$.

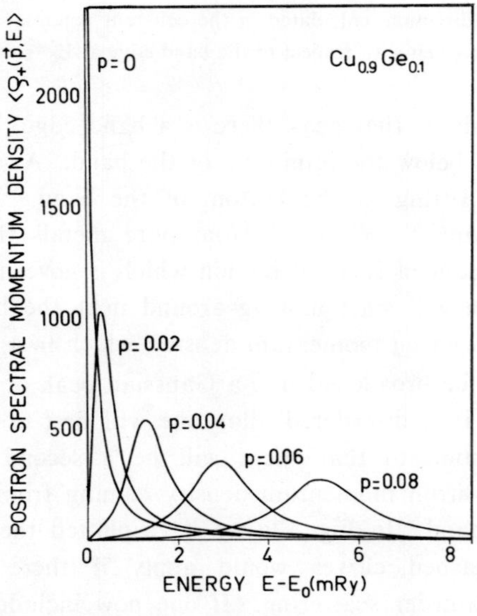

Fig. 14. Same as Fig. 12, but for $Cu_{0.9}Ge_{0.1}$. From ref. 25.

positron band is small in Cu-Ni alloys. The reason for this is that the positron near the bottom of the band sees only s scattering phase shifts. For copper and nickel these s phase shifts are very similar, i.e. to a positron copper and nickel look very much the same. Since the dashed curves are Gaussian and the solid curves look very nearly Gaussian, we have treated both types of curves as Gaussians. By subtracting the halfwidths quadratically we have separated the disorder component from the thermal component (Fig. 13b). As one expects, the thermal component goes to zero as $T \rightarrow 0$, but the disorder component retains a finite (although very small) value at $T=0$.

In CuNi alloys the damping of the positron band is small, but there are other alloys where one can expect a much larger damping because the scattering properties of the atoms differ more than in the case of CuNi. The next figure (Fig. 14), for instance, shows the positron spectral momentum density in $Cu_{0.9}Ge_{0.1}$.[25] Here the positron band is much more heavily damped and the spectral density becomes very wide as soon as one goes slightly upward in the band. Hence, in this alloy disorder scattering will be a much more important effect. Indeed, we have calculated the positron momentum distribution and have found a much larger difference between thermal motion only and thermal motion plus disorder scattering (Fig. 15a). Again we have separated the thermal and disorder parts and one sees that the disorder contribution is much more important (Fig. 15b). The slight increase of the disorder contribution with temperature of course comes from the fact that by raising the temperature one increases the positron thermal motion. The positron is then able to sample states higher up in

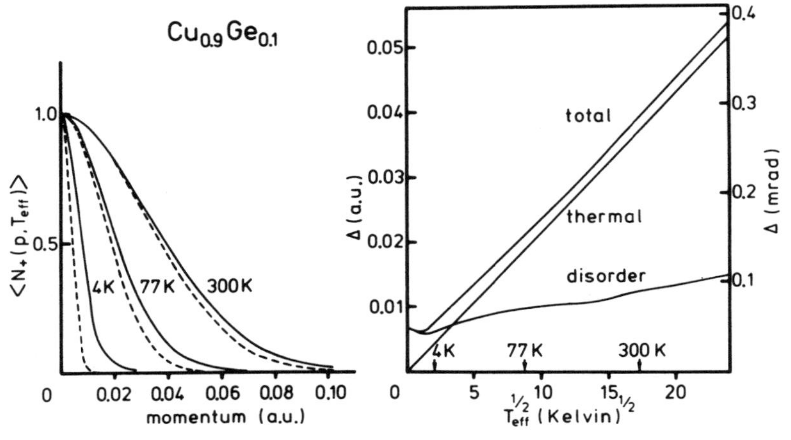

Fig. 15. Same as Fig. 13, but for $Cu_{0.9}Ge_{0.1}$. From ref. 25.

the conduction band which are more heavily damped than the states close to the bottom of the band. We have also done calculations for other alloys such as $Cu_{0.9}Al_{0.1}$ and $Cu_{0.7}Zn_{0.3}$.[25] The effects are very much the same in all these alloys.

There is a moral to this story. If one tries to measure the electron lifetime at the Fermi surface by looking at the slope of the Fermi break, one must be aware of the existence of the disorder contribution from the positron. First of all, one has to correct for this contribution and secondly its existence means that there is a kind of material-dependent lower limit to the resolution of such an experiment. Thermal motion may always be lowered by cooling the sample but disorder scattering cannot be reduced below a certain limit.

Of course one can also look at this in a different way. Positron disorder scattering is an effect of interest for its own sake, and hence one would be very happy with an alloy where the effect is so large that with the angular resolution of current 2D machines one can measure it. If we look at the size of the effect in $Cu_{0.9}Ge_{0.1}$, half the full width at half maximum is approximately .06 milliradians, so the full width of the positron momentum density at T=0 K comes to say .1 − .15 milliradians, which is a little less than the resolution of current high-resolution machines. However, since these are expected to improve in the future I think this effect may well become observable.

This brings me to the end of my talk and I hope that you have found some things of interest in it.

I wish to thank Professor A. Bansil, Dr. L.P.L.M. Rabou and K.E.H.M. Hanssen for their collaboration and for many interesting discussions. Thanks are also due to G.J. Langedijk for preparing the drawings. This work is part of the research program of the Stichting voor Fundamenteel Onderzoek der Materie and was made possible by financial support from the Nederlandse Organisatie voor Zuiver-Wetenschappelijk Onderzoek.

K. Lynn

Do you have theories for some of these Heusler alloys, on the question of spin density enhancement versus charge density enhancement? Say for example there is some polarization of the positrons; so you might have to include effects of spin density enhancement which may not be the same as a charge density enhancement.

P. E. Mijnarends

The calculations we have done did not take into account any type of enhancement.

Lynn

But in the end do you think you would eventually have to include that?

Mijnarends

I think one will have to include that in some way in order to get a good agreement with the experimental results. One will have to do that.

P. Platzman

If you wanted measurements of small effects like lifetime effects and so on, how big do you estimate are all the things you leave out, like shakeup and things like that due to correlation effects between the positrons. It's a little bit like Kelvin's question. There are a whole host of effects involving the correlation between the positron and electron and the shakeup connected with the fact that the final state is not really an eigenstate anymore. They lead to broadening effects and so on. How big do you think those are?

Mijnarends

These enhancement effects do not affect the position of the break.

Platzman

I'm interested more in shakeup.

Mijnarends

I would not know.

A. Bansil

The point is, one is comparing say copper with copper aluminum. Whatever effects were there in copper are still there in copper aluminum, but there is the additional effect of smearing. Presumably one could be reasonably safe in observing the composition dependence.

Question

What about the other kind of disorder, vacancies?

Mijnarends

If one is interested in the electronic structure of the alloy as such one tries to make the sample as much vacancy free as possible but in a case where one has a sample with vacancies, yes one will see smearing due to the scattering of the vacancies.

Question

Could you decouple that from the smearing that you get from the electron bands? Could that be a parameter that one could vary to look at defects?

Mijnarends

My feeling is that the smearing due to the vacancies will probably take over. It will drown the other effects.

REFERENCES

1. A. T. Stewart, Can. J. Phys. **35**, 168 (1957).
2. S. Berko and J. S. Plaskett, Phys.Rev. **112**, 1877 (1958).
3. Peggy Lee, for example on EMI 054-85001.
4. S. Berko and J. Mader, Appl. Phys. **5**, 287 (1975).
5. S. Berko, M. Haghgooie and J. J. Mader, Phys. Lett. **63A**, 335 (1977).
6. P. E. Mijnarends, in *Positron Solid-State Physics*, edited by W. Brandt and A. Dupasquier (North-Holland, Amsterdam, 1983), p. 146.
7. K. Watanabe, Trans. Jpn. Inst. Met. **17**, 220 (1976).
8. K. H. J. Buschow, P. G. van Engen and R. Jongebreur, J. Magn. Magn. Mat. **38**, 1 (1983).
9. P. G. van Engen, K. H. J. Buschow, R. Jongebreur and M. Erman, Appl. Phys. Lett. **42**, 202 (1983).
10. R. A. de Groot, F. M. Mueller, P. G. van Engen and K. H. J. Buschow, Phys. Rev. Lett. **50**, 2024 (1983); J. Appl. Phys. **55**, 2151 (1984).
11. K. E. H. M. Hanssen and P. E. Mijnarends (submitted for publication).
12. M. Methfessel and J. Kübler, J. Phys. F **12**, 141 (1982).
13. P. E. Mijnarends and L. P. L. M. Rabou, J. Phys. F, in press.
14. B. Williams, *Compton Scattering* (McGraw-Hill, New York, 1977).

15. We are indebted to Dr. R. A. de Groot for providing the electron and positron crystal potentials from his augmented-spherical-wave calculation (Ref. 10).

16. A. Bansil, Phys. Rev. B **20**, 4025 and 4035 (1979); Phys. Rev. B **23**, 3107 (1981).

17. B. L. Györffy and G. M. Stocks, in *Electrons in Disordered Metals and at Metallic Surfaces*, edited by P. Phariseau, B. L. Györffy and L. Scheire (Plenum Press, New York, 1979), p. 89.

18. P. E. Mijnarends and A. Bansil, Phys. Rev. B **13**, 2381 (1976); Phys. Rev. B **19**, 2912 (1979).

19. A. Bansil and P. E. Mijnarends, Phys. Rev. B **30**, 628 (1984).

20. R. Harthoorn and P. E. Mijnarends, J. Phys. F **8**, 1147 (1978).

21. R. Prasad and A. Bansil, Phys. Rev. Lett. **48**, 113 (1982).

22. A. Bansil, R. S. Rao, R. Prasad, H. Asonen and M. Pessa, J. Phys. F **14**, 273 (1984).

23. A. T. Stewart and J. B. Shand, Phys. Rev. Lett. **16**, 261 (1966).

24. Positron-phonon interaction which distorts the Gaussian shape is neglected here. See e.g. H. J. Mikeska, Z. Physik **232**, 159 (1970).

25. P. E. Mijnarends and A. Bansil, in *Proc. 7th Int. Conf. on Positron Annihilation*, edited by R. M. Singru and P. C. Jain (World Scientific Publ. Co., Singapore), in press.

Positron Studies of Simple Metals

Alec Stewart

Department of Physics
Queen's University
Kingston, Canada K7L 3N6

In the beginning it was always surprising that a positron could collect around it a big pile of charge, this polarization charge, and yet when it annihilated, it seems to annihilate with unperturbed electrons. This always was a really pleasant and startling thought. I suppose with hindsight, its easy to see that when a positron and an electron approach one another their interaction becomes so fierce that they are really independent of the rest of the universe while they are close together and that since they have attracted one another, their total momentum reflects their original states, before they came together.

Beyond this simplistic view Kahana[1] showed that the heaping up of the polarization cloud around the positron had slightly more than a proportional amount of electrons near the Fermi surface because those electrons were more perturbable. This was the beginning of the subject of the enhancement in the annihilation rate and its momentum dependence. There are at present two or three different theoretical versions of this enhancement and the experiments are almost conclusive enough to select among these theories.

The other thing that is interesting is to watch the motion of the positron inside the little polarization cloud. It was noticed many years ago[2] that the effect of temperature on the Fermi surface, Δk, was negligible compared with the effect of temperature on the motion of the positron, k_+.

For electrons:

$$\frac{\hbar^2}{2m}(k_F + \Delta k)^2 = E_F + k_B T$$

$$\text{whence} \quad \frac{\Delta k}{k_F} = \frac{k_B T}{2E_F} \sim \frac{1}{300} \quad \text{at} \quad T \sim 300\text{K}$$

For positrons:

$$\frac{\hbar^2}{2m} k_+^2 = \frac{3}{2} k_B T$$

$$\text{whence} \quad \frac{k_+}{k_F} = \left(\frac{3k_B T}{2E_F}\right)^{1/2} \sim \frac{1}{10} \text{ at } T \sim 300\text{K}$$

The thermal smearing of the Fermi surface is not observable, whereas the thermal motion of the positron as we shall see is easily observed. Experiments to measure this effect are shown in Fig. 1. These are old results[3] taken some years ago. They show that the positron has a small thermal momentum distribution so that when you measure the annihilation momentum you are measuring a kind of convolution of the electron momentum distribution with the positron momentum

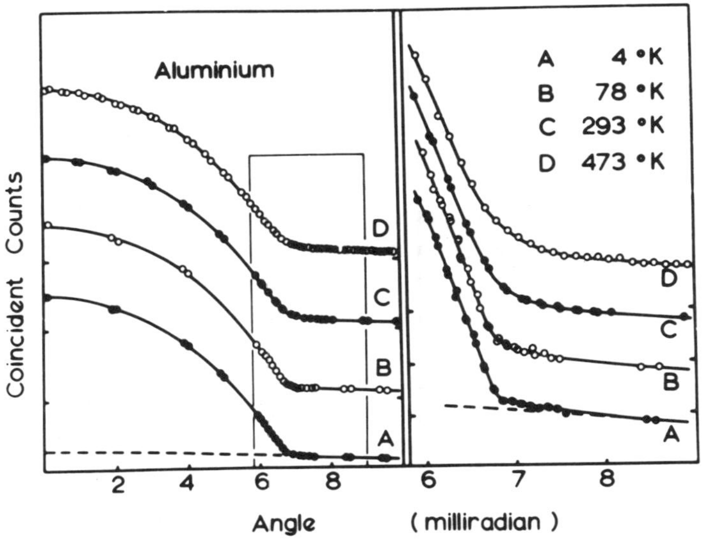

Fig. 1. Angular correlation of photons from positrons annihilating in $A\ell$. The solid line is a fit to the data. Ref. 3.

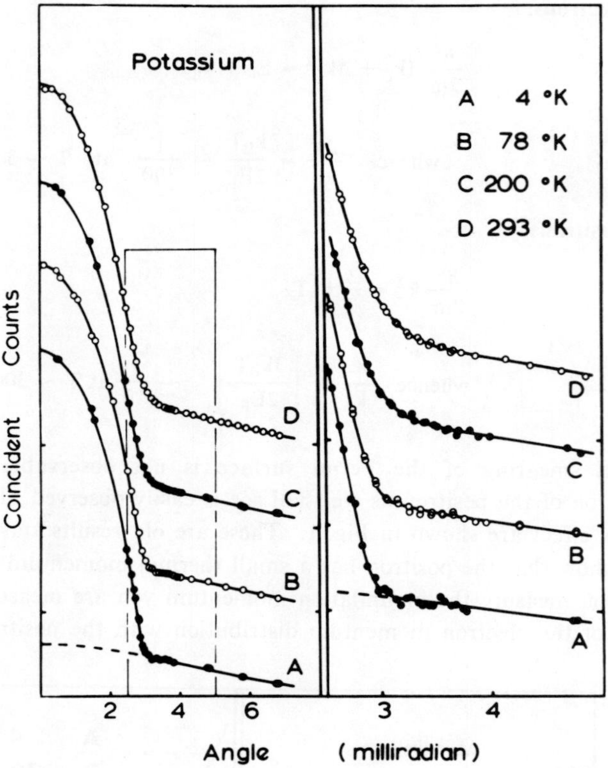

Fig. 2. Angular correlation of photons from positrons annihilating in potassium. Ref. 3.

distribution. You can see this positron motion at the position where the electron momentum has a cutoff, that is, at the Fermi surface. These corners at the end of the parabola show clearly that the positron motion is measurable. Figure 2 shows similar results for potassium. When one does the simplest and crudest gaussian fit of this little curvature at the cut-off, that is, assuming the positron is a free particle with a gaussian momentum distribution, one obtains the temperature of this gaussian distribution as a fitting parameter. Plotting the temperature of the positron thereby obtained against the temperature of the specimen, one observes the results shown in Fig. 3.

Two things are immediately obvious here: first, the linearity proves thermalization. That is when you double the temperature of the environment you double the motion, that's what you mean by thermalization.

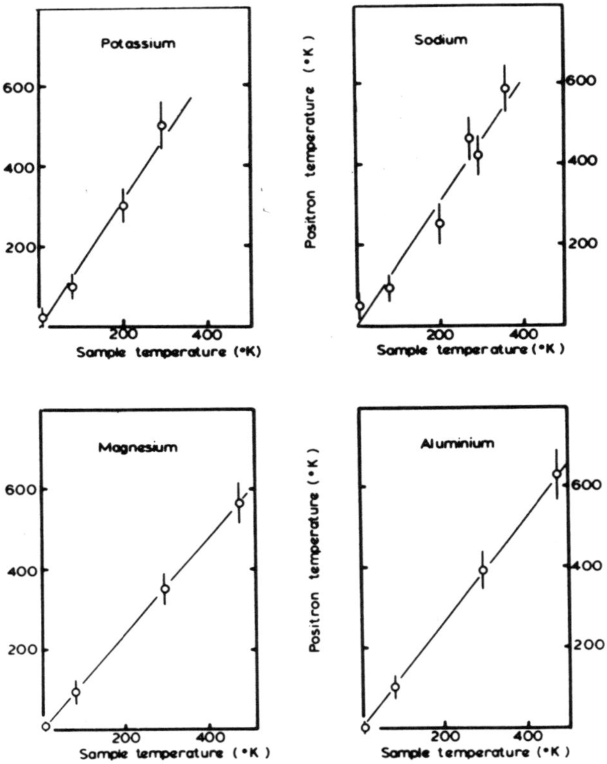

Fig. 3. Plot of positron apparent temperature vs. specimen temperature. Most of the points shown are the average of two independent experiments at the same temperature. Ref. 3.

Secondly, since the straight line says that they are thermalized then of course the positron must be at thermal temperatures so that the slope of this line then cannot measure temperature but must measure something else. And so following the traditions of solid state physics, an effective mass is declared to exist. That is

$$e^{-\frac{p^2}{2mk_BT_{eff}}} \rightarrow e^{-\frac{p^2}{2m^*k_BT}}$$

and the problem is solved. It would have been solved if a couple of theoreticians hadn't gotten into the business and started to calculate the effective mass of the positron. That created a certain amount of chaos as the table shows (Fig. 4). The first column is the effective mass measured by fitting the gaussian to those corners. The second is a band mass from Kubica and Stott.[4] The next columns

			m_e/m	
Metal	m^*/m (experimental)	m_B/m ref. 4	ref. 5	ref. 6
K	1.60 ± 0.1	1.15	1.19	1.45
Na	1.55 ± 0.1	1.07	1.15	1.30
Mg	1.20 ± 0.1	1.10	1.10	1.15
Aℓ	1.30 ± 0.1	1.15	1.07	1.08

Fig. 4. Positron effective mass m^* determined from present work, the band effective mass m_B, and the effective mass arising from positron — electron correlation. Ref. 3.

are electron clad positron effective masses, one column calculated by Hamann[5] and the last column calculated by Baldo and Pucci.[6] Perhaps one should take the product of the band mass with the electron mass and see if it fits but it doesn't very well, so that is a problem.

There was another problem besides the fact that we couldn't get a decent fitting. I always suspected that this potassium data and perhaps the sodium data was not really a straight line, that it was a curve, that if you like, the effective mass varied slightly with temperature. Therefore we decided we should measure it more accurately and more carefully and we should also learn a little bit about the work the theoreticians were doing.

The chief work[7] was done many years ago by a colleague who was working with Dick Farrell at Maryland named H. J. Mikeska who showed two effects resulted from putting a positron in a lattice with an electron gas. Of course there is always the electron polarization cloud around it but with the phonons of the lattice two things happened. First intense scattering effectively broadened the gaussian distribution. Secondly, the real momentum distribution was not quite gaussian, the principal differences being below 1% intensity where the tail decreases as p^{-4}. We had ignored these long tails, especially since they are way down at the 1% level. However, when you integrate p^{-4} tails on the long slit apparatus you get p^{-2} and this makes a big difference. That means the integral of the momentum distribution as seen in the long slit high resolution apparatus has a p^{-2} tail which can extend to overlap the big parabola. The convolution of these two distributions is then changed. Figure 5 shows[8] the difference between these functions. The curve marked G is a gaussian momentum distribution; you see I've plotted it logarithmatically now, and F is a realistic momentum

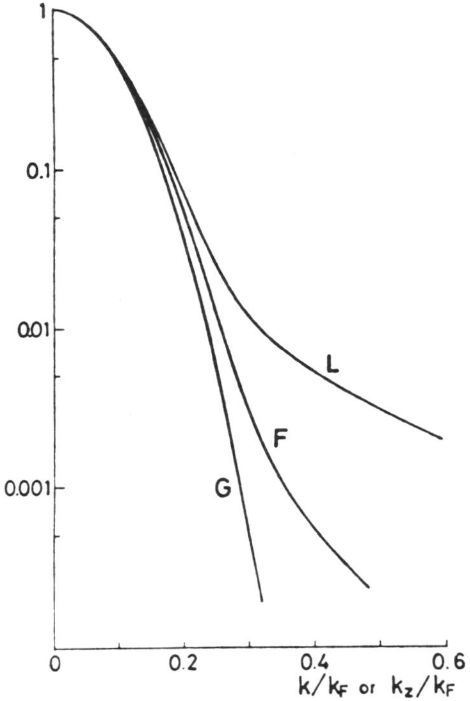

Fig. 5. Momentum distributions for a positron of quasi-particle mass $m^* = 1.2\ m_e$ at temperature $T = 0.01T_F$. G: Boltzmann distribution (no phonon effect). F: realistic distribution F(k) for $\tilde{\chi} = 1.0$. L: long slit projection $L(k_z)$ of the distribution F(k). Ref. 8.

distribution in a real metal. In addition, we have plotted here curve L. It is obtained by integration over two directions in the plane corresponding to long slit measurements. You see there is a big tail at less than 1% intensity but it is very important when you do the convolution of that with the big central parabola. In fact, in our best data recently we can detect the difference between using the proper momentum distribution and fitting an ordinary gaussian. Figure 6 shows some very precise recent data[9] on potassium for different temperatures and a couple of different crystal orientations. In Fig. 7 these data are cut off at about 20% height and show only the bottom parts. We see that the parabola is going way up to the next floor and we notice that the corner which measures the motion of the positron is still not a large feature. In order to get this tiny little bit we have to know all about this parabolic portion, all about the ion core and something at least about the higher momentum components which can be seen

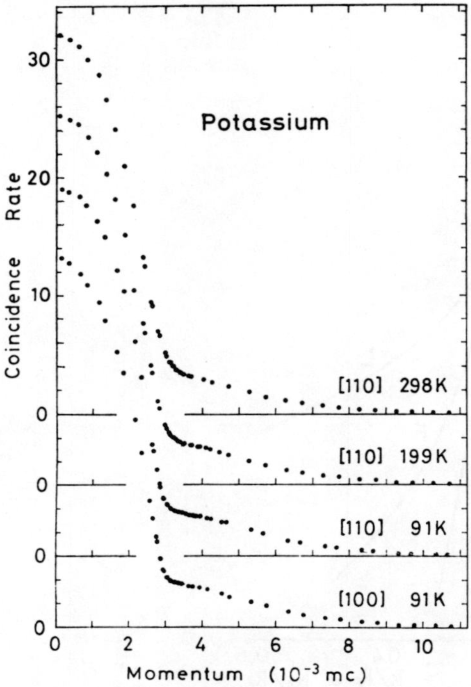

Fig. 6. Long slit projections onto the [110] and [100] directions of the momentum distribution of photon pairs from positrons annihilating with electrons in potassium single crystals at the temperatures idicated. Ref. 9.

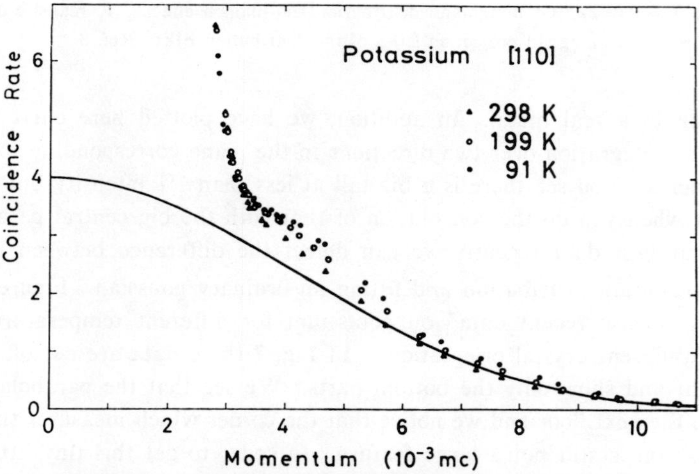

Fig. 7. Expanded portion of the data along [110] in Fig. 6. The solid line shows the experimentally observed momentum distribution for argon with an intensity suitable for an estimate of the core contribution to the data at 91 K. Ref. 9.

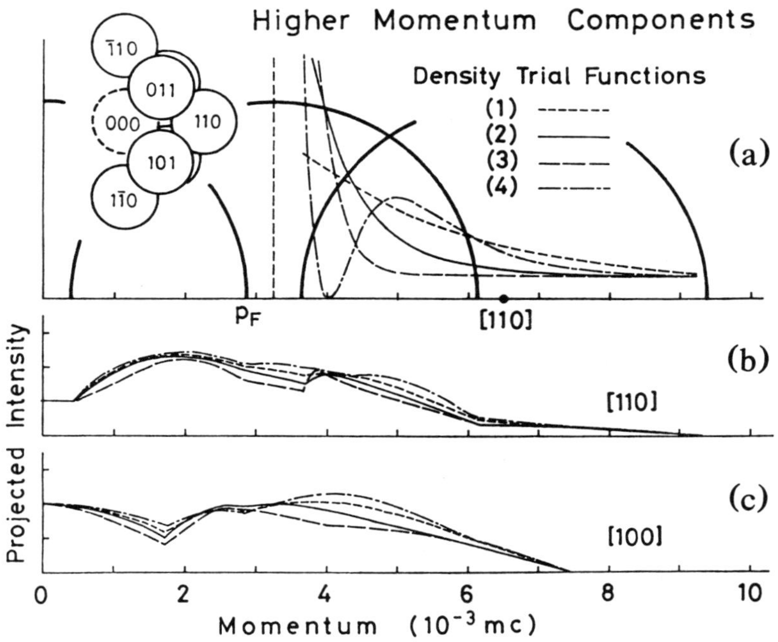

Fig. 8. Long slit projection onto the (b) [110] and (c) [100] directions of the (110) higher momentum components with various hypothetical momentum density functions (a). Large semicircles in (a) indicate the projections of the Fermi sphere centered at each (110) reciprocal lattice point [insert in (a)] onto the [110] direction. Ref. 9.

beyond the Fermi cut-off. (Notice how they vary with temperature.) We engaged in a fitting process which can be called fitting by "reasonableness", invented by my colleague Toshio Hyodo. This is a tedious process but it means that we used for this parabola a spherical momentum distribution and tried different enhancement factors to see whether or not when subtracted from the data the remainder would look like core plus higher momentum components. That is, we know that the parabola is smooth and the core is smooth and the only parts with structure are the higher momentum components which we tried to simulate with a few crude models. (I showed this at the Texas meeting). We assigned the higher zones with a momentum distribution that is not physics but is just a guessed trial function. We then calculated the projection in the long slit data. The results are drawn in the (b) & (c) parts of Fig. 8 which shows that topology is the chief effect determining h.m.c. shape. With this kind of calculation of the higher momentum components we then did fitting by the

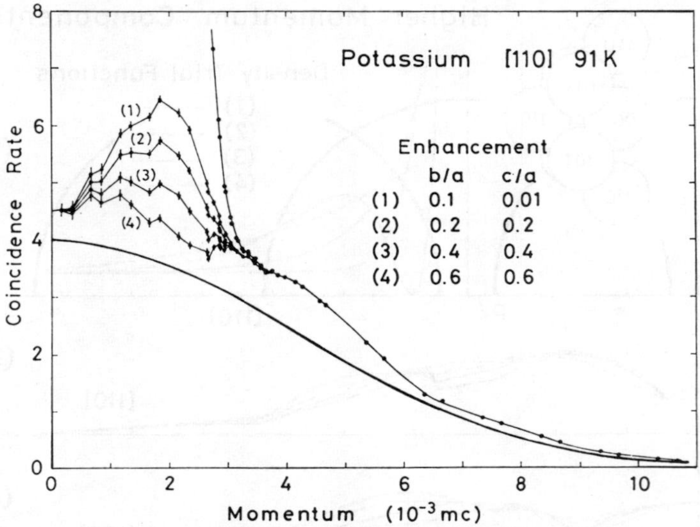

Fig. 9. An expanded portion of the data along the [110] direction at 91 K. The curves (1)-(4) are examples of the remainder after the parabolic part with various enhancement factors was subtracted from the data. Other parameters are fixed at $p_F = 2.87\ 10^{-3}$ mc, $m^* = 1.4$, and $\tilde{\chi} = 1.3$. Ref. 9.

"reasonable remainder" technique. For example, Fig. 9 shows the way we fitted to determine what enhancement should be used.

When the enhancement is selected we then vary the effective mass of the positron, again choosing which of the underlying remainder really does look like the higher momentum components. (Fig. 10) You notice that we now are fitting separately the electron enhanced effective mass of the positron m^*, and this parameter marked $\tilde{\chi}$ which is a measure of the positron-phonon interaction. To show the effect of separating them completely we have to look at the data at different temperatures. Fig. 11 shows the various pairs of values, since they're not orthogonal. Out of this we conclude that since the real effective mass m^* is of course independent of temperature over this small temperature range, the dot in the center is the constant one yielding an effective mass of 1.4 for the electron clad positron in potassium, and simultaneously yielding 1.3, for the reduced interaction between the positron and the lattice.

K. Lynn

Alec, is there a physical feel for $\tilde{\chi}$, I mean the size of that?

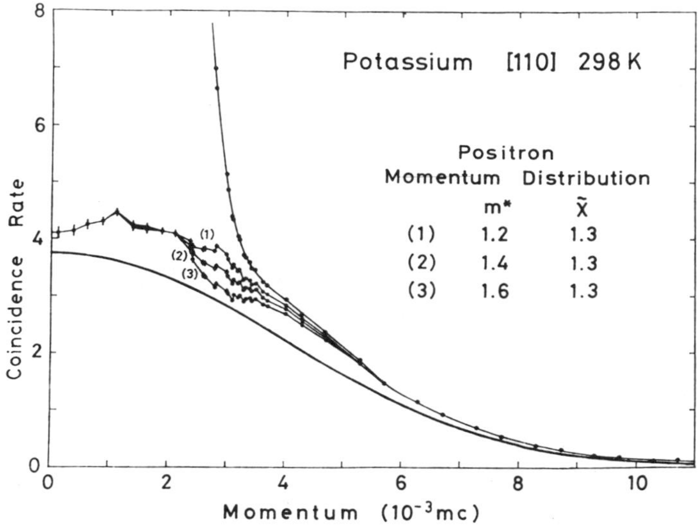

Fig. 10. An expanded portion of the data along the [110] direction at 298 K. The curves (1)-(3) are examples of the remainder after the parabolic part with various smearing functions was subtracted from the data. The value of $\tilde{\chi}$ is fixed, as is $p_F = 2.81\ 10^{-3}$ mc and $b/a = c/a = 0.35$. Ref. 9.

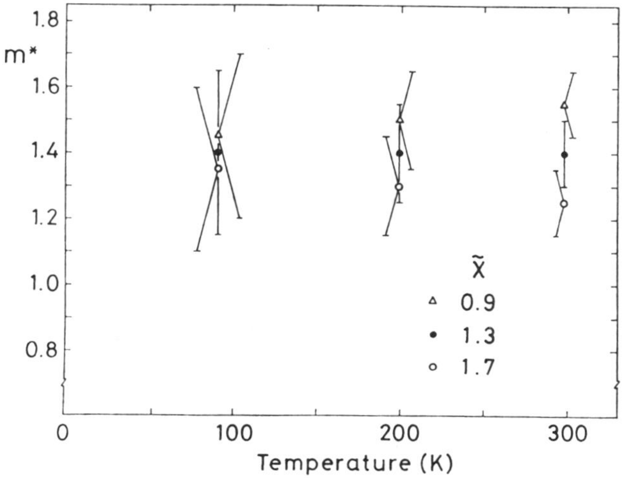

Fig. 11. Values of m* obtained from fitting data at various temperatures keeping $\tilde{\chi}$ fixed. Requiring temperature independence leads to $\tilde{\chi} = 1.3$ and $m^* = 1.4\ m_c$. Ref. 9.

Stewart

Not for me. There may be for my colleague Tom McMullin who helped me through this, but I don't know what it means in terms of how much the positron pushes on the lattice, no. In fact, I'm trying to understand and to get some feeling for it.

Finally, having obtained an enhancement, a positron effective mass and a positron-lattice interaction we put all of this together and synthesized the result we would get in another crystallographic direction of potassium. Figure 12 shows the synthesis for K (100) and it agrees well with the data.

So, finally, here are our net results for the enhancement, for the effective mass and for some kind of lattice interaction. The accuracy of that effective mass is about ± 0.1. This will be published soon.

I said that we have to know three things, the positron motion is the smallest part of this kind of momentum distribution and to get it with any accuracy at all you need to know very well the underlying core annihilation. There is no way of knowing what that is, so for a first guess for the potassium ion core we use the observed argon momentum distribution, (positron annihilation in liquid Argon).

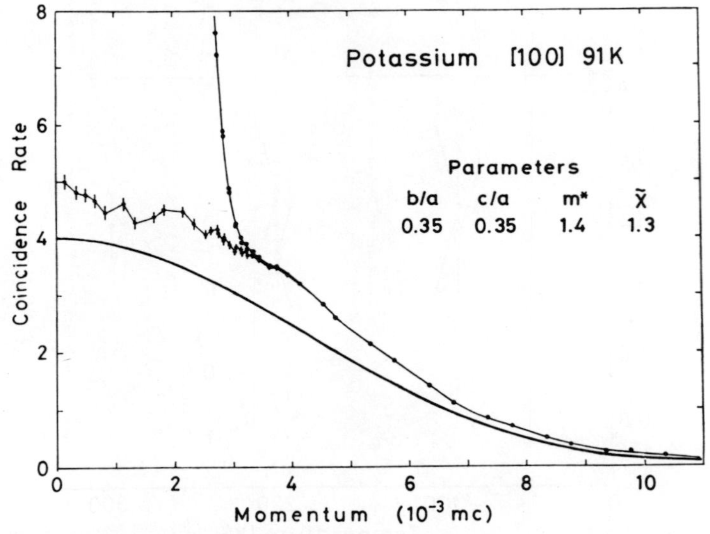

Fig. 12. An expanded portion of the data along [100] at 91 K, and the remainder after the parabolic part with the parameters determined by the analysis of the data along [110] has been subtracted. Ref. 9.

This is in fact quite accurate. You would expect two reasons for change from argon; the charge on potassium is higher so it will pull the electrons in but it will also repel the positron so you will sample slightly slower electrons. These are certainly compensating tendencies. Also, with Peter Mijnarends, we did a calculation of the momentum distribution for an ion core of potassium and compared it with a momentum distribution calculated for argon and they were essentially identical though, unfortunately, neither was quite like the data. It proved that the wavefunctions were almost alike. So we hope that our method of using the argon core is correct for Potassium. But the thing that is most unknown is the higher momentum components because at this time they had not been calculated. So with crystal in hand, I crossed the ocean to Roy West's place about a year or so ago. Unfortunately his machine was being rebuilt or something and wasn't running while I was there and so then off to Geneva where with that very hospitable group I did some direct measurements of these higher momentum components in potassium, which I'll show you. I'm showing this for

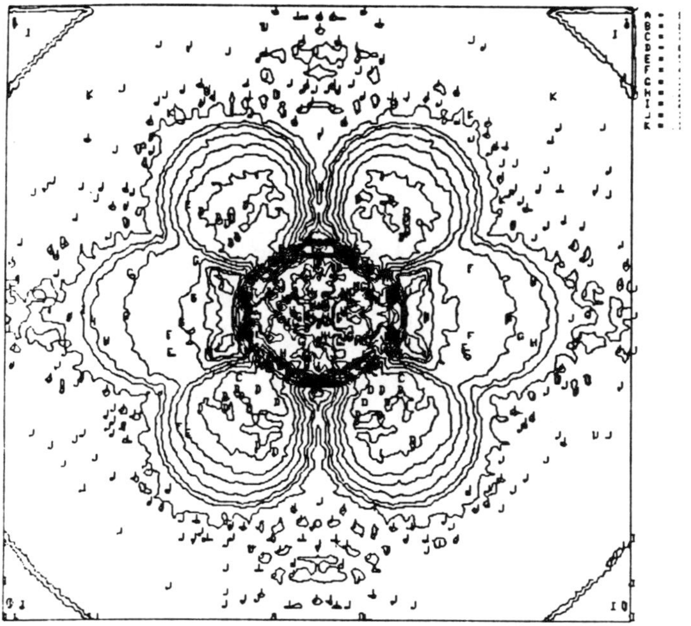

Fig. 13. Two-dimensional projection of the momentum distribution of photons from positrons annihilating in a single crystal of potassium. The central region has had a spherical pattern subtracted to leave revealed the low intensity (1-2%) higher momentum components from the [110] Brilluoin zones. Ref. 10.

another reason. Steve is not really interested in things like the positron mass, stuff like that, but 2-D; 2-D he likes. So here is 2-D. Fig. 13 is for potassium at 98°K. The central parabolic part, or rather in 2-D, the spherical part of the main distribution of momentum in the first zone has been subtracted to look at the low intensity higher momentum components. Thus the central zone shows mostly noise and has little meaning, but all of these other spheres have meaning. The two on the horizontal axis are the nearest (110) spheres in the reciprocal lattice of potassium. Each of the other four corresponds to two (110) spheres above and below the plane seen superimposed in projection so that the intensity of these is twice the intensity of the other two.

Further you can see the nearest (200), at the top and bottom. Similar data, nothing to do with our effective mass experiment, is shown in Fig. 14 for lithium. The shape of these higher momentum components is very different from those in K. You notice that again we subtracted a sphere and subtracting a sphere from the lithium Fermi surface is not quite correct, so that there remain bulges in the (110) directions where the anisotropies of the lithium Fermi surface appear.

Fig. 14. Two-dimensional projection of the momentum distributions of photons from positrons annihilating in a single crystal of lithium. As in Fig. 13, a central spherical distribution has been subtracted to leave revealed both the anisotropy of the Fermi Surface of Li and the higher momentum components from the several [110] Brillouin zones. Ref. 10.

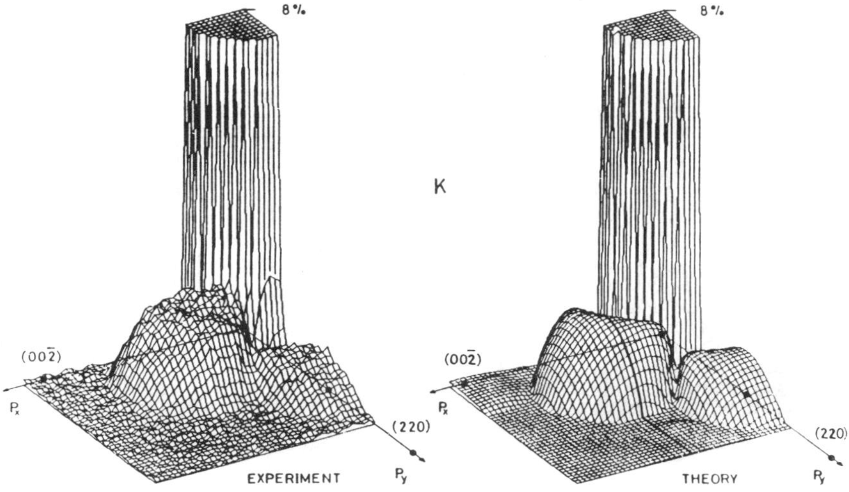

Fig. 15. Higher momentum components of the experimental angular correlation of annihilation photons from K compared with the KKR calculation. The vertical scale is shown by the truncation at 8% of the conduction electron intensity at $p_x = p_y = 0$. Ref. 11.

Another interest of these momentum distributions is that at many positions there is no higher momentum component at all because you can see all the way through the reciprocal lattice without being in contact with any Fermi spheres. So the intensity in these places can be used to normalize the core contribution.

The same data in isometric projection is shown in Fig. 15 for potassium. It is compared with a calculation done by the Mijnarends group in Holland. These slides show 1/4 of the data I showed before. I should have mentioned with those data that there has been no folding done at all. The first thing we noticed was that we had oriented the crystal correctly! This is one quadrant comparing the experiment and Peter Mijnarends theoretical calculation. You'll see the single (110) sphere on the P_y axis nearest the main central sphere and the larger bump is the two (110) spheres that are above and below the plane here superimposed in projection. You can just see the (200) on the P_x axis. The experiment on the left is in delightful correspondence.

In a later slide I will show you a profile of the density distribution along the solid line to compare experiment and theory. The equivalent data for lithium is in Fig. 16 and you will notice the great difference in the shape of the momentum distribution in the higher momentum zones of lithium than in potassium. The

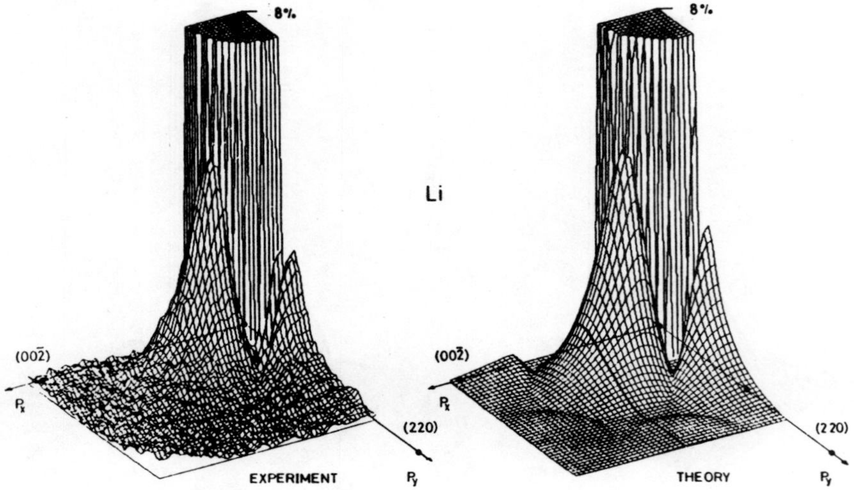

Fig. 16. Higher momentum components of the experimental angular correlation of annihilation photons from Li compound with the KKR calculation. The vertical scale is as in Fig. 15. Ref. 11.

reason for this is understood and if we had more time Peter would tell you why.

Finally, to compare the intensities directly, which I promised you, we show in Fig. 17 profiles through that cut shown by the heavy line. You notice that it goes down to zero at the axis. This is a direct comparison with the data and the experiment with no normalization other than to the intensity at (000) for both calculation and theory. The agreement is quite spectacular. The theoretical, solid, line has not been smeared for the instrument resolution which you can see in the experimental data as it drops down to zero.

In conclusion, two principal deductions: a) The observed h.m.c. in K are sufficiently like some of the simple models postulated in our analysis of the data for the effective mass to lend credence to the analysis and the results. b) The effect of enhancement in the higher zones of K and Li seems close to the enhancement at (000).

K. Lynn

You said that you could use the holes in the higher momentum components to get the core enhancement or the core rate. Did you in fact do that, I didn't hear the number? How much above the ipm?

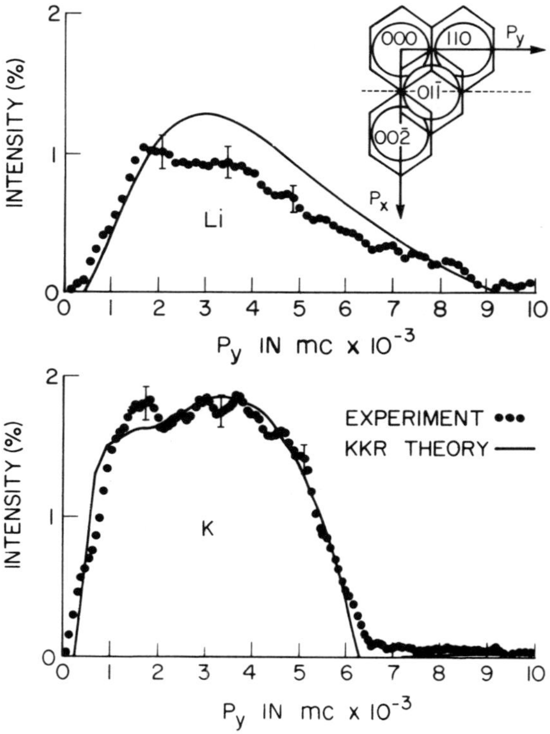

Fig. 17. Profile through the projection of $(01\bar{1})$ and $(10\bar{1})$ HMC's as shown in the inset and by the heavy lines in Figs. 15 and 16. Typical statistical errors are shown, but the scatter of points has been reduced by a smoothing procedure. The vertical scale is as in Figs. 15 and 16. Ref. 11.

Stewart

No, we have no number yet for that.

Lynn

That would be interesting.

S. Berko

First of all you said that Berko is interested in 2-D but not the effective mass. I want to make a correction. As a very polite Hungarian, I do not usually like to do experiments which my esteemed friends from Canada are known to be involved in and so when it came to effective mass in alkali metals I kept my hands off desperately. We did do one effective mass measurement which I thought was different than yours, namely in a semiconductor, and to make sure of that, quite clearly the semiconductor's effective mass is smaller than in a metal on the other hand it was a much harder experiment because one doesn't have the very sharp break that you see. But the thing that I want to say is that, as I understand it, if you normalize the theory at (000), you get the right height for the umklapps. Which means that the umklapps are enhanced the same way, more or less, as the central.

Stewart

Which conclusion we've put in the paper.

Berko

Which is exactly what we got with aluminum and copper. Nature starts saying something, namely that the umklapps seem to be very much along the lines of this (000).

Stewart

The data is not good enough to see the momentum dependence in the higher zone. Not quite yet.

Berko

In the reconstructed 3-dimensional regrown copper, studied at Brandeis, the data just starts barely being good enough to see that there is a slight deviation in the first umklapp close to the Fermi surface. This may also be reflecting the increase of the enhancement as you approach the Fermi surface from inside.

Given your enhancement parameters are they good enough to now rule out already or do you have to go to Cesium to completely rule out the Finnish group?

Stewart

The answer is probably good enough now.

Berko

For those who are not initiated in this thing, Kahana's theory and the Finnish group (Arponen and Pejanne theory) agree in the high density metals and therefore also in aluminum and copper and so on. They disagree violently however for Cesium.

Stewart

And the newer theory was allegedly designed to pick up many of the Feynman diagrams that Kahana and Carbotte had left out so in principle the newer theory should be right but it doesn't look that way.

P. Platzman

In this line I want to issue a little word of caution. The fact is that the electrons in a metal are a rather complicated many-body system, and one doesn't even understand them exactly in the absence of the positron. All the theories you're talking about are theories of that electron gas in the absence of the positron. Now you add the positron, that adds a complication. Basically the point is you don't even understand the electron gas. If I asked questions about the momentum density of the electron gas in the absence of the positrons, so that while it's very nice to get some of these trends and so on and try to get masses out and measure them very precisely in things like potassium and sodium and the alkali metals where the electron correlations are becoming important, you would have to treat things with a little grain of salt. It's almost better, in many cases, to think of the theory as experiment in some sense. In other words, you have an array of theoretical calculations all of which involve uncontrolled approximations. Really, what you should say is: here's the experiment, and it's in rough agreement with this terrible theoretical estimate, because you are looking at very tiny effects on a system which is not, while it's understood pretty well, is not understood extraordinarily well.

Stewart

In many parts of this field the theory is changing about as often as the experiment and in some parts the experiment is not nailed down very well.

Platzman

What you have to remember is that it is just impossible to do calculations for an electron gas with $r_s = 3$ or 4 with an accuracy of 1%. You know roughly what happens but you are not going to know exactly.

Stewart

I should add that I cheated a little here because I represented to you that the motion of the positron in potassium electron gases is well understood, it is pretty well understood. However, when we then went back and said what about aluminum, because in aluminum, for which I showed you some decent data, there should be a greater phonon interaction and yet the apparent mass measured by the old criteria of fitting a Gaussian is lower so this is wrong. Thus recently we've accumulated a great stack of precision data on aluminum and will process it in the same manner because it seems to look wrong at present.

Berko

There's one thing I'm really worried about here, assigning things that have very small effects, like outside of the smearing, to necessarily phonons because as you know the Kahana theory really has no electrons, not electrons but momentum density, spilling outside of the Fermi surface; but any better theory does. It was always a miracle to me that the bottom of the electron correlation doesn't have a long tail outside of the Fermi surface. So you can see these things you have to be careful of, because maybe they are positron-electron correlations not entering into Kahana's diagrams of this stuff.

Can I say one more thing? It finally dawned on me why Allen Mills has suddenly gone and studied alkali halides. I never quite understood why Bell had all this interest in semiconductors and metals including alkali halides, and that is because Platzman must have warned Mills all the time "you have to take all these theories with a grain of salt".

REFERENCES

1. S. Kahanna, Phys. Rev. **117**, 123 1960.
   J. Carbotte, and S. Kahanna, Phys. Rev. **139**, 213 1965.
2. A. T. Stewart, J. B. Shand and S. M. Kim, Proc. Phys. Soc. **88** 1001 1966.
3. P. Kubica and A. T. Stewart, Can. J. Phys. **61** 971 1983.
4. P. Kubica and M. J. Stott, J. Phys. F **4** 1969 1974.
5. D. R. Hamann, Phys. Rev. **146** 277 1966.
6. M. Baldo and R. Pucci, Nuovo Cimento B **23** 202 1974.
7. H. J. Mikeska, Z. Phys. **232** 159 1970.
8. T. Hyodo, T. McMullen and A. T. Stewart, Can. J. Phys. **62** 297 1984.
9. T. Hyodo, T. McMullen and A. T. Stewart, Phys. Rev. B 1986.

10. Data and data reduction with L. Oberli, A. A. Manuel, R. Sachat, P. Descouts, and M. Peter.

11. L. Oberli, A. A. Manuel, R. Sachot, P. Descouts, and M. Peter; L. P. L. M. Rabou and P. E. Mijnarends; T. Hyodo and A. T. Stewart, Phys. Rev. B **31** 1147 1985.

Positron Studies of the Electronic Structure of Metals and Alloys

R. N. West

School of Mathematics and Physics
University of East Anglia
Norwich NR4 7TJ, U.K.

Introduction

I would like to start this talk by saying how pleased and honored I am to be able to contribute to this very important occasion and also how nervous I am. Not only, you should understand, because of the overwhelming presence of the man we're all here to honor, but also because this gathering probably includes more experts in the field of 2D angular correlation than have ever been assembled together before.

The first thing that I'd like to do is change my title. The programme refers simply to metal alloys but this talk, like those of the two previous speakers, is concerned with electronic structure. I would also like to insert a strategic "and" and redefine my subject as "Positron Studies of Electronic Structure in Metals and Alloys". Today, thanks to Steve Berko[1], that must mean 2D angular correlation.

I am glad that the previous speakers have done some of my work for me in reminding you that what we are talking about in this session is the measurement and interpretation of the electron-positron momentum density ρ_{ep} ($\underline{p}$). This, as is now well known[2–5], can be regarded as reflecting two very important aspects of an electronic structure. From it we can get, on the one hand, information about the nature and form of the electron and positron states. Secondly, we get information about the occupancy of those states in terms of some suitable quantum number like $\underline{k}$; i.e. in metals, the Fermi surface. Of course, that in itself should be enough to establish the claims of positron annihilation over those alternative techniques that just give information about the Fermi surface or about the energy density of states, or, indeed quite often, just about the density of states at the Fermi surface. But that perhaps is not quite enough and we also have to remember that one of the major justifications for plunging money into these positron measurements is that they can work, where many of those other techniques cannot work, in the presence of comparatively short electronic mean-free paths. So we can, if we want, study metals at elevated temperatures where there are interesting structural or magnetic phase transformations. We can also look at concentrated disordered substitutional alloys.

When I talk about a disordered substitutional alloy I'm talking about a system in which it is just the potentials that are disordered. We still have a well-defined crystal structure, be it fcc, bcc or whatever, but on any particular site we may find (in a binary system for example) either an A-type or a B-type atom. Then, as Peter Mijnarends has already told us, we shall get all sorts of complex scattering effects on both the positrons and the electrons. But even with these scattering effects the Bloch wave vector $\underline{k}$—concept can still be useful. Alloy theorists[6] use $\underline{k}$ as a quantum number and look, in general, at situations where it still has some claim to be regarded as a good quantum number. Those are the systems that we can most usefully study.

The Electron-Positron Momentum Space Density

We have already seen in the experimental and theoretical data of the two previous talks the enormous amount of information that is implicit in the 2D angular correlation or electron-positron momentum density. The first talk developed its theme in terms of some of the most sophisticated concepts and representations of electron energy-band theory but I would like to go back to some relatively simple and familiar representations.

In the independent particle model (IPM) we can express the three-dimensional (3D) electron-positron density in terms of the real space wavefunctions of the electrons, $\psi_{\underline{k}}^{j}(\underline{r})$, and the thermalized ($\underline{k} = 0$)[7] positron, $\psi_{+}(\underline{r})$. In atomic units ($e = \hbar = m = 1$, $c = \alpha^{-1} = 137$),

$$\rho_{ep}(\underline{p}) = \text{const.} \sum_{\text{occ}\underline{k},\, j} \left| \int_{V} d^3r\, e^{-i\underline{p}\cdot\underline{r}}\, \psi_{\underline{k}}^{j}(\underline{r})\, \psi_{+}(\underline{r}) \right|^2 \qquad (1)$$

where, in a crystal of volume V comprising N unit cells of volume Ω, $\psi_{\underline{k}}^{j}(\underline{r})$ and $\psi_{+}(\underline{r})$ are (for later convenience) normalized to N and to V respectively. j is an electron energy-band index and k is the reduced Bloch wave vector. Thus far the crystal periodicity is only implicit but we can make it more explicit if we follow the seminal paper of Berko and Plaskett[8] and write

$$\psi_{\underline{k}}^{j}(\underline{r})\,\psi_{+}(\underline{r}) = \sum_{\underline{G}} C_{\underline{G}}^{j}(\underline{k})\, e^{i(\underline{k}+\underline{G})\cdot\underline{r}} \tag{2}$$

where the G's are reciprocal lattice vectors.

The inverse relation

$$C_{\underline{G}}^{j}(\underline{k}) = \Omega^{-1} \int_{\Omega} d^3r\, e^{-i(\underline{k}+\underline{G})\cdot\underline{r}}\, \psi_{\underline{k}}^{j}(\underline{r})\,\psi_{+}(\underline{r}) \tag{3}$$

will later be useful. Now if we substitute Eq. 2 into Eq. 1, and sum over doubly occupied spin degenerate states, we obtain a new expression for ρ_{ep} which, aside from the usual density of states factor, $V/8\pi^3$, is[8]

$$\rho_{ep}(\underline{p}) = 2\Omega \sum_{j} \theta\,(E_F - E_{\underline{k},j}) \sum_{\underline{G}} |C_{\underline{G}}^{j}(\underline{k})|^2\, \delta_{\underline{p}-\underline{k},\underline{G}} \tag{4}$$

Here θ is the Heaviside unit step function and E_F is the Fermi energy.

I should remind you that this last equation is particularly useful because it shows us in a very transparent way the difference between p and k which arises from the so-called higher momentum components (HMC) in the particle wavefunctions. It also shows, in a nice simple way, that we are probing the k–space occupancy and that we are looking, in principle, at all the occupied states.

We now have many experimental results to confirm the essential correctness of this picture. In the simple metals we can see directly the Fermi surface[5] and the HMC[5,9]. Both here and in the results for less-simple metals the agreement between experiment and refined theoretical calculations is often impressive (Fig. 1). But when we come to still more complex multi-band systems like the transition or rare earth metals, where the effects of many small Fermi surface sections are redistributed by the delta function (Eq. 4) throughout the momentum density, it is often difficult to separate the Fermi surface effects from those of the electron wavefunctions. There is a tremendous amount of information in such 2D spectra and I think that sometimes, at least from the point of view of understanding, there is a little too much information. Even in

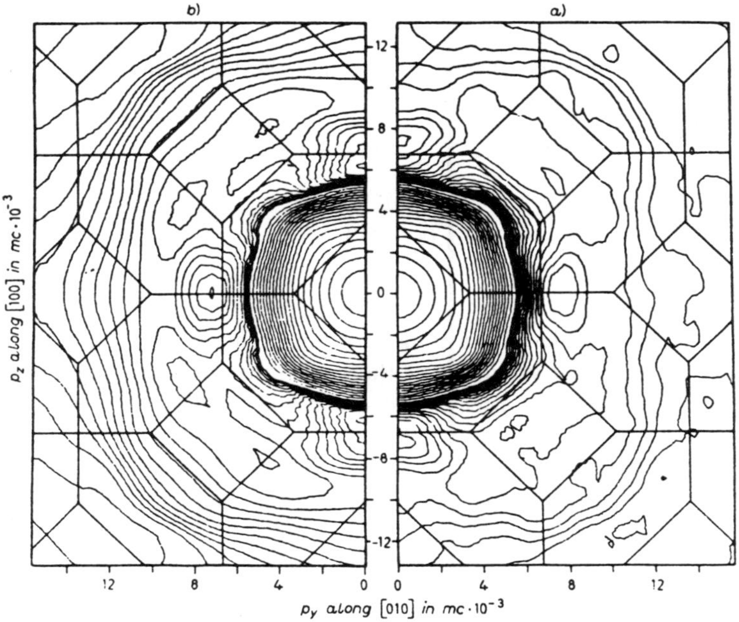

Fig. 1. Contour maps of experimental vs theoretical 2D distributions for Cu with integration direction p_x along [001]. The presentation is in the form N (p_y, p_z) − R (p_y, p_z), where R (p_y, p_z) is an isotropic and approximately gaussian distribution chosen so as to display most clearly the anisotropies in the 2D distribution, N (p_y, p_z). The contour intervals are changed at p ~ 5 · 5 m rad, again to emphasize anisotropies.[5]

the case of data for comparatively simple metals it is very nice to be able to compare an experimental 2D spectrum with its theoretical equivalent. But it's not always easy to say much more than "Oh, that's nice" or "No, that doesn't seem to agree too well". It is for that reason that the alternative treatment of angular correlation data, by the folding procedure,[10,11] for which Beardsley *et al*[12] a few years ago, so generously coined the name LCW, has come so much into its own.

The LCW Transformation

In a LCW transformation[10–12] one takes the momentum density (or its projection) and forms the superposition

$$\sum_{\underline{G}'} \rho_{ep}\,(\underline{p} + \underline{G}') = 2\Omega \sum_{\underline{G}'} \sum_{j} \theta\,(E_F - E_{\underline{k},\,j}) \sum_{\underline{G}} |C^{j}_{\underline{G}}\,(\underline{k})|^2\, \delta_{\underline{p} - \underline{k} + \underline{G}',\,\underline{G}} \qquad (5)$$

In this new expression the double sum over reciprocal lattice vectors allows us to regroup the various terms so that the effect of the delta function, which originally threw events from the first Brillouin Zone into the higher momentum regions, is completely removed. That is,

$$\sum_{\underline{G}'} \rho_{ep} (\underline{p} + \underline{G}') = 2\Omega \sum_{\underline{G}'} \sum_{j} \theta (E_F - E_{\underline{k}, j}) \sum_{\underline{G}} |C^j_{\underline{G}} (\underline{k})|^2 \delta_{\underline{p}-\underline{k}, \underline{G}'}$$

$$= \sum_{\underline{G}'} \rho_{ep} (\underline{k} + \underline{G}') \tag{6}$$

What we have now achieved is a transformation from a single center distribution, which just has the point symmetry of the system, to a new distribution which the translational symmetry has been included and which is in effect a distribution of Bloch wave vectors in the repeated zone scheme. The importance of this $\underline{k}$–distribution is that in it the wavefunction effects are to a large extent suppressed, but those of the Fermi surface are made more clear. Within the first Brillouin zone we may write

$$\rho_{ep} (\underline{k}) = \sum_{j} \rho^j_{ep} (\underline{k}) = \sum_{j} \rho^j_e (\underline{k}) \chi^j (\underline{k}), \tag{7}$$

that is simply a sum of the individual band occupancies in $\underline{k}$ each multiplied by the factor $\chi^j (\underline{k})$ that contains all the wavefunction effects. We can calculate $\chi^j (\underline{k})$, although more often in the past we have elected to ignore it or at least its $\underline{k}$–dependence.

So what we have said is, if we fold $\rho_{ep} (\underline{p})$ we get a $\rho_{ep} (\underline{k})$ which is virtually the electron $\underline{k}$–space density $\rho_e (\underline{k})$ aside, that is, from some factors that for the moment we'll choose to ignore. That has been the basis for many comparisons between theory and experiment[13–16] and, very quickly, I'd like to illustrate the consequent situation with the aid of Fig. 2.

This is a plot of information versus effort. On the left hand side of the figure, I have put in a point marked E for experiment. There is only one point on that curve for me. Different people may want to argue different points but that is my representation of the information contained in $\rho_{ep} (\underline{p})$ and a rough measure of the experimental effort involved in obtaining one or more of its 2D integrals. On the right hand side I have inserted a corresponding theoretical point and in setting it so far to the right I am trying to emphasize that, not only does a calculation of ρ_{ep} (p) involve considerable effort, even when you have finished the calculation

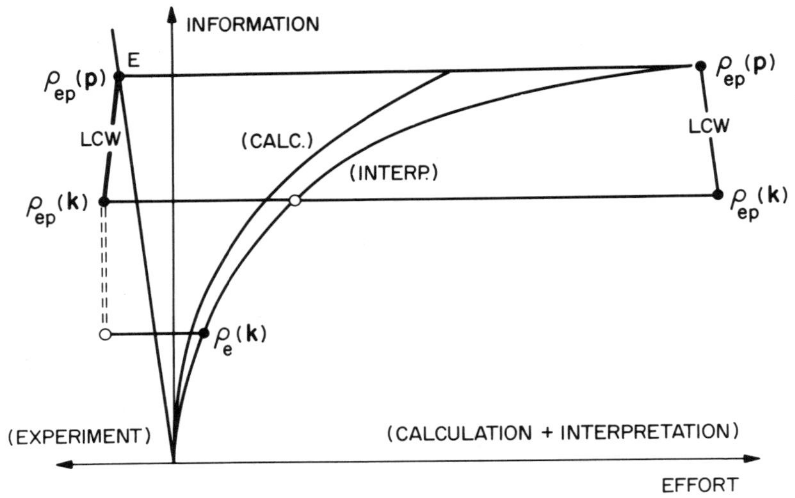

Fig. 2. A graphical representation of the author's view of the cost-effectiveness of the different possible paths in the interpretation of electron-positron momentum densities.

you still have a long way to go before you really understand the results and, equally importantly, the comparison with experiment. That is the reason why we so often take the easier path via the LCW theorem to the $\underline{k}$–space distribution, then ignore the little subscript p in ρ_{ep} ($\underline{k}$), and compare it with the electron $\rho_e(\underline{k})$ which is so much easier to calculate and interpret.

Of course there is a lot of information being thrown away when we run down those "slippery slopes" from ρ_{ep} ($\underline{p}$) to ρ_e ($\underline{k}$) and the basis for experiment-theory comparison is much more secure at the level of ρ_{ep} ($\underline{p}$). However there is also a lot more information available at the ρ_{ep} ($\underline{k}$) level if we calculate it rather than ρ_e ($\underline{k}$). We can already do that[17] by LCW folding a theoretical ρ_{ep} ($\underline{p}$) but then we just lose information with no reduction at all in effort.

I would like, in the rest of this talk today, to fill in the open circle. I'm not suggesting that my colleagues and I are the first to do this. Others[17,18] have calculated ρ_{ep} ($\underline{k}$) but I want to describe a route to ρ_{ep} ($\underline{k}$) in which we can achieve a valuable compromise between the information we get and the effort we have to expend. But, before I do that I would like further to set the scene and at the same time indulge myself by showing you some of our recent data[16] for Nb_{50} Mo_{50}.

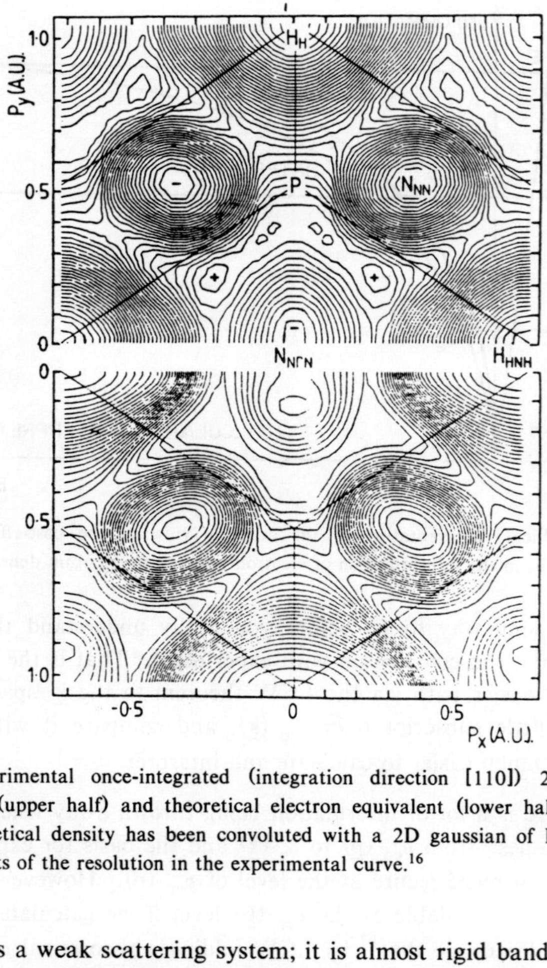

Fig. 3. Experimental once-integrated (integration direction [110]) 2D electron-positron $\underline{k}$–space density (upper half) and theoretical electron equivalent (lower half) for a Nb_{50} Mo_{50} alloy. The theoretical density has been convoluted with a 2D gaussian of FWHM 0.09 AU to simulate the effects of the resolution in the experimental curve.[16]

Nb – Mo is a weak scattering system; it is almost rigid band. It is a favorite of ours because it's alloys are almost as simple to handle as a pure metal. In Fig. 3 (bottom) I show an integral of ρ_e ($\underline{k}$) in the form of a contour plot. The large and well defined minima around the N points represent well-known hole pockets. There are other pockets around Γ and H and various other structures all of which are relatively well understood in terms of the electronic structure of groups V and VI transition metals. In the upper half of the figure I show the corresponding measured integral of ρ_{ep} ($\underline{k}$). I think you will see that there is a remarkable qualitative, and almost quantitative, agreement, which in theory[19] should be exact in respect to the Fermi surface singularities in the infinite resolution equivalents of these curves. These particular experimental data were obtained with a relatively modest resolution of slightly less than 0.1 AU or 0.65 milliradians but the statistics are so good (the data of Fig. 3 represents ~300 individual datum points each involving $\sim 5 \cdot 10^6$ counts) that we can sensibly generate first and second derivatives. By comparing the derivatives of the experimental and theoretical spectra we have been able to fix some major Fermi surface parameters to a precision of order of 0.02 AU, i.e. ~5–fold improvement in the effective resolution. This, I would assert, is more than enough to test the claims of competing theories. However, there is a problem when we go from a nice rigid-band system like this to a strong scattering system and that indeed is where we have to go sooner or later. In Fig. 3 we have Fermi surface sections that are large and well-defined and the comparison of ρ_{ep} $(\underline{k})$ with ρ_e $(\underline{k})$ is valid. But, if we go to strong scattering alloys, we will start to see significant difficulties arising from the residual differences between ρ_{ep} $(\underline{k})$ and ρ_e $(\underline{k})$ and if we want information about such effects of disorder as Fermi surface blurring we do need to be able to handle ρ_{ep} $(\underline{k})$ on the theory side.

The Electron-Positron $\underline{k}$-Space Density

There has never been any problem in principle about a calculation of ρ_{ep} $(\underline{k})$. The quantities of interest are well defined and one can describe them, at least in the IPM, quite realistically. One merely needs to calculate the various $\chi^j(\underline{k})$ (equation 7). We can see from Eq. 6 that each χ^j $(\underline{k})$ is just a sum involving the pair momentum eigenfunctions $C^j_{\underline{G}}$ $(\underline{k})$. Its $\underline{k}$ dependence arises from the cross terms in the equation that defines the $C^j_{\underline{G}}$ $(\underline{k})$'s in terms of the corresponding positron and $\underline{k}$ dependent electron momentum functions[10].

A few years ago, Dennis Lock, a student of mine, and I searched around for alternative representations for χ^j $(\underline{k})$ which might give us more insight into the probable form and strength of the $\underline{k}$ dependence.[11] One thing we realised was that, because the χ^j $(\underline{k})$ had to be periodic in $\underline{k}$–space, we could always represent them as a Fourier series involving the real-space lattice vectors, i.e.

$$\chi^j\ (\underline{k}) = \sum_{\underline{G}} |C^j_{\underline{G}}\ (\underline{k})|^2 = \sum_{\underline{R}_j} a^j\ (\underline{R}_j)\ e^{i\underline{k}\cdot\underline{R}_j} \tag{8}$$

This is fine but, on its own, it doesn't actually help you very much in arguing the form of χ^j $(\underline{k})$ for a particular electron energy band: the relationship between such reciprocal and real space Fourier coefficients is rather obscure. Only recently I was reminded by Nobu Shiotani that we have an alternative real space representation for χ^j $(\underline{k})$ which I had come across earlier and then completely forgotten because I thought it so trivial and obvious. It is,

$$\chi^j(\underline{k}) = \int_\Omega |\psi^j_{\underline{k}}(\underline{r})|^2 |\psi_+(\underline{r})|^2 \, d^3r. \tag{9}$$

You can derive this result either directly from Eq. (1) by invoking the Bloch theorem and working through some lengthy real space algebra[20] or by substituting Eq. 3 into Eq. 8 and making use of the identity[21]

$$\sum_{\underline{G}} e^{i\underline{G}\cdot(\underline{r}-\underline{r}')} = N\delta(\underline{r}-\underline{r}').$$

Thus $\chi^j(\underline{k})$ is quite simply the IPM annihilation rate with the state $\psi^j_{\underline{k}}(\underline{r})$. It is just an integral of the product of positron and electron densities. Because these are densities and because the phase is gone we should consider this latest formula (Eq. 9) most carefully because it is potentially much easier to calculate. Unlike the $C^j_{\underline{G}}(\underline{k})$ it has no cross terms in it. It is much simpler.

The first thing to see is that to take full advantage of Eq. 9 it is no use resorting to plane wave expansions; that just takes you back to Eq. 8. You have to stay in $\underline{r}$–space and one very good way of staying in $\underline{r}$–space is with an orbital angular momentum ℓ–based representation.

A recently graduating student of mine, Jerry Kaiser, has set up the problem in KKR.[22,23] Now I don't want to obscure my discussion with formulae. I myself am easily put off by complex bandstructure formulae and the KKR bandstructure method is well described elsewhere.[24–26] Let me just remind you that we set up the problem in terms of non-overlapping muffin tin potentials and write the electron wavefunctions as,

$$\psi^j_{\underline{k}}(\underline{r}) = \sum_{\ell\mu t} a^t_{\ell\mu}(\underline{k}; E_{\underline{k},j}) i^\ell \, Y^t_{\ell\mu}(\hat{r}) \, R_\ell(r; E_{\underline{k},j}). \tag{10}$$

Here, the $a^t_{\ell\mu}$ are appropriately symmetrized wavefunction coefficients that normalize $\psi^j_{\underline{k}}(\underline{r})$ to one in the unit cell. The $Y^t_{\ell\mu}(\hat{\underline{r}})$ are the similarly symmetrized angular harmonic functions of argument $\hat{r} = (\theta, \phi)$ and corresponding to angular momentum ℓ and row μ of the irreducible representation t. $R_\ell(\underline{r}; E_{\underline{k},j})$ is a solution of the radial Schrödinger equation inside the muffin-tin ($r < r_{mt}$) which is matched, at $r = r_{mt}$, to an analytic form

($r > r_{mt}$) which is completely defined by a scattering phase shift and the energy E.

If we substitute expressions like that in Eq. 10 into Eq. 9, some very unpleasant non-diagonal matrices appear. But, if we are only interested in s, p, and d electrons, as we so often are, we can terminate the electron ℓ − sum at $\ell = 2$. If we do that we shall almost certainly terminate the positron wavefunction at $\ell = 0$ because the next non-zero term is $\ell = 4$ and that becomes very expensive in computing time. So $\psi_+(\underline{r}) = R_+(r)$, the original spherical Wigner-Seitz cell type of positron wavefunction[8] which has served us well for so many years. If we do restrict to $\ell = 2$ it turns out that, in the case of a cubic crystal, a lot of the original and non-diagonal matrices simplify[27,28] and the expressions for $\chi^j(\underline{k})$ and thus $\rho_{ep}(\underline{k})$ assume a much nicer form. Specifically,

$$\rho_{ep}(\underline{k}) = 2\sum_j \theta(E_F - E_{\underline{k},j}) \sum_{\ell t} T_\ell^t(\underline{k}; E_{\underline{k},j})\, \rho_\ell^t(E_{\underline{k},j}) \tag{11}$$

where,

$$\rho_\ell^t(E) = \int_\Omega d^3r \left(\sum_u |Y_{\ell\mu}^t(\hat{r})|^2\, |R_\ell(r;E)|^2\, |R_+(r)|^2\right) \tag{12}$$

Now $T_\ell^t(\underline{k}; E)$ is determined by the structure and normalization alone and the wavefunction information is all in the $\rho_\ell^t(E)$ in the form of a modest number of diagonal matrix elements. We can get considerable interpretational value from studying these functions.

In Figs. 4-8 I illustrate our approach with some recent results for Niobium. Fig. 4 shows some sections through the band structure obtained by Kaiser[27] which agree well with the results of other calculations.[29–30] We are interested in three bands. We have a completely full band 1 which starts out s-like at Γ then hybridizes and becomes progressively more d-like in the outer parts of the zone. Bands 2 and 3 both cross the Fermi level creating various hole sections. With our new representation we can study, as a function of energy, the annihilation rate of the different s, p and d contributions making up these various bands.

In Fig. 5, I show, as a function of energy, a parameter, $\zeta_\ell^t(E)$, which is just the ratio $\rho_\ell^t(E)$ (Eq. 12) to the same function but with $R_+(r) = 1$; i.e. I'm factoring out everything except the effect of the positron-electron overlap. I have also included the energy range of interest which is from the bottom of band 1 up

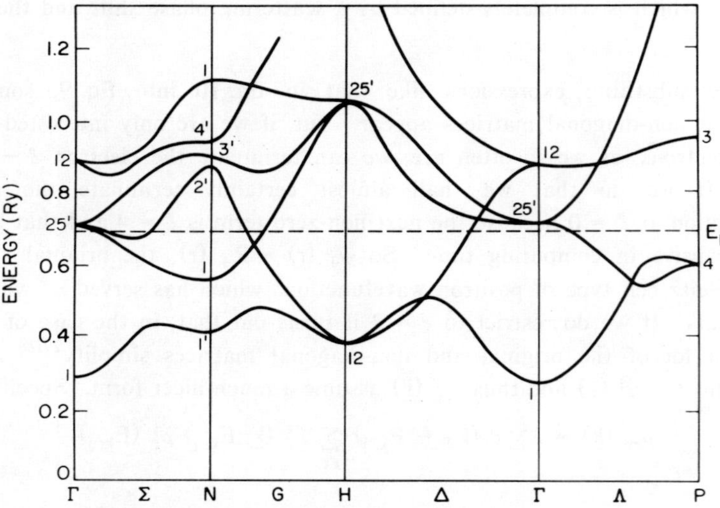

Fig. 4. The band structure of niobium as obtained by Kaiser.[27]

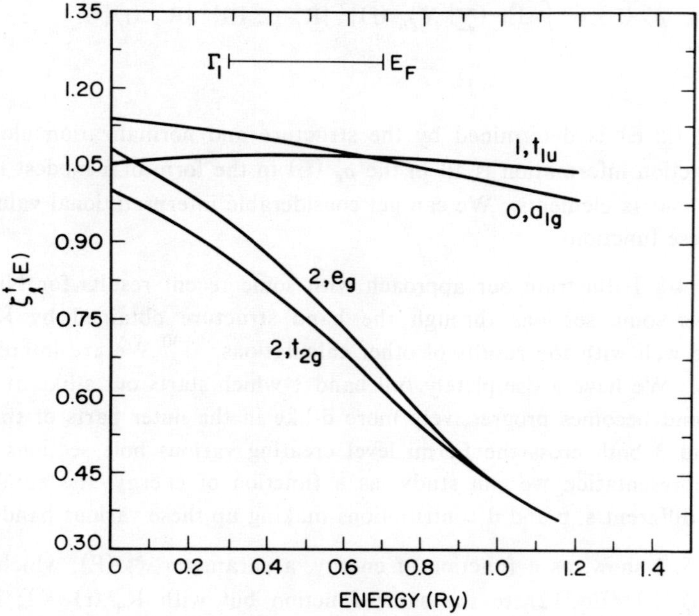

Fig. 5. The ratio ζ_ℓ^t (E) of the electron-positron to the electron only matrix elements for niobium ζ_ℓ^t (E) = ρ_ℓ^t (E)/ρ_ℓ^t (E; $\psi_+ = 1$) for the various ℓ, t as noted. The inset horizontal bar defines the energy range spanned by the occupied conduction electron states.

to the Fermi level. We can see, first of all, that for the $\ell = 0$ and $\ell = 1$ partial waves there is comparatively little E dependence. On the other hand, as soon as we go to $\ell = 2$ there is a large decrease in the annihilation rate and a stronger E dependence. The small difference between the two sets of d states of different symmetry can be explained quite easily if you just look at the lobes in the corresponding electronic charge density. In e_g these lobes are directed along the (100) interstitial directions where the positron density is greatest. t_{2g} is directed along (110) towards a region in which there is a saddle point and the positron density is smaller. Even so, there is not much difference between these two contributions. What is much more important is the big difference between $\ell = 0,1$ and $\ell = 2$. The strong E dependence of the d contributions also has a relatively straightforward explanation. As E increases the $\ell = 2$ electronic charge density, within the muffin tin, grows at the expense of that outside where the bulk of the positron density resides. This E dependence implies, through the E vs $\underline{k}$ dispersion curves, a related $\underline{k}$ dependence in the individual d band contributions to ρ_{ep} ($\underline{k}$). However, although this effect appears, the combination of hybridization and the large difference between the s, p, and d contributions is still the most significant factor.

The resulting $\underline{k}$–dependence of some individual band contributions is shown in Fig. 6. Here I show four different curves for each Niobium band along the $\Gamma - N$ direction. Curves (a) and (b) are ρ_e ($\underline{k}$) and ρ_{ep} ($\underline{k}$) respectively as obtained from our earlier described analysis. Curves (c) and (d) are the same quantities but as obtained from the corresponding momentum densities ρ_e ($\underline{p}$) and ρ_{ep} ($\underline{p}$) with an LCW sum truncated at $p_{max} = 3.712$ AU ($\underline{G} = 2, 2, 2$), which is about the momentum limit of most contemporary 2D machines. You will see that if you truncate the LCW sum you do get some losses and the losses are more profound in ρ_e ($\underline{k}$) than in ρ_{ep} ($\underline{k}$). This is natural since any ρ_{ep} ($\underline{p}$) will fall off more rapidly than ρ_e ($\underline{p}$) because of the positron's repulsion from the ion cores where the average momentum is greatest. However, it is interesting to note, at least in this example, that the loss in ρ_{ep} ($\underline{k}$) is almost $\underline{k}$–independent. If you truncate you just lose sensitivity to the Fermi surface breaks. What is much more important is the difference between ρ_{ep} ($\underline{k}$) and ρ_e ($\underline{k}$) in each case.

If we look at band 1 around Γ, where the band is s-like, the contribution is large. As $\underline{k}$ increases we move into the region of hybridization (Fig. 4), the band turns over becoming increasingly d-like and its contribution goes down. If we put together the data of Figs. 5 and 6 we can deduce directly the relative proportions of sp and d in the corresponding $\underline{k}$–states. If we look at the next band we see, as we pass from Γ to N (once we are past the hole that is!), that the annihilation

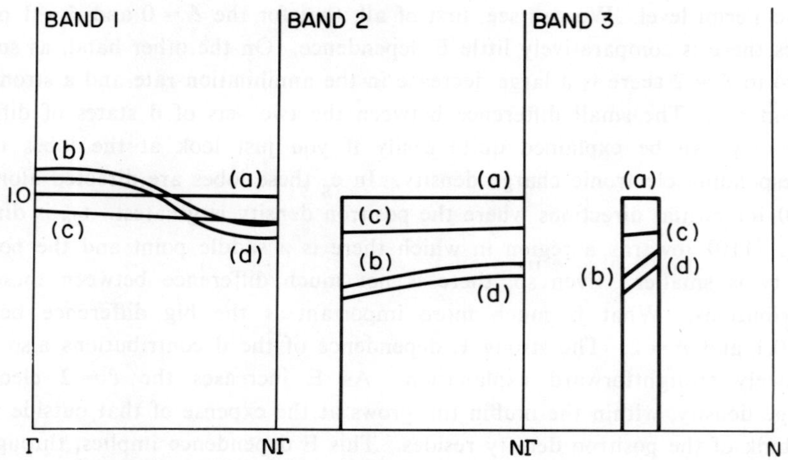

Fig. 6. The individual niobium conduction band contributions to the k−space densities along Γ − N. (a) ρ_e, (b) ρ_{ep}, both from Eq. 12; (c) ρ_e, (d) ρ_{ep} from a truncated LCW sum.

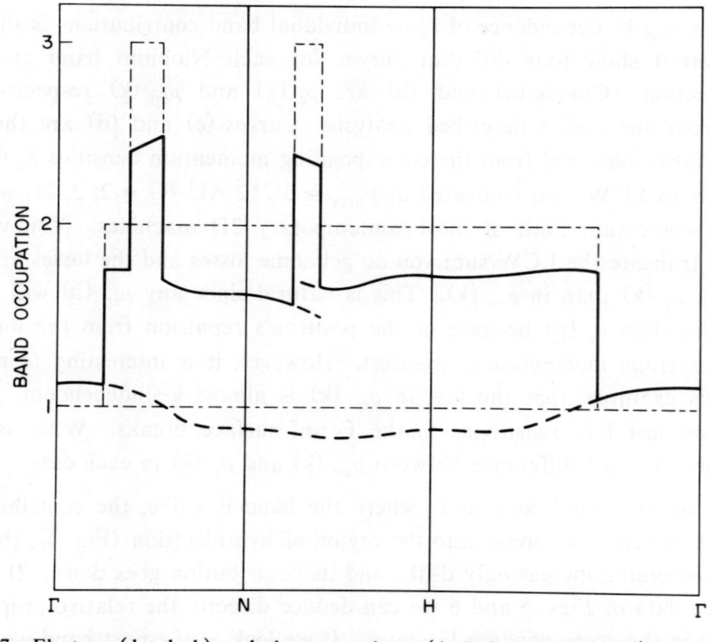

Fig. 7. Electron (dot-dash) and electron-positron (dashed and solid lines) k−space densities along some high symmetry directions in niobium.

rate is increasing. This is because, although this band is essentially d throughout, it is going in the other direction, i.e., $\partial E/\partial k$ is negative. Finally, in the third band, we see the effect, in the short occupied region, of a change from d to sp in the direction of k increasing. Although this is a very small sample I think you can already see, in the data of Figs. 4 through 6, that we have a useful framework for quickly and efficiently seeing the essential factors that are determining the matrix elements χ^j ($\underline{k}$).

If we now put all the individual band contributions of ρ_{ep} ($\underline{k}$) together and do the same for ρ_e ($\underline{k}$) we can look at the overall distortion in ρ_{ep} ($\underline{k}$) that arises from the positron wavefunction (Fig. 7). Were we to do the same comparison of an experimental ρ_{ep} ($\underline{k}$) with a theoretical ρ_e ($\underline{k}$) we would probably normalize each distribution to the same overall integrated area. We then might say, "OK, things look quite good except that we don't seem to have enough of a hole around Γ". We now know that the reason we don't have enough of a hole is because, around Γ, the s-like band 1 contribution is coming through loud and clear and everywhere else the d contributions are being suppressed.

So far I have been talking about 3D sections; now I would like to show some experimental results and theory for the 2D case. Figure 8 shows some curves for

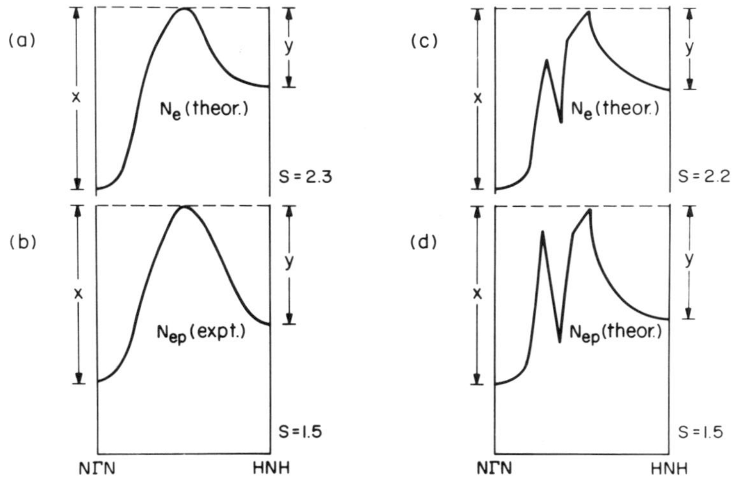

Fig. 8. Sections through the 2D electron (N_e) and electron-positron (N_{ep}) distributions for niobium along NΓN – HNH (see text for explanation). The theoretical curve in (a) has been convoluted with a 2D gaussian of FWHM 0.1 AU^{-1} to simulate the effect of experimental resolution.

Niobium corresponding to integration along (110). I denote the 2D distributions by $N_{e(p)}$ (k_y, k_z) or, in short, by $N_{e(p)}$. In Fig. 8 the original 3D plane of integration had $\Gamma - N$ as a central perpendicular (to the integration direction) line: the notation is as in Fig. 3 and in Bull et al.[16] In Figs. 5(a), (b) I show an experimental N_{ep} and the corresponding section of the theoretical N_e convoluted with a 2D gaussian of FWHM 0.1 AU to mimic the effect of the experimental resolution. You will see immediately that although these two curves are qualitatively similar there is more skewness in N_e. If you like, there is a bigger hole pocket at Γ in N_e. That, of course, is exactly what we were seeing a moment ago in Fig. 7 and we would expect to see a similar picture result from a further calculation of N_{ep}. However, I have to confess that so far we have not done enough calculations to complement fully the curves (a) and (b) (Fig. 8) which include the effects of convolution with a 2D resolution function. To do that you have to calculate not only the section of interest but also a lot of other neighboring sections. Thus far we have just done the principal (unconvoluted) section (d) which we compare with the corresponding N_e (c). Now it is interesting to see that the difference in skewness apparent in curves (a) and (b) is almost exactly reproduced in their "infinite resolution" theoretical equivalents, curves (c) and (d). It looks as if, in this 2D section for Niobium at least, that most of the difference between our folded experimental distribution and the N_e that we have been comparing it with can be understood in terms of independent particle effects. It seems, since they are the only thing now neglected, that many-particle correlation effects[31] in these $\underline{k}$–space distributions are almost $\underline{k}$–independent.

So far our results are only preliminary and most of our calculations have been performed in relatively brute-force fashion. However we can see ways of doing them very much more efficiently and rapidly, although still not so easily as those for ρ_e $(\underline{k})$. I would point out that one can calculate ρ_e $(\underline{k})$ without even a complete band structure calculation. You can simply search for the Fermi energy, find the corresponding $\underline{k}$'s and by that route produce a complete ρ_e $(\underline{k})$ most cheaply. If you want to calculate ρ_{ep} $(\underline{k})$ you must do a band structure calculation because you have to evaluate matrix elements. The important point here however is this: since those matrix elements involve densities rather than wavefunctions, which means that their $\underline{k}$–dependence can always be represented as a simple Fourier series (Eq. 8), all you need to do is a small number of $\underline{k}$ values within a symmetry element of the first Brillouin zone. You don't need large numbers; you can probably get away with 40 or 50 $\underline{k}$ values in a complete calculation of ρ_{ep} $(\underline{k})$ taking no more than 1.5 to 2 times the time required for the primary bandstructure calculation. Seen in those terms, both ρ_{ep} ($\underline{k}$) and its intermediate by-products such as the ρ^{t}_{ℓ} (E) (Eq. 12) could be of enormous value. They might also give us additional insight into many-particle correlation effects. Who knows, it may well be that in complex many band systems, a description of correlation enhancement effects may be more understandable if the associated enhancement factors[31] are viewed as a function of $\underline{k}$ rather than $\underline{p}$.

In conclusion let me remind you of what I have been arguing in general. It is that perhaps we could sometimes compromise a little on the information vs effort curve (Fig. 2). Instead of having a very tough time at the complex and interpretationally difficult level of ρ_{ep} ($\underline{p}$), or coming right down and making a qualitative or at best semi-quantitative comparison of folded experimental data with ρ_e ($\underline{k}$), we could more often focus on ρ_{ep} ($\underline{k}$). It is a function which on the one hand is fairly cheap to produce and which, on the other hand, still contains an enormous amount of information.

Acknowledgments

I would like to thank all those whose work and ideas have contributed to this paper and particularly Dr. J. H. Kaiser on whose Ph.D. thesis much of it is based. Both he and I would also want to acknowledge many valuable discussions with Dr. B. L. Gyorrfy, W. D. Temmermann, Z. Szotek and last, but by no means least, with Dr. N. Shiotani whose resurrection of Eq. 9 provided the starting point.

P. E. Mijnarends

You mentioned the modeling you've done with a truncated sum over the reciprocal lattice vectors. If you wanted to eliminate or minimize the truncation effect how large a range of momentum would you have to measure? Infinite?

R. West

We are seeing in these results, I think, between 5 and 10% losses at worse and that's going out to about 25 milliradians. The actual loss obviously depends upon the metal. There must be a scaling problem somewhere in there and certainly if we went to the 3d's rather than the 4d's I suspect we would need to go further. I suspect that with 25 milliradians we would be losing much more let's say in vanadium or copper. So I can't really answer the question quantitatively; we haven't done enough yet.

P. E. Mijnarends

I think the matter of convergence depends very much on the concentratedness of the wavefunction. In fact one can explain Fig. 4 just by thinking in terms of a radial wavefunction which is rather spread out at the bottom of the d-band and which becomes more concentrated as you go up in energy in the d-band.

R. West

But no, you also have the nodes moving and you have to look at the r-space wavefunction in all it's glory and watch the nodes moving about and the corresponding changes in the overlap of the positron wavefunction. I would argue that that is a little more complicated, and I would say that my matrix element, which just looks at the annihilation rate as a function of energy into a principle contribution or component of any electron wavefunction, is perhaps a little bit easier to read. But I agree with you about what's happening.

Question

What frightened me a little bit, in your method of calculation, concerns the radial wavefunctions which you obtain: do you obtain them up to the muffin tin radius or do you obtain them through a Wigner Seitz sphere?

R. West

Yes, at the end of the day you have to do a Wigner Seitz integral. You don't stop at the muffin tin, you do a complete Wigner Seitz cell integral. As you know, the codes for those calculations are available and I think those integrals can be done with relatively high precision.

R. Nieminen

Maybe I can make one, possibly two very brief comments. One is that I think that one of the big advantages of positron and high resolution Compton measurements is that they do contain the higher momentum components which of course as you have emphasized is a great consolation. The process of folding back which also you have emphasized in some sense is related to the Bloch wave. The second point which I think perhaps because of time you didn't bring out is that really at this point in time one can carry out very reliable calculations of electron positron momentum distributions in concentrated alloys, KKRCPA. I think that the next few years could see for the first time really serious comparisons of theory and experiment. Because concentration is an independent parameter and hopefully many of the other complicating things don't change rapidly, I think that should be an exciting sort of a possibility in the next few years.

R. West:

Can I just answer that one? I always find myself in danger of perhaps suggesting to some people that I am for throwing away the momentum density. I'm quite prepared to offer another talk on the uses of the momentum density, and without folding, on any occasion. But I think we have to be realistic, and I would argue that if we go to a situation where we have strong scattering transition metal alloys or transition metal — simple metal alloys, we are going to get into a difficult situation. A situation where there are going to be so many bubbles around the baby that unless we blow them away we're not even going to see it at all. There really is, I think, a problem here. You can get into a situation where there is just too much complexity; there is too much information and this is why, I think, if you have a alternative intermediate route, with some very clear concepts you should make full use of it. For example — just a last observation; if we had not got the concept of Fermi surface where would any of us be?

A. Bansil

A very interesting punch-line for the session.

REFERENCES

1. Berko S, Haghgooie M & Mader J J 1977 Phys. Lett. **63A** 335.

2. Berko S in Compton Scattering, e. B Williams (McGraw-Hill, 1977).

3. West R N 1973 Adv. Phys. **22** 263.

4. Mijnarends P E in Positrons in Solids ed P Hautojärvi (Springer 1979).

5. Berko S in Positron Solid-State Physics eds W Brandt & A Dupasquier (North-Holland 1983).

6. Faulkner J S 1982 Prog. in Mat. Sc. **27** 1-87.

7. Bergersen B & Pajanne E 1974 Appl. Phys. **4** 25.

8. Berko S & Plaskett J S 1958 Phys. Rev. **112** 1877.

9. Oberli L, Manuel A A, Sachot R, Descouts P, Peter M, Rabou L P L M, Mijnarends P E, Hyodo T & Stewart A T, 1985 Phys. Rev. **B30**, 1147.

10. Lock D G, Crisp V H C & West R N 1973 J. Phys. **F3** 561.

11. Lock D G & West R N 1975 Appl. Phys. **6** 249.

12. Beardsley G M, Berko S, Mader J J & Shulman M A 1975 Appl. Phys. **5** 375.

13. West R N in The Physics of Transition Metals ed Rhodes P (IOP 1981).

14. Farmer W S, Sinclair F, Berko S & Beardsley G M 1979 Sol. St. Commun. **31** 481.

15. Manuel A A, Samoilov S, Sachot R, Descouts P & Peter M, 1979 Sol. St. Commun. **31** 955.

16. Bull C R, Kaiser J H, Alam A, Shiotani N & West R N 1984 Phys. Rev. **B29** 6378.

17. Singh A K & Singru R M 1984 J. Phys. **F14** 1751.

18. Rabou L P L M & Mijnarends P E, 1984 Sol. St. Commun. **52** 933.

19. Majumdar C K, 1965 Phys. Rev. **140** A237.

20. Shiotani N 1984 Sol. St. Phys. (Japan) **19** 526.

21. Ziman J M 1969 Elements of Advanced Quantum Theory (Cambridge: CUP).

22. Korringa J 1947 Phys. **13** 392.

23. Kohn W & Rostoker N 1954 Phys. Rev. **94** 1111.

24. Ham F S & Segall B 1961 Phys. Rev. **124** 1786.

25. Faulkner J S, Davis H L & Joy W 1967 Phys. Rev. **161** 656.

26. Bulter W H, Olsen J J, Faulkner J S & Gyorffy B L 1976 Phys. Rev. **B14** 3823.

27. Kaiser J H, West R N & Shiotani N 1985 - submitted to J. Phys. F: Metal Phys.

28. Mattheiss L F 1970 Phys. Rev. **B1** 373.

29. Elyashar N & Koelling D D, 1976 Phys. Rev. **B13** 5362.

30. Wakoh S, Kubo Y & Yamashito J, 1974 J. Phys. Soc. Japan **38** 416.

31. Arponen J & Pajanne E, 1979 J. Phys. **F9** 2359, 1979 J. Phys. **C12** L161, 1979 Ann Phys. (N.Y.) **121** 343.

Low Energy Positron Measurements of Defect Profiles

K. G. Lynn

Brookhaven National Laboratory
Upton, L.I., New York, 11973

I want to thank the people who organized this symposium for asking us to present these results. It should be noted that this paper assumes that the reader is familar with slow positron research as this is not intended to provide a general review but an up-to-date presentation of recent results. After the positron moderation process had become efficient enough ($\sim 10^{-5}$), a few people like me jumped in to do material and surface science experiments. At present this fast to slow positron moderation is typically 10^{-3}.

One interesting outcome is the observation of the diffraction of positrons by the lattice.[1,2] I'll show you the Brookhaven version of positron and electron diffraction which I think is a nice picture. Recently Bill Frieze, Dave Gidley and myself constructed a brightness-enhanced electrostatically focused beam (BEEF),[3] an idea that Allen Mills had promoted a few years ago.[4] We had good indication, during some reemitted positron-energy loss measurements,[5] that this idea could be realized in the back reflection mode. Figure 1 shows multiple spot electron and positron diffraction pattern for W(110). The top one is for LEED and you can see the specular beam (00), the (01), the (10), and the $(1\bar{1})$. These patterns were taken on a 3" channel electron multiplier array (CEMA) with a resistive anode read-out and took about 200 seconds for a LEED beam and about 2000 seconds for a low energy positron beam. The beam that generated this picture has a phase space which is a little bit less than a millimeter-degree.[3] The electron gun and the positron gun phase space characteristics were designed to be

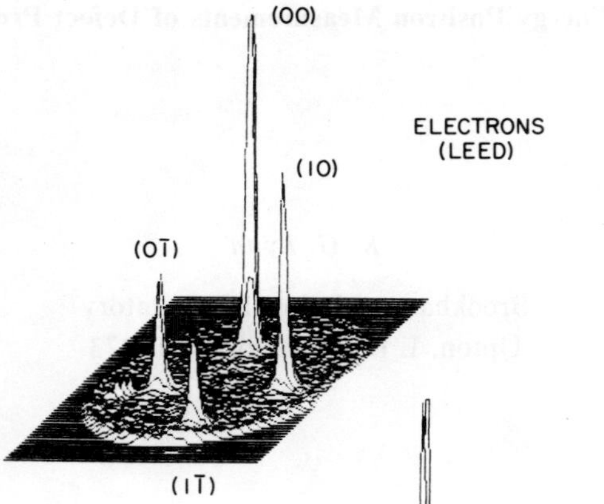

Fig. 1. Isometric plots of low-energy diffraction results for electrons and positrons on W(110). The specular and three diffracted spots are shown in each case. The plots are normalized to give equal heights of the specular spots; absolute scattering probabilities into the specular are 1% and 2% for electrons and positrons, respectively. Beam energy was 250 eV and beam orientation was along the [110] bulk direction at $\alpha = 10^\circ$ from the surface plane.

the same. We are now finishing a study where we have compared the detailed I-V characteristics for positrons and electrons on a Cu(100) sample. At present we find both results at large angles to be consistent with the accepted multi-layer relaxation on the Cu(100). It turns out, with brightness enhancement as pointed out by Mills, that you can potentially produce a better phase space characteristic positron beam than you can a LEED beam. This allows one to do some very nice experiments, e.g. examine critical exponents on the surface, owing to large transverse coherence lengths.

In a different area Peter Schultz, Bent Nielsen, and myself[6] re-examined some old Brookhaven data. I'm not convinced yet that the positron in a "perfect metal" is really understood all that well nor do I believe that the positron-phonon-electron coupling strengths are understood. There was a theory,[7] put forth approximately ten years ago in which the positron-acoustic phonon coupling was the dominant scattering mechanism. Of course you can put in impurity scattering and positron-electron interactions, but for the perfect, well annealed, high purity metal, the acoustic phonon-positron coupling is the dominant term. This theory predicted a $T^{-1/2}$ temperature dependence of the diffusion coefficient.

Now, as Risto Niemenen and the Finnish group have done better Monte-Carlo calculations,[8] and Allen Mills and Bob Wilson[9] at Bell Labs determined the positron implantation profiles, we can start to understand more about the diffusive characteristics of the positron as well as its temperature dependence. Figure 2 shows our data for the square log of the diffusion length versus the log of the temperature. The dotted lines in the high temperature region indicate where vacancies begin to trap positrons. The presence of vacancies will change the diffusion coefficient by a significant amount because they will remove positrons from the diffusion process. Thus, you can see vacancies easily with this technique.[13]

The lower temperature dots or lines indicate what we believe are shallow defects or impurities which are still present in well annealed crystals. Whether these shallow traps are actually present near the surface and do not remain in the bulk is not known. However, there is certainly a diffusion limited mobility temperature dependence at low temperatures in aluminum and in cadmium. The data shown in Fig. 2 represent the most negative slopes obtained after successive heat treatments. For example, the cadmium was annealed so well that we left, probably, a mil of Cd inside our vacuum chamber. The aluminum was also very well annealed, and the Mo and Nb were taken up to within 50°K of their melting temperatures. To get the temperature dependence of the diffusion coefficient we have fit the best line through the data. The logarithmic slope of the square of the diffusion length in Mo and Nb is about −0.8, $A\ell \sim -0.6$, Cd ~ -1, and for the other direction of Cd, it is something like −2.5. I don't believe these are absolute numbers, however I do believe that we will have to rework the theory on positron diffusion coefficients which might provide some interesting new information about positron scattering inside a bulk crystal.

All these data have been taken with the positronium yield method where we measure the back-diffusion of positrons to the surface as a function of positron incident energy. The main requirement of this method is that the surface conditions do not change during one set of energy measurements. We verified that the Ps fraction did not change during the measurement. In addition we have

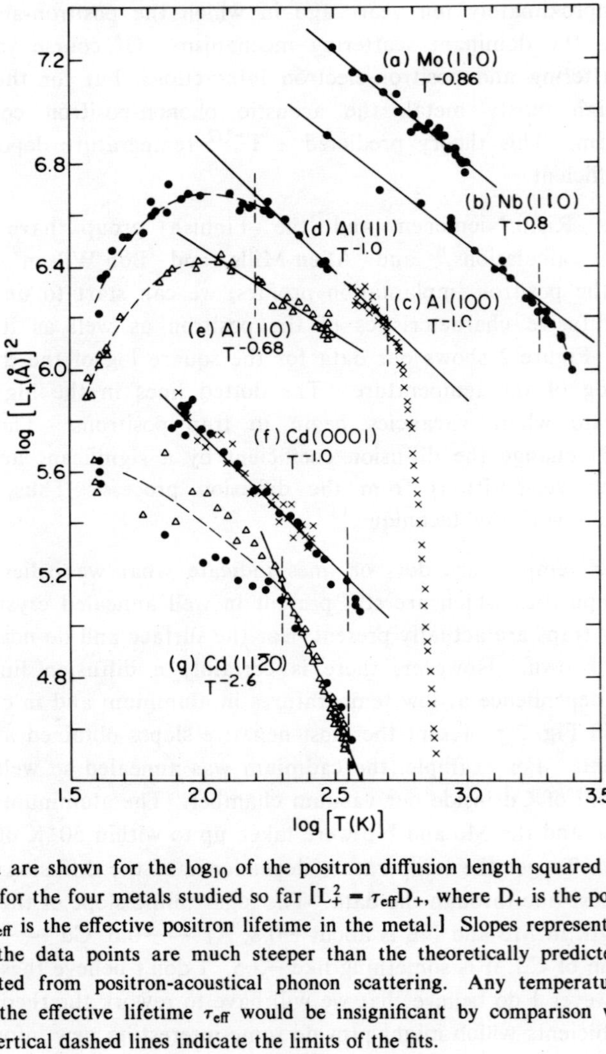

Fig. 2. Data are shown for the $\log_{10}$ of the positron diffusion length squared L_+^2, vs $\log_{10}$ of the temperature for the four metals studied so far [$L_+^2 = \tau_{eff} D_+$, where D_+ is the positron diffusion coefficient and τ_{eff} is the effective positron lifetime in the metal.] Slopes represented by the lines drawn through the data points are much steeper than the theoretically predicted $T^{-1/2}$ which would be expected from positron-acoustical phonon scattering. Any temperature dependence associated with the effective lifetime τ_{eff} would be insignificant by comparison with the slopes observed. The vertical dashed lines indicate the limits of the fits.

checked the systematics of the measurement; positron implantation profiles, systematics in determining the positronium fraction, scattering of gamma rays, and detector characteristics have been discussed in reference 10.

P. Platzman

You may be interpreting the data wrong. The diffusion constant is a bulk property, so what does this bulk property have to do with the surface orientation? When you measure an anisotropy for Cd, aren't you measuring some reflection coefficient dependence or some other thing?

Lynn

The diffusion coefficient is a bulk property, but the positron wavefunction is very anisotropic in Cd. Therefore it is not too surprising that there would be a difference in the positron diffusion characteristics. In particular the positron wavefunction is much different in the $\bar{c}$ direction than the $\bar{a}$ direction in Cd. Figure 2 shows a bulk crystallographic direction not in the plane of the sample. The potential will be squeezing, forcing a positron wavefunction to be anisotropic, and certainly Cd exhibits a significant anisotropy in the thermal expansion.

A. Stewart

At what temperature does this low temperature fall-out occur?

Lynn

Well, that depends slightly on how well you anneal your crystal. For aluminum it's like 100-125°K. In cadmium it would change a little bit, depending on the sample history. So it's very critical how well you anneal your crystal. Another potential problem is that oxygen may diffuse into an aluminum crystal, which could cause shallow traps. Sputtering the sample could also cause some problem. However, our work and that of the Finnish group show that you can anneal out the damage at 300°C. You may leave a residual number of argon bubbles inside the aluminum that provide shallow traps (i.e. low positron binding energy) at low temperatures.

Question

And the temperature dependence of positronium coming off the surface, how do you know that?

Lynn

We measure the quantity f_0 (the Ps fraction at $E \approx 0$) in every experiment. That's the branching ratio at the surface and it does change as a function of temperature. However, as long as our temperature does not vary during the f vs. E measurement it does not change the curvature of the f or S vs. energy.

Figure 3 shows a plot of the parameters determined in ref. 8 where we plot the positron stopping profile and projected range of positrons for three different energies in $A\ell$. In fact, a very nice property of positrons which I think

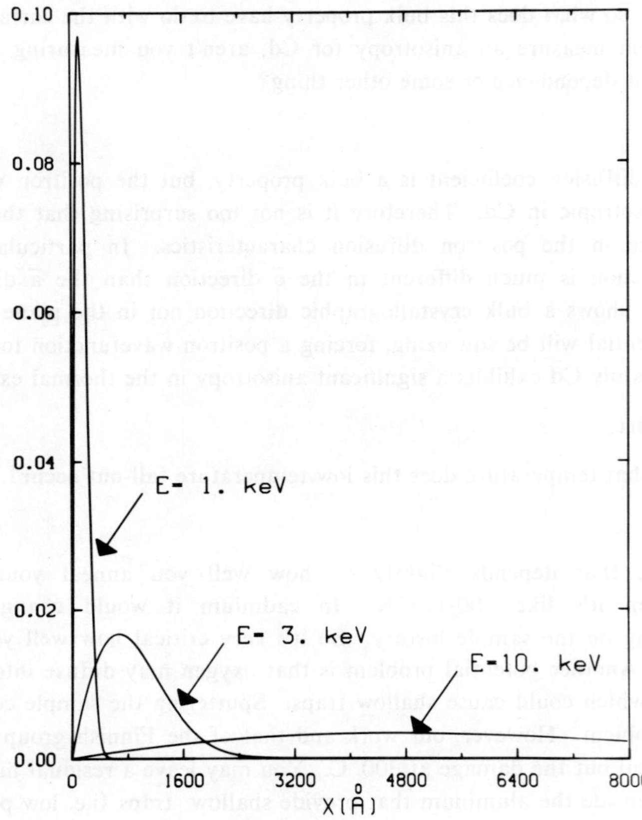

Fig. 3. Expected positron implantation profiles (ref. 8) for Aℓ at three different incident positron energies (1, 3, 10 keV).

Allen Mills and I initially missed, is that the profiles are not exponential. They're kind of a derivative of a gaussian and that turns out to provide you with a greater capability to perform defect profiling measurements. This includes interfaces, profiling ion beam damage, ion beam mixing; whatever you want to do, you will get some defect profiling information.

These profiles are from a theoretical calculation,[8] or rather, it's the theorists trying to do experiments. It's a Monte Carlo technique and one must deal with statistical problems correctly. Certainly the parameterization of the data has some statistical errors which have not been treated properly.

If you have lattice vacancies or disorder you will get positron trapping in such open volume regions. You can start to get new information by measuring the number of particles that diffuse back to the surface, or the momentum or lifetime distribution of the trapped particles. To probe these defects you need to inject positrons with some implantation profile. In addition, the disorder in the lattice can be characterized by a defect profile, which, in our case, is typically peaked a few thousand angstroms in from the surface. With sufficient implantation energy, positrons pass through the defect layer and you find a significant amount of defect trapping. This has a great many interesting applications, especially in providing new information on the defect profile near the surface or at interfaces.

In one case nickel samples were implanted with high energy He atoms (30 Kev, 90 KeV and 180 KeV). We then used positrons to probe the damage and He^4 bubble profile as a function of energy via a Doppler broadening "S" parameter.[11] This measurement is simply made by using a high resolution solid state detector (Ge in this case) to detect the energy spread of the annihilation photons emitted from the sample. We have also performed a parallel set of

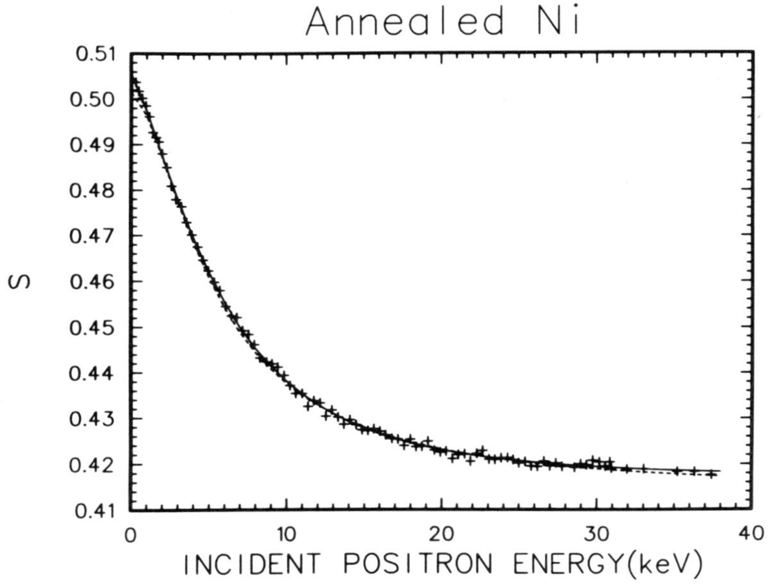

Fig. 4. Data of S-parameter as a function of incident positron energy measured for a unirradiated Ni sample annealed at 1100°C in situ. The solid curve is a fit to the data with a Makovian representation of positron implantation profile and the dotted line is a fit using an exponential stopping profile.

positronium fraction measurements which provides positron diffusion length information.

The S parameter versus incident positron energy for the unimplanted Ni sample, well-annealed at 1100°C, is shown in Fig. 4. The dashed curve shows the fitted result for an exponential positron stopping profile and the dotted lines represent the derivative of the Gaussian positron profile, which in other experiments provides a better representation of the data.[12] In this case the exponential profile results in an equally good fit (dashed curve). A comparison between the two fits suggests the reasonableness of using our slightly modified profile (m = 2.0). One can observe that the positron diffusion length is smaller for the exponential fit, as has also been observed in Si.[12] This is associated with the strong coupling of the fitting parameters.[13] The values of the diffusion length determined from this S-parameter measurement are in general agreement with previous positronium (Ps) fraction versus incident positron energy measurements for Ni and Cu.[14] It should be noted that one expects some difference between these two measurements which depends on the annihilation characteristics of Ps and e^+ in materials. Either S_s (surface S-parameter) or S_b (bulk S-parameter) has the same fitted value for both implantation profiles, which is expected

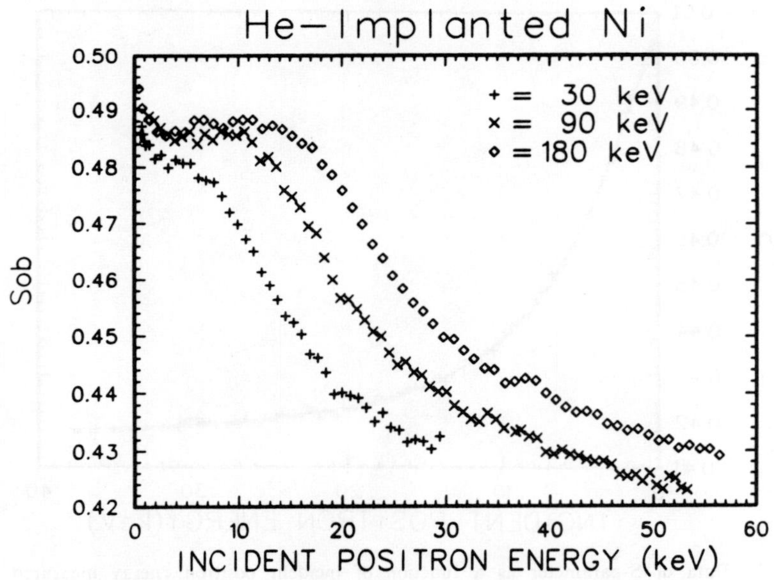

Fig. 5. Observed S-parameter versus the incident positron energy for He-implanted Ni samples with different He-implantation energies as indicated in the Figure.

because they only depend on the material and not the positron profiles. In either case, S_s contains ≅17% more in the central region in the annihilation gamma spectrum compared with S_b (i.e. a narrower energy spectrum).

Figure 5 displays the measured S-parameter versus incident positron energy for three He-implanted Ni samples which have been studied. The scatter in the data points is associated with statistical fluctuations. The effects upon the observed S-parameters, S_{ob}, due to the introduction of open volume defects via He-implantation are clear when compared to the reference sample (Fig. 4). These changes arise from the fact that the annihilation at a defect site produces greater S-parameters (narrower lineshape) than the bulk as previously stated, and that as the incident positron energy increases, more and more implanted positrons are removed from F_b (fraction that annihilate in the bulk), and F_s (fraction that annihilate at the surface) and contribute to the fraction trapped in the defects F_d. As this energy increases further, positrons are implanted beyond the defect region. F_d decreases and eventually, at sufficient energy, most of them decay in the bulk region F_b as in the case of a defect free sample.

To see the consequences of these created defects in a more direct way, ΔS ($S_{measured}-S_{annealed}$) vs. incident positron energy, are shown in Fig. 6, for the

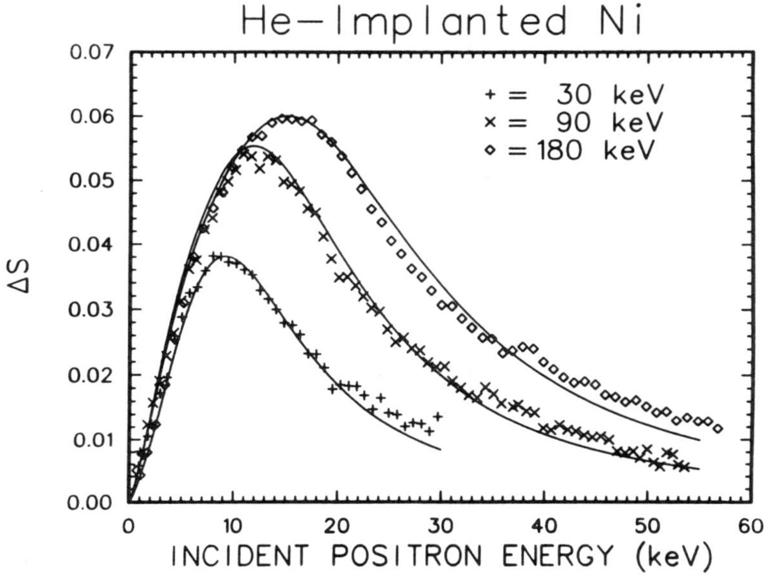

Fig. 6. Difference data obtained by subtracting the fitted curve of the reference data (Fig. 4) from the S_{ob} (Fig. 5). The solid curves are the results of fitting.

three irradiations. The adjustment, $\Delta S \equiv 0$ at $E = 0$, has been constrained in these data. As one would expect, the peak position shifts to increasing positron energy at the higher He implantation energy. If one defines the change as the ratio of ΔS data to the bulk S-parameter S_b, the maximum changes in the S-parameter in these samples were approximately 9.1%, 13.2%, and 14.4% for the 30, 90, and 180 keV He implantation energy, respectively. The maximum change in the S-parameter measured for single vacancies in Ni is approximately 6%.[21] This is compelling evidence that the trapping sites under study are void-like owing to this large percentage change measured in the S-parameter for these samples compared to vacancy defects.

Figure 7 is displayed in an attempt to provide a physical picture of the positron, ^4He, and damage profiles on the same depth scale. Fig. 7a shows the positron implantation profile at the incident positron energy where the peak maximum occurs in the ΔS curve shown in Fig. 6. The He profile results are obtained from a Monte-Carlo code TRIM[15] program for 1000 projectiles and are shown in Fig. 7b. The average depth of the ^4He implanted 10° off axis at 30, 90, and 180 keV is calculated to be ≈1100Å, ≈2980Å, and ≈5135Å, respectively. The damage energy deposited by the ^4He, namely the damage profile calculated from the same program is shown in Fig. 7c. It should be noted that the ^4He stopping profile does not directly represent the defect profile introduced when the ^4He is losing energy, nor does the damage profile itself. The defect profile is the combined result of these two with the damage profile as the dominant factor. If the types of defects are essentially the same throughout the region, it is then reasonable to think that the ΔS peak should occur at the incident positron energy at which the positron implantation profile has maximum overlap with the defect profile. On the other hand, if the large volume defects are likely to be created in the maximum deposited damage energy region, the ΔS peak would appear at the energy such that the associated positron implantation profile has the same peak position in depth as the damage profile. By comparing Fig. 6 and 7, one notes that neither case is in good agreement. One possibility is that the trapping does not occur instantly after thermalization, therefore diffusion still plays a role. In the peak region of the defect profile, the trapping probability is much higher (i.e. in this case one would expect some pre-thermalized positron trapping), resulting in a steep slope in the positron density distribution, and the net flux would diffuse toward the central region of the defect profiles. The sample surface can also be regarded as having a high trapping rate. Therefore, for the 30 keV He implanted Ni sample, the mean damage depth is approximately within a diffusion length, thus the positrons have to be implanted slightly deeper so that trapping by defects

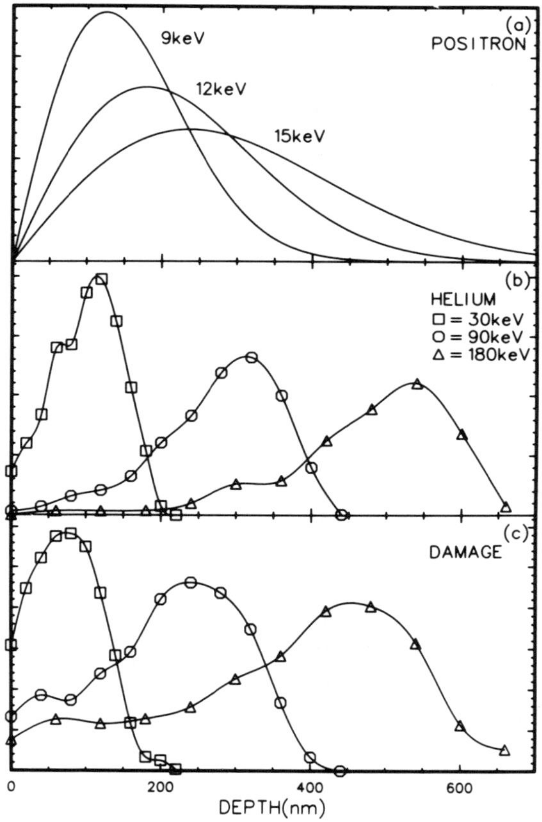

Fig. 7. Positron (a) and helium (b) implantation profiles and the associated damage profile are plotted with the same horizontal scale for comparison. The vertical scales are in arbitrary units. The positron profiles are generated from a Makovian profile (8) with m = 1.9, n =1.62, and E_0 = 1.6keV. The selected energies are corresponding to the peak positions of the ΔS curves in Fig. 7 (b) and (c) are results from a TRIM calculation (15) with different He-implantation energies shown in the legend. The solid curves joining the symbols are the spline-interpolation.

can dominate surface trapping. However for the other two samples with higher He implantation energies, the central region of the defect profiles are much deeper than one diffusion length, as are the positron mean implantation depths at 12 keV and 15 keV. Therefore, surface trapping is no longer as important, trapping by defects automatically dominates, and the positron flux will diffuse deep inside the metal until they are trapped. For still higher positron implantation energy, the positron profile broadens rapidly and the overall trapping is reduced.

A theory has been fit to the data and the resulting fits are shown as solid curves. The details at this theory and results will be published later,[16] however we have assumed no pre-thermalized trapping into these defects. The agreement between the model and data is quite good as can be seen in the figure. The most successful aspect of this simple model is that the assumed defect profiles terminated at the depths where the calculated Helium or damage profiles are well attenuated. The starting point of square defect distributions are, however, not as consistent, especially for the result of the 30 keV Helium implanted sample. The agreement between the model and data is not unique because the fitting parameters are strongly correlated with one another. Most of the positrons become localized at defects. The situation is further complicated by knowing that there are a variety of defects present in the sample, all of which have different trapping rates and S parameters.

In an attempt to clarify the situation, isochronal anneals were performed to find out if specific defects would anneal out at various temperatures. We found that little change had occurred after annealing the sample to 500°C. Significant increases in ΔS were found by further annealing as can be seen by the 700°C data. Annealing beyond this temperature showed a subsequent decrease in ΔS and is consistent with the thermal He release data of Jäger and Roth.[17] The thermal He release showed a large change at 1100–1300°K with essentially all of the He released at $T_a \approx 1400K$. We interpret the increase in ΔS peak as a coalescence of the He bubbles where the larger defects grow at the expense of smaller ones into faceted bubbles. This lowers the He density and thereby the electron density.

The dose used for this work was 10^{17} He^4 atoms per cm^2. It was an ill-chosen number because that's a high dose for positron studies and we were in saturation. We've now installed a small helium ion source in our beam that we can raster with a known ion current and watch when the implanted atoms are thermally released. We still have sensitivity down to 10^{12} helium atoms or argon atoms per cm^2 at energies of 3 keV on a surface of Mo, where the type of defects are relatively well known. In this Ni case we have multi-vacancies, platelets, and He bubbles and it is a very complicated microstructure. Now we are reducing our in situ ion currents and have a sensitivity approaching 10^{11} He^4/cm^2.

With Tom Van Veen, who was visiting from Delft University, we started doing some low dose measurements. We took an annealed sample and measured the Ps fraction (the log of the number of positrons that diffuse to the surface and form Ps) as a function of energy. Figure 8 shows the annealed curve, and three curves taken with different doses of 3 keV He ions. It is still a high dose region,

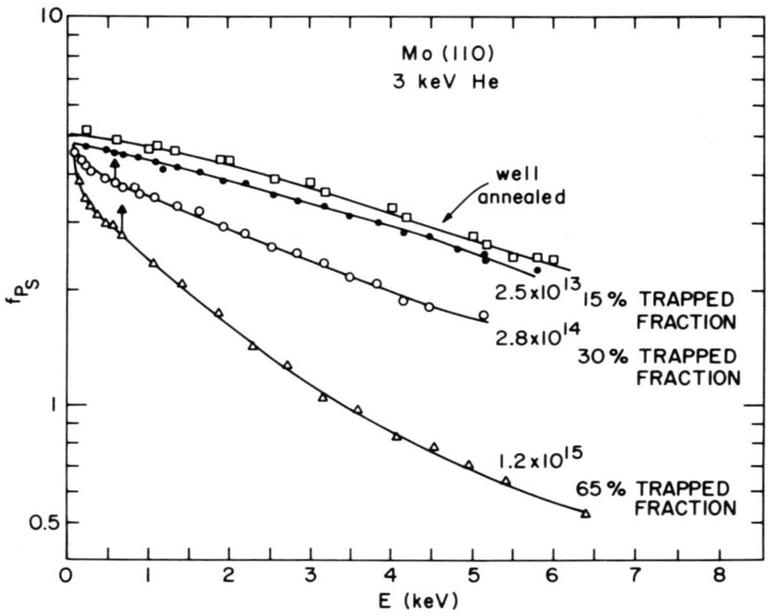

Fig. 8. Fraction of the positrons that return to a Mo(110) surface and form Ps plotted versus incident positron energy. The individual curves are for different 3keV He^+ fluences and corresponding different fractions of trapped positrons, at those defects generated by the He implantation.

but it's getting to the point where it's interesting for those people who want to know about helium interactions in metals.

Now you can start to determine the concentration of defects in the sample, or the fraction of positrons that are actually trapped. In Figure 8, we have plotted the Ps fraction on a log scale vs. energy. The way we can determine the fraction of positrons that are actually trapped is to implant the positrons beyond the damaged region. The Finnish group has done something very similar using Ar^+ as the implanted species. With annealing studies we couple the positron measurements with the helium desorption measurements. For example, if we damage the sample with helium ions, we generate and fill the defects with helium. Then you can determine from the emitted helium atoms, using a very accurate quadrupole mass spectrometer, the defects that are annealing out of the crystal. Using the annealing studies coupled with both helium desorption spectroscopy and positron S-parameter and f measurements, one can determine the type and concentration of the defects that reside in the crystal. I think it's a

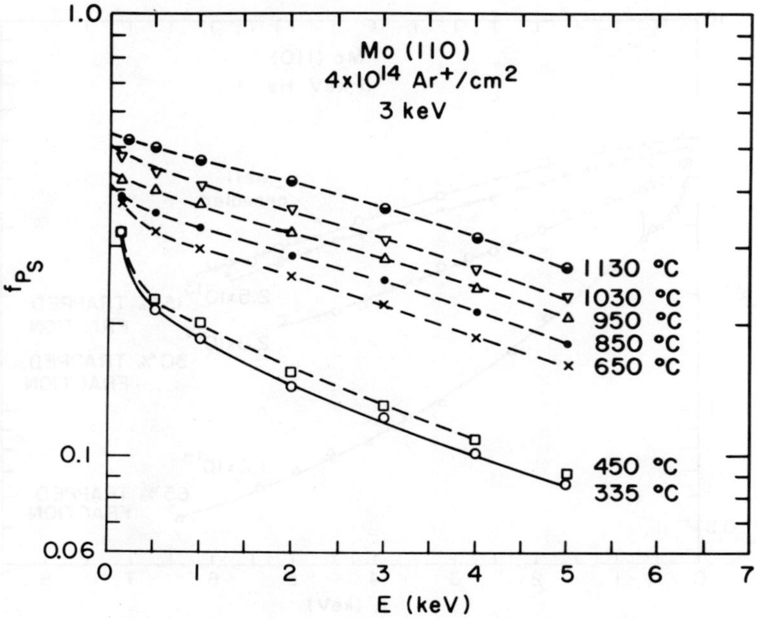

Fig. 9. Logarithmic plot of the returning positrons that form Ps while escaping a Mo(110) surface versus incident positron energy. The various curves shows the effect of ischronal annealing (twenty minutes at temperature) after irradiation with 3keV Ar^+ with a total fluence of $4 \times 10^{14} Ar^+/cm^2$.

very nice coupling of probes to look at radiation damage for ions in metals. This approach can be made to be a very broad based program in ion interactions with materials.[18,19]

Figure 9 shows the annealing characteristics of a Mo(110) crystal implanted with 3 keV Ar^+ ions to a dose of 4×10^{14} Ar^+/cm^2. These were isochronal anneals where the sample was held at temperature for 20 minutes. The Ps fraction was measured versus the incident energy at room temperature after the anneals were performed. All irradiations were carried out at room temperature under ultra-high vacuum conditions.

The irradiation time, and sample temperature are important and there are debatable parameters in the theory on how these bubbles form, so annealing times are critical. Certainly to carry out these irradiations at different sample temperatures would be very useful in resolving these theories.

So, in conclusion we have reviewed new studies on both the diffusion characteristics and trapping of positrons at defects created by ions in metals. I think there's a bright future in using these beams as a material science tool to probe interfaces, for example $Si - SiO_2$, but I think somebody has to put a significant effort into these studies to resolve many unanswered questions. I expect that this method will prove to be useful in interfacial science studies.

A. Stewart

I thought that the metallurgists had a nice problem of the wearing qualities of steel after it had been nitrogen bombarded. The problem is that the durability or the wearability of the surface exists long after the implantation depth has been worn off. Have you ever filed away the surface and seen if your implant goes further in than the implanted ion?

Lynn

The answer is no, I haven't done it, but I think it's a nice experiment. I really hope that somebody takes the effort and the time to try develop these variable energy techniques into a useful material science technique that will have impact on interfacial science. I think the information is here in these measurements, however it will no doubt be difficult to extract this information unambiguosly. This study of helium in metals took an investment of something on the order of 2 to 3 months, so I think that you have to have a strong interest and want to continue along these lines.

Question

I just wanted to comment that if you did go to low doses, 10^{12} per cm^2 or something like that, you could compare it with very detailed defect structures that are known from other techniques and you can compare your annealing curves. Those two annealing curves that you showed look like they were composite things.

Lynn

Yes and we have compared them to helium desorption studies.

Question

If you look at those, you have very simple defect structures, di-vacancies. There is some very high concentration tetrahedral tri-vacancies for example in nickel that you should see. When they anneal them, they occur in a very characteristic stage and they decay in a very characteristic stage, and so one should look for those signatures.

Lynn

That's why we're using the helium desorption spectroscopy, HDS. We use the helium to create, as well as occupy, the defect. For example if you have a di-vacancy, you have in most cases two helium atoms in an open volume region. These He atoms come out at a certain temperature allowing you to determine the di-vacancy fraction, which anneal out at a different temperature than mono-vacancies. The HDS is a complementary probe. Positrons give you defect-specific information, and the HDS tells you the concentration of each fraction.

REFERENCES

1. I. J. Rosenberg, A. H. Weiss and K. F. Canter, Phys. Rev. Lett. *44*, 1139 (1980); A. H. Weiss, I. J. Rosenberg, K. F. Canter, C. B. Duke and A. Paton, Phys. Rev. B. *27*, 867 (1983).
2. A. P. Mills, Jr., and P. M. Platzman, Solid-State Commun. *35*, 321 (1980).
3. W. E. Frieze, D. W. Gidley and K. G. Lynn, Phys. Rev. B, *31*, 5628 (1985).
4. A. P. Mills, Jr., Appl. Phys. *23*, 189 (1980).
5. D. Fischer, K. G. Lynn, and W. Frieze, Phys. Rev. Lett. *50*, 1199 (1980).
6. P. J. Schultz, K. G. Lynn and B. Nielsen, Phys. Rev. B *31*, (1985).
7. B. Bergersen, E. Pajanne, P. Kubica, M. J. Stott and C. H. Hodges, Solid State Commun. *15*, 1377 (1974).
8. S. Valkealahti and R. M. Nieminen, Appl. Phys. *A32*, 95 (1983); S. Valkealahti and R. M. Nieminie, Appl. Phys. *A35*, 51 (1984).
9. Allen P. Mills, Jr. and Robert J. Wilson, Phys. Rev. A *26*, 490 (1982).
10. P. J. Schultz, K. G. Lynn and H. H. Jorch in Proceedings of Workshop on Slow Positrons in Surface Science, Helsinski University of Technology, Finland (1984) - unpublished.
11. Positrons in Solids, Topics in Current Physics 12 Ed. P. Hautajarvi 12 Berlin Heidelberg Spruiger-Verlag 1979; W. Triftshäuser and G. Kögel, Phys. Rev. Lett. *48*, 1741 (1982).
12. Bent Nielsen, K. G. Lynn, A. Vehanen and Peter J. Schultz, Phys. Rev. B (in press).

13. K. G. Lynn, Positron Solid State Physics, School of Physics "Enrico Fermi" (North-Holland, Amsterdam 1983).

14. K. G. Lynn and D. O. Welch, Phys. Rev. *22*, 99 (1980).

15. J. P. Biersack and L. G. Haggmark, Nucl. Instr. and Meth. 174, 257 (1980).

16. K. G. Lynn, D. M. Chen, B. Nielsen, R. Pareja and S. Myers, unpublished.

17. S. Jager and J. Roth, Journal of Nucl Mat. 93/94, 756 (1980); Nucl. Instr. and Meths. 182/183, 975 (1981).

18. B. Nielsen, A. VanVeen and K. G. Lynn, to be published in Seventh International Conference on Positron Annihilation, World Scientific Publishing Co.

Surface Positrons and the Many Positron, Many Electron System

P. M. Platzman

AT&T Bell Laboratories
Murray Hill, New Jersey 07974

It would be preferable to have lunch but I'll go ahead and try to make it short. Coming up here this morning, (everything is very clear at 6:00 a.m.), I had the opportunity to think a little bit about my connections with Steve Berko over the years and I decided that it's been a very long time that I've known Steve Berko.

I got started in positron physics as an undergraduate working with Martin Deutsch. I first came to visit Brandeis at about the same time that Steve came here. Ultimately, I didn't join Brandeis, instead I went to Bell Labs and did other things. However, over the years Steve and I seem to meet at almost every meeting. He would always tell me that positrons were good for doing physics and I kept saying x-rays were better. We have had these discussions many times. Then he really did it to me. In 1975 he sent me Allen Mills, presumably to work with me, but now I'm not so sure, which brings me to the subject of my talk.

When Allen came and said would you like to give a talk at Steve's "festschrift", or whatever you choose to call a 60th birthday gathering, I said fine, I'll just do something easy and talk about something old. That was, by the way, Positrons at Surfaces, which is really very old, at least 6 months old and most of you have heard about it. Allen said, 'No no, you shouldn't talk about that. We want you to talk about something really interesting like many electron, many positron systems". I said, "well, I don't know anything about that," and he said, "oh that's all right, you should talk about it anyway". So here I am and of course I have "decided" that I will talk on Many Electron, Many Positron Systems.

I want you all to know, before I begin, that for the first time in my career, thanks to Allen Mills, I'm talking about a subject which I know absolutely nothing about. Steve Berko told me, when I talked to him on the phone recently, that he was going to get everybody to sign a disclaimer which said that anything that was said here, he could work on. He wanted to make it clear that nobody was going to keep him from working on anything. In a real sense I would like to go one step further. Please Steve Berko, or anyone else, work on this subject, because really very little is known about it, and I'm not going to do it all.

Since Allen Mills got me to give the talk, I got him to make the first figure. (Fig. 1) The fact is that what we're talking about here is pretty far-out. We are going to discuss what happens when you start to get positrons and electrons at sufficiently high density that you can start to think of them almost like you think of electrons in a metal. We will ask what kind of properties that system has.

Allen has told me how I'm going to get such a system. We're going to have an accelerator at Bell Labs and it's going to produce a big burst of fairly relativistic (20 MeV or so) electrons. These electrons are going to slam into a target which is, for example, tungsten. From the tungsten moderator is going to come some number of slow positrons. I'm told that about 10^7 slow positrons will

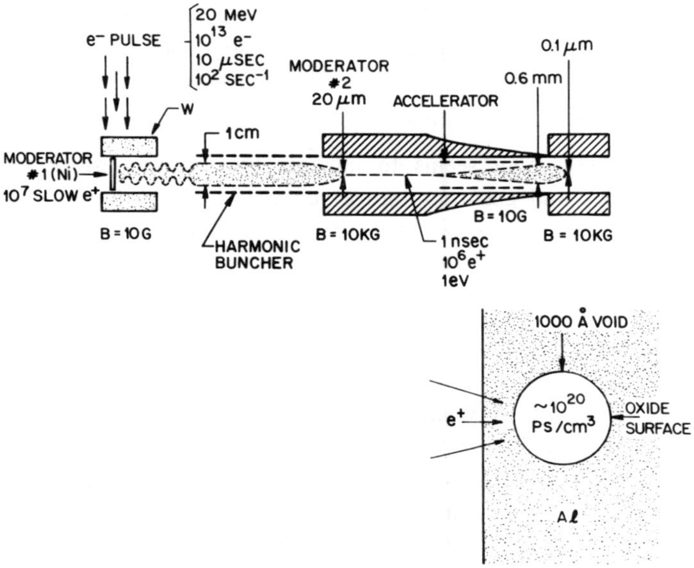

Fig. 1. Apparatus for production of positronium at high density.

be produced, assuming a conversion efficiency of about one part in 10^6, for each pulse of 20 MeV electrons.

As an aside let me tell you a story which you in particular should appreciate. This beautiful free-electron laser is being built at some cost approaching 1 million dollars. However, because they (Bell Labs management) didn't really spend enough money on it, it may be that in the long-run it will only function as a positron source. So at least we'll do positron physics with the waste product of the free-electron laser.

These 10^7 slow positrons can now be focused. All of the junk between the tungsten target and the sample in Fig. 1 is meant to indicate that Allen Mills, Kelvin Lynn etc. can focus positrons by using various kinds of lenses, meaning magnetic fields which squeeze the positron beam. Avoiding all the details, what it amounts to is that you start out with 10^7 slow positrons in a 1 cm^2 spot and wind up at a sample with a 1000Å diameter beam of 10^6 1 keV positrons.

This focused beam can then be stopped into some kind of a solid target in which you're going to form your high density $e^+ - e^-$ plasma. That target is also shown in Fig. 1. It will be some type of porous medium i.e. one with lots of 1000Å holes in it. It could be one of the beautiful bubbles that Kelvin had, or one that we can perhaps make by lithography. Actually it's not clear that you absolutely need the cavity because when these 1 keV positrons stop, they stop fairly close to the surface. Most of them come out, typically as positronium. If they all come out in some very short period of time there's going to be a high density very near the surface on a subnanosecond time scale.

This type of reasoning seems to indicate that it may be possible to put $10^6 - 10^7$ electrons and positrons into a volume which is of the order of 10^{-15} cubic centimeters i.e. a density $n \leqslant 10^{22}$ cm^{-3} of positrons and their associated electrons. The question I want to ask in the rest of the talk after a slight digression is: is this 3-D system interesting?

There is a possibility that we can form a dense 2-D Ps liquid near a surface. The point is that Ps near a surface, at least in some pictures, is typically confined to the surface by a Van der Waals potential, and a repulsive correlation potential with the electrons.[1] (This was to have been my entire talk.) In Fig. 2 we show the surface of the metal and schematically a physisorbed Ps atom. Since it is known that positronium is bound to a metal surface by something of the order of a few tenths of an electron volt, the uncertainty principle would imply that the distance scale in Fig. 2 is about 4Å. The density in the metal tends to fall off on

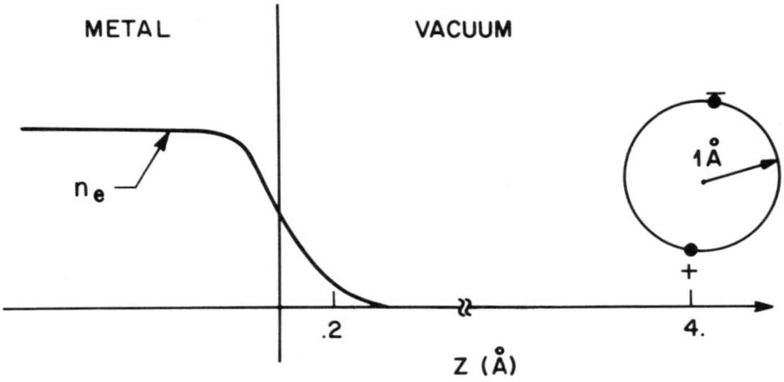

Fig. 2. Distance scale for physisorbed positronium.

a distance scale of .2Å, or so. Thus, the Ps is essentially going to float out in the vacuum bound to the surface by a weak Van der Waals potential. They will remain in this 2-dimensional bound state if the surface density (n_{2D}) is such that the kinetic energy along the surface, their 2-D Fermi energy,

$$\epsilon_F = \hbar^2 \pi \, n_{2D}/m \tag{1}$$

is less than their binding energy. In this case they will behave like a 2-D liquid. Thus if I go to $n_{2D} \leqslant 10^{14}$ per square centimeter I should be able to consider a 2-dimensional positronium liquid, or whatever you want to call it, trapped at a surface. Now this kind of density is fairly easy to get, and such experiments might be possible. I'm not going to tell you too much about it because the theory of 2-D liquids is even vaguer than that of 3-D liquids.

Let's go back and stay for the rest of this talk in 3-D. Suppose we had such a degenerate 3-D liquid of e^+ and e^- and I wanted to understand its properties theoretically. It turns out that a great deal of work, (and this is where I stole everything for the talk) has been done on the problem of electron-hole liquids in things like silicon or germanium.[2] The electron-hole liquid is analogous to the $e^+ - e^-$ liquid. Theoretical and experimental solid state physicists have spent a lot of time trying to sort out such problems because they appear to exhibit a very interesting array of phase transitions.

So why electron-positron when you can work with electron-holes in semiconductors? Well here, for once, we have a big advantage. In a semiconductor, when one looks at electron and holes, one has many additional complications. The primary problem is band structure. In addition, one has the problem of what the underlying lattice is doing to the problem, phonon interactions, maybe even impurities, if you don't have a very pure material. So, any really detailed theory for electron-hole liquids starts out by saying: I can't solve the problem of electron-hole liquids with bandstructure and electron phonon interactions etc., so, let me consider a perfectly ideal system of isotropic electrons and isotropic holes with equal masses. The problem with unequal masses and more complicated band structure is very difficult. So they say let's try it with equal isotropic masses and use that as a model for the semiconductor system.

There exists no such ideal system in solid state physics. The electron positron plasma is such an object. The Fermi surfaces are perfectly matched, they're just spheres, and the electrons and positrons have the same mass and there's no bandstructure. What's more, because the electrons and positrons annihilate with one another one might be able to get some very interesting information about what's going on inside this structure.

So let me tell you now about calculations that were primarily done by Nozières and Comte.[3] They considered the problem of an electron hole plasma. They were particularly interested in the ground state. I want to show you how enormously rich even the question of what the ground state looks like is, and how basic it is to a lot of many body problems in solid-state physics.

The ground state properties of such a system are purely a function of the density, n per unit volume. A typical length in the problem is the Bohr radius $a_o = \hbar/me^2 \cong .5 \times 10^{-8}$ cm which is also the radius of the Ps atom. The one parameter governing the ground state calculation is $r_s \equiv n^{-1/3}\, a_o$. Now I think experimentally we're always going to be in a dilute limit $r_s >> 1$, although one is in general interested in the question of what happens at arbitrary densities. To get an r_s of order one, we would have to get to densities of the order of 10^{24} or so.

In the dilute limit which is the one that we'll primarily be concerned with, at least for the next 10 years anyway, there are a number of possibilities. One can ask, do these positronium atoms at $T \cong 1\,^\circ K$ crystallize into some kind of solid? Well I think the answer to that, in any realistic regime, is no. An estimate of the energy of interaction ΔE_I between positronium atoms due to their Van der Waals interaction is

$$\Delta E_I \cong \left(\frac{e^2}{r}\right) \left(\frac{a_o}{r}\right)^5 \tag{2}$$

You find that on a distance scale of about 100Å this is a very small number, about 10^{-9} eV. That means you don't need very much temperature, only about 10 microdegrees, to melt this solid at densities which are of the order of 10^{18} per cubic centimeter. Of course as we go up to 10^{22} we're going to wind up with bigger numbers, but still it's very probable that we will see no crystallization happening here because the energy scale is too small.

One other very interesting question, once we rule out crystallization is, what about Bose condensation of positronium? Now that sounds a little bit crazy but it's true. The positronium atom is made up of two fermions, and it's a Boson, so if I cool it down to low temperatures, as everybody knows, it should condense into some kind Bose liquid. The details are tied up with questions of the spin and I really don't want to get involved with that except to say that the spin adds some additional degrees of freedom. It's a little bit like the problems of spin polarized hydrogen.[4] One of the things that makes this system experimently interesting is that because positronium is so light, the Debye wavelength $\lambda_D = \pi\hbar/\sqrt{mk_BT}$ is about 1000Å at $T = 1°K$. We expect, independent of the details of the interparticle forces, that a Bose transition will occur when[5]

$$n\,\lambda_D^3 \cong 1\,. \tag{3}$$

We really have only one example of Bose condensation in nature, and that is helium which is a strongly interacting system and not susceptible to theoretical analysis. So it is an interesting thing to look for another such system. At present, there is a vast army of people working on polarized hydrogen, and asking about a Bose condensation of that. It's a new system with a different set of physical parameters. They will have to go to millidegree temperatures. So far it hasn't been successful. However, dense positronium at ($n \cong 10^{18}cm^{-3}$) may Bose condense at $T = 10°K$ which will give us some new information about many body phenomena.

At high densities, the system (assuming we could get $n > 10^{24}$ cm^{-3}) begins to look more like a metal. In this metallic regime, we might have the analog of crystallization, i.e. some kind of a charge density wave. No one has done any calculations on this state. In addition there is the question as to whether we can even have a metal. It turns out that when one calculates the properties of a metal made up of e^+ and e^- they find out it's not really a metal. They find out that it's something called an exitonic metal. This means that if you look at the excitation spectrum of this crazy metal, this positronium metal, it doesn't look at all like the excitation spectrum of an ordinary metal. It does not seem to have a continuum of excitations down to zero energy. As we shall see in a moment, it looks very much like the bound exiton, meaning our old positronium atom. There's a gap for the excitation of a positron-electron pair, even at high densities.

In order to treat both of these limits, since they both look a little bit alike, i.e. a Bose condensation of excitons at low densities and an excitonic metal at high densities, Nozières and Comte (NC) had a very nice approximation to the ground state wave function.[3] It is an approximation that is meant to go continuously from the dilute to the dense state. What they did was point out that Bose condensation and an excitonic metal are a lot like superconductivity. It's not exactly superconductivity because the positrons and the electrons have opposite charge so they don't really carry a current very well, however they said 'Let's try a BCS-like wavefunction[7] and try it in both of these cases and see if I can get some reasonable results for the ground state energy'. They tried such a wavefunction;

$$\phi_\lambda^o = \prod_k (u_k + v_k a_k^+ b_k^+) |0> \qquad (4)$$

In Eq. (4) a_k and b_k are the annihilation operators for electrons and positrons. The state $|0>$ is the vacuum state and u_k and v_k are the amplitudes for the zero and one pair states respectively. This wavefunction is meant to show you that the ground state contains an arbitrary number of electron-positron pairs just as in BCS there is an arbitrary number of electron hole pairs. In the dilute limit the pair is meant to be like a positronium atom, i.e. $|v_k|^2$ is the Fourier transform of the Ps atom wave function. The fact that you have an arbitrary number just means you have a different kind of canonical ensemble and you're not really conserving particles. The average number of particles is fixed at the density of the system.

Using this wavefunction, NC approximately calculated what the energy per particle was as a function of the density. Even within the NC calculation there are a number of different results which have very interesting physical properties depending on further approximations to calculation of the energy.

First of all, all of these states that they calculate are in some sense Bose condensed states. I'm not talking about finite temperature but only the ground state, so every one of these states is Bose condensed. The energy per particle at very low densities is the energy of positronium. In the high density limit all theories agree that everything looks like the energy of a degenerate electron-hole plasma. We simply add up the Fermi energy of the electrons and the positrons, i.e. the energy goes like $n^{2/3}$. Now, in the calculation that they did with the BCS kind of wavefunction that I wrote down, they essentially take that wavefunction, calculate the expectation value of the Hamiltonian, and they find a nice smooth curve like the one labeled (a) in Fig. 3. Then they try to improve that calculation by including some screening into the problem. When they do that they find curve (c) in Fig. 3, which is somewhat controversial. This curve gives a very different behavior from (a). If I take the derivative of the energy with respect to the density I find out that (a) describes a system which is always under pressure. In other words, if I fix the density it remains a gas all the time. If I look at curve (c), I see immediately that there's a minimum which is below the minimum energy for free Ps. This means that if I have a dilute gas of this stuff it will condense at 0°K and make a liquid at the density of the minimum. So the pressure is negative in region I and positive in region II. As I increase the density, if I keep it under pressure I can get to region II. It is a big debate as to whether the minimum lies below free Ps or whether the minimum even exists.

NC go on to argue that there can be a situation, if I treated the screening self-consistently, where I could get a curve that looks like Fig. 3(b). What this means is that if I start out with an exiton gas and increase the density when I get to point B_2 the whole thing phase separates into two different kinds of liquids at

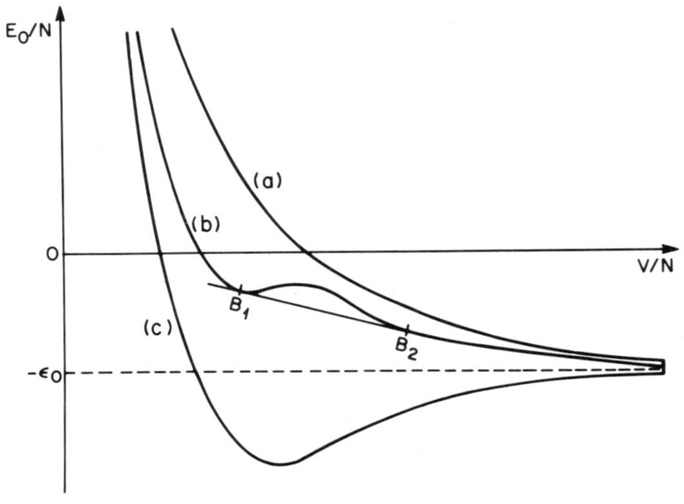

Fig. 3. Various possibilities for the liquid-gas phase diagram of an electron-hole system. a) homogenous state; stable at any density, b) liquid-gas equilibrium, c) pure liquid.

finite pressure. The tangent construction as shown in Fig. 3 determines the equilibrium situation. You don't have to worry about where it comes from. Basically you take the derivative and when the two tangents are equal you would get a system which has two different densities forming at finite pressure.

This is very close to what Mott[7] predicted for the metal-insulator transition in a metal. He asked what happens to a gas of hydrogen atoms as you squeeze them together. The reason you get such a crazy behavior in this more complicated theory is that with screening the dependence of the energy on the density becomes a very cooperative phenomenon. As the excitons get closer together there is some screening which is involved because they're not really excitons. Then, as the screening gets a little bigger the binding energy goes down and therefore the screening goes up, so at some point you get a catastrophe, i.e. you get what amounts to a first order transition. The question of whether such a transition occurs is interesting theoretically and would be interesting experimentally if you can get a handle on it. Unfortunately, I think at 10^{22} you're going to be at too low a density. To get to B_2 you're going to have to be at 10^{24} cm^{-3}.

Ok, now I want to show you some of the more detailed results of the NC calculation just to give you a little feel for what we've been talking about. In addition to the ground state energy curves, they calculated a number of things about the system. In Fig. 4 we see a series of plots of the quantity v_k^2, our variational parameter for varying densities. It is is like a Ps atom wavefunction, i.e. it's a pair wavefunction. What you see in curve 4(a) at low densities is

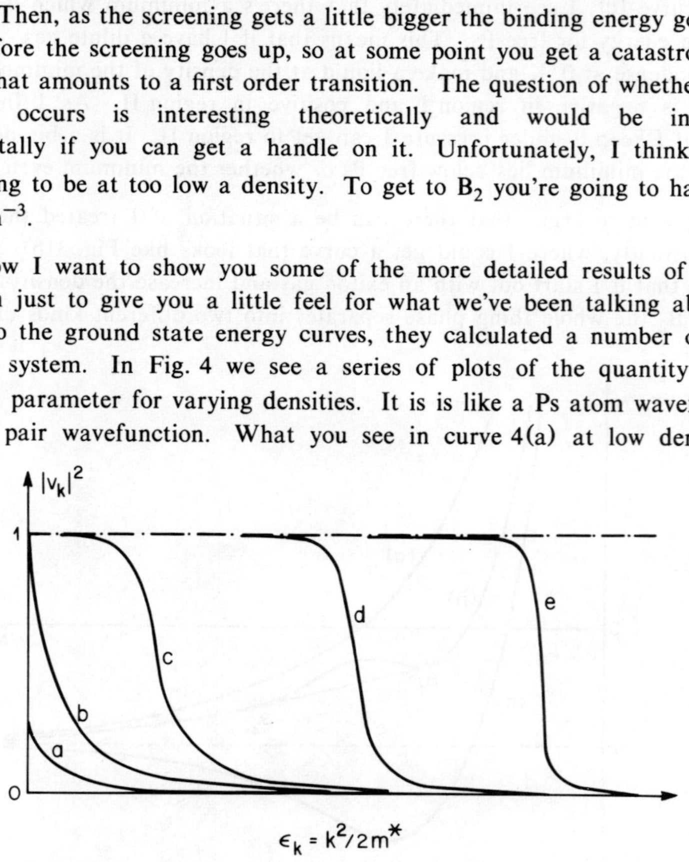

Fig. 4. The ground state fermion distribution, $|V_k|^2$, for an electron-hole system. The curves a) through e) are for increasing density.

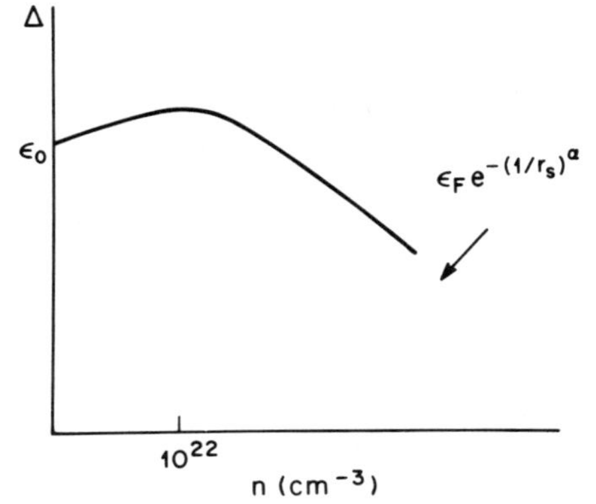

Fig. 5. Gap in the excitation energy versus density for an electron-hole liquid. E_o is the positronium binding energy.

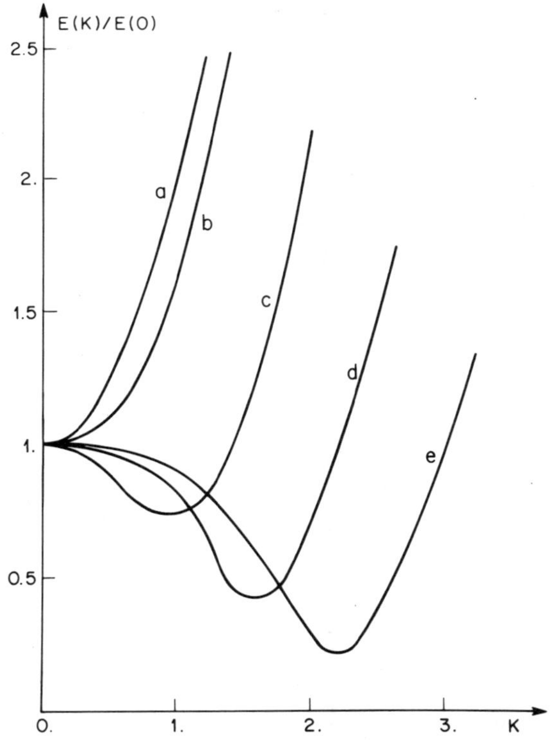

Fig. 6. Dispersion relation for pair excitation in an electron-hole liquid.

basically Ps, and as you crank up the density (b)–(e), $|v_k|^2$ goes continuously to a Fermi distribution as we would expect if the system were a degenerate metal. They also calculated what the gap energy was for the excitation of a particle and a hole as a function of density. Figure 5 shows you that there is a finite gap even at high density. In Fig. 5, the dilute limit is just the positronium binding energy ϵ_0. As you crank up the density, it starts to increase a little bit and then it decreases exponentially almost like a BCS formula for the transition temperature.

If they calculate the energy ϵ_k to produce a pair with some momentum k, (Fig.6) they find some rather interesting things. At very low densities ϵ_k increases like $\epsilon_k \cong \sqrt{k^2/2m^* + \epsilon_g}$ where m^* is an effective mass. As you go up in density, what you find is that the minimum energy to produce a pair is out at something which approaches what you would think would be the Fermi momentum for the almost metallic liquid. That really means that you're getting this gap appearing because you have two Fermi surfaces which have identical shapes. If you looked at this closely you would find that the electrons and positrons that are pairing are the ones right out near the Fermi surface, exactly as in the BCS theory. This gap is the analog of the BCS gap. So you go from something which looks a little bit like the positronium gap, to a gap at the Fermi surface, at finite wave vector k, that is still roughly the Ps binding energy.

I thought I would also show you how the ground state value compares with more sophisticated theories, just to show you there really are a lot of theories and nothing is certain. In Fig. 7 the nice smooth dotted curve labeled NC is the

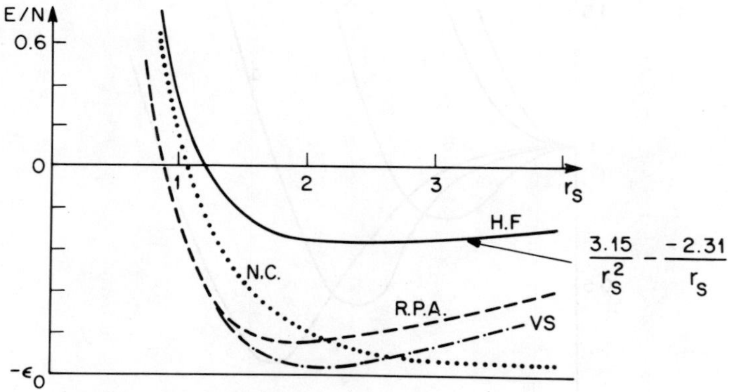

Fig. 7. The actual ground state energy per particle as computed in various approximations. (H.F.=Hartree-Fock, N.C.=Nozieres-Comte, RPA=random phase approximation, V.S.=Vashita-Singwi).

same as in Fig. 3. The solid curve is a Hartree-Fock approximation for this gas which is expected to work well in the high density region. The dashed curve is RPA which takes into account screening much better than Hartree Fock. It includes some dynamic screening. You see that they tend to develop minima but none of them go to the right low density limit.

I thought I would say one or two words about transition temperatures. Although no one has a microscopic theory of transition temperatures it is possible

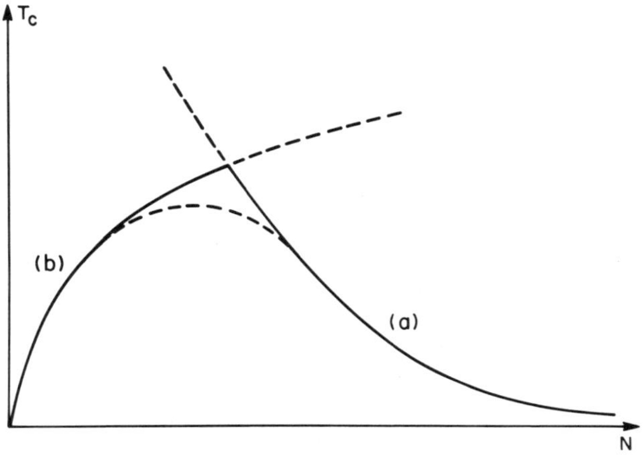

Fig. 8. Transition temperature vs. density for an electron-hole liquid.

to make some reasonable estimates. At high densities one's physical intuition is that there will be a transition temperature which is of the order of the gap in the system. In other words, the transition temperature $T_c \sim E_F \exp -(1/r_s)^2$ (Fig. 8(a)). At low densities we just have positronium atoms. We expect this curve to look like $T_c \sim n^{2/3}$, which is just the statement that you get a Bose condensation of this gas when the thermal DeBroglie wavelength equals the interparticle spacing (Fig. 8(b)). For a density of $n \cong 10^{20}\ \mathrm{cm}^{-3}$ which is near the intersection of the two solid curves the transition temperature is of the order of 50°K for positronium.

Even less is known about collective modes, but one can speculate. In Fig. 9(a) we have sketched (guessed at) the collective excitation spectrum in the dilute limit. It's actually quite interesting. If the positronium atoms interact with one another rather weakly, we have a real theory for calculating what the collective modes are in this system: the so-called Bogolubov approximation for a weakly interacting Boson gas. Of course one of the main reasons for doing spin polarized hydrogen is that it may provide us, for the first time, with a system where you can really do perturbation theory. Helium, on the other hand, is very complicated and strongly interacting. The curve in Fig. 9 (a) is qualitatively what one would get from a Bogolubov approximation to the excitation spectrum.[6] It shows that there is a collective mode which goes to zero frequency, even though the single particle mode has a gap. This mode is a little bit like sound, although it is collisionless. It's not quite hydrodynamic sound. Ultimately as k increases it goes over to some type of single particle excitation which looks a little bit like the center of mass motion of Ps;

$$\omega = k^2/2m^* \ , \ m^* = 2m_e \ . \tag{5}$$

Now for helium, you remember, you do get sound but then you get a maximum and a minimum, the so-called Roton minimum.[5]

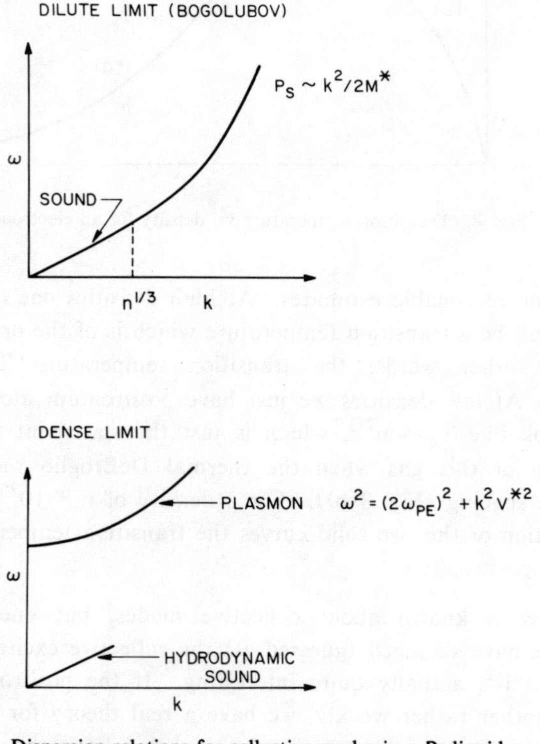

Fig. 9. Dispersion relations for collective modes in a Ps liquid.

In the dense limit the situation is very different (Fig. 9(b)). In this region, as far as I can tell, we have a plasmon branch where the k=0 frequency is given by twice the plasma frequency of either the positron or the electron. In addition there is some dispersion which depends on the Fermi velocity i.e.

$$\omega^2 = (2\omega_{pe})^2 + \beta k^2 v_F^2 \qquad (6)$$

with β of order one. As far as I can tell, there are no other collective modes except hydrodynamic sound, which is very different from these collisionless collective modes.

Of course, this is all speculation. Nothing is known quantitatively but it should give you an idea that it would be very interesting to look at such things because these questions are not simple questions to answer. Somebody once said that there are only two kinds of people; the ones that get confused or the ones that get stimulated. So I hope I have stimulated you a little bit. I think the only conclusions we can really draw from all of this preliminary theory is that if we get there, it surely will be interesting. Thank you.

Question

I have two questions or comments. You were talking about crystallization of positronium, how about crystallization of ionic positronium, Ps^- and Ps^+? You could imagine three positronium atoms rearranging into two ions and then being electrostatically attached to one another. Depending on the density, that might be a lower energy configuration.

Platzman

All those things are possibilities, in fact, people talk about positronium molecules also. You make a molecule and then they could be Van der Waals coupled; but Ps^- and Ps^+, I don't think so because it costs too much energy to ionize the Ps.

Comment

These are polar.

Platzman

The Ps minus has a radius typically of say, 10Å or something like that? A little smaller than that. And so, you could get the polarization energies quite large. The electrostatic energies could be quite significant, but probably not enough to compensate for the Ps binding energy.

Question

And the other question was, well it's not a question, to point out to you that at higher densities, people have talked about the annihilation laser existing and that's been calculated using some model of an e^+ e^- metal.

Platzman

Supposedly some of the gamma-rays lines coming from space were talked about in the context of stimulated emission of .5 Mev γ-rays coming from the recombination of electron-positron pairs. Chandra Varma of Bell Labs did a little bit of work on that and even published a short note on it in Nature. However I think it's rather speculative. This is on a little firmer footing than that.

A. Bansil

Are the lattice effects really that small?

Platzman

There is no lattice present in what I was talking about.

Bansil

You're sitting on top of it?

Platzman

Well, actually I was thinking in the three-dimensional case where it's kind of floating out in space. Now for the two-dimensional case, the surface will play an important role. Two-dimensions is even more confusing than three. Just to give you an idea how confusing it is, which is why I didn't want to discuss it, there is no such thing as an ideal Bose condensation in two-dimensions. So the whole question of whether there is even a dilute limit, or if I could think of some ideal problem of a real Bose condensate and what it meant other than at T = 0, is a completely open question. Such questions are being actively thought about by people who think about physi-absorbed spin polarized hydrogen, floating around on the surface of liquid Helium. But again, the temperatures of interest with Ps at surfaces would be much higher than for hydrogenic systems. However, the surface would play a big role for two dimensions. Nevertheless, it would be interesting anyway.

E. P. Gross

In the last slide you had a dispersion curve for what you called hydronamic sound and a plasmon mode. I assume you mean collision-less?

Platzman

No, I really mean hydrodynamic. As far as I know there is no collisionless ion acoustic mode for this system with two equal species. What I mean by hydrodynamic, is that I squeeze the positrons and the electrons together. In the collision-free regime there is no mode down here but there must be hydrodynamic sound.

Gross

I find it very difficult to believe. Classically there certainly are two modes for the collisionless Vlasov equation?

Platzman

The Vlasov equation does *not* give two modes because when the two species have the same mass, you'll find out that there is no analog to the ion acoustic mode. It's overdamped by Landau damping. If you simply use the Vlasov equation you discover that the modes are always overdamped. The only time you get an underdamped mode is if you think about hydrodynamic sound, real honest-to-God hydrodynamic sound.

Gross

Collision dominated?

Platzman

Right, collision dominated. There is no collision-free mode in the sense of the Vlasov equation. There is only the phonon mode. If the mass ratios are a little bit different, you'll get such a mode, ok. What are the modes in such a dense limit? Even neglecting all of the problems with gaps and so on which may be delicate, what does this thing really look like? I could be wrong, I just drew that curve on a dark night.

Gross

I'm even more embarrassed because that thing was worked out by Krook and myself in a paper for the equal mass case and I can't remember.

Platzman

Well I'd like to see the reference. I have a reference to some comments in the journal by Bert Fried, but I don't think they refer to anything by Gross.

Gross

The two-component plasma is in our paper of 1956, [E. P. Gross, M. Krook, Phys. Rev. 102, 593 (1956)]. Because the dispersion relation is easy to solve for equal masses we worked it out in detail, but I've forgotten.

Platzman

And you didn't find it was overdamped?

Gross

I just can't remember.

Question

I just wondered if there is a timescale associated with Bose condensation and if there is, how does it compare with the annihilation of this quasi-solid?

Platzman

Good question. When the positrons go into a solid, they will tend to reach the temperature of the solid in some time which is probably picoseconds. It may be longer if the lattice temperature is 1 °K, i.e. it may be as long as 10^{-10} second. Thus when they come out of the surface, they should be quite cool. They will equilibrate among themselves in some time which is very short compared to the nanosecond annihilation lifetimes that we're talking about. Thus we can hope that the stuff will come to some kind of equilibrium. However you're right, it's an absolutely essential point. A similar situation arises, for example, in experiments being done in CuCl. People pump copper chloride with a laser then they want to say things about Bose condensation of the excitons in copper chloride. They turn off the laser and they want to know whether their metastable state really is in some type of equilibrium or is it in some state which was prepared by the laser, in which case you're not really looking at Bose condensation.

Question

Is the nanosecond still the timescale that you want to work on when you have this kind of a system?

Platzman

Yes, I think so because it's a Ps kind of lifetime.

Question

It's still the same thing?

Platzman

Yes, it's typically a nanosecond.

Berko

Before the positron community is going to reach your densities there's a very high likelihood that not one but two or three positronium atoms can indeed be formed simultaneously before they clump and annihilate. You will surely be doing experiments in this regime with these voids. Now the question is, suppose you have two or three instead of the very large number of positronium atoms; is there any signature that one should look for, for molecular formation?

Platzman

That's a little bit like people who are interested in clusters of atoms. The answer is that three will be an interesting six body problem and I don't know what the signature would be, but maybe there is something in the angular correlation profile or the lifetime.

REFERENCES

1. P. M. Platzman and N. Tzoar, "Positronium at or Near Surfaces" Proceedings of the Int. Conf. on Positron Physics, Delhi (1985).
2. T. M. Rice in "Solid State Physics", vol. 32 (H. Ehrenreich, F. Seitz, D. Turnbull, eds.) Academic Press (1977). (For experimental work see J. C. Hensel, T. G. Phillips, and G. A. Thomas in same volume.).
3. P. Nozières and C. Comte, J. Phys. (France) **43**, 7, (1982) p. 1083.
4. See for example I. Silvera, Physica **109-110B**, 1499 (1982).
5. E. M. Lifshitz and L. P. Pitaevskii, "Statistical Physics" v. 5 of "Course of Theoretical Physics", part 2, Pergamon (1980).
6. J. R. Schrieffer, "Theory of Superconductivity", Benjamin (1964).
7. N. F. Mott, "Metal-Insulator Transitions", Taylor and Francis, London (1974).

Slow Positron Optics

K. F. Canter

Department of Physics
Brandeis U. Waltham, MA 02254

There's been a change in my topic from what was earlier advertised. If anyone is interested in positrons in helium, which is a very exciting topic, I'll be glad to talk to them later. I don't want anyone to feel that they're missing out. However, Allen recommended I talk about something that we could sell to people who want to make a positron beam. So what I'm going to talk about, to begin, is the knowledge of electron optics I've acquired over the last six years. The second half of the talk will concern the future of positron optics and the type of experiments you can do with electrostatically focused positron beams. By "the future" I mean the next ten years.

In addition to this talk, there will be published in these proceedings two papers[1,2] which present documentation and the description of how to design the key features of an electrostatically focused positron beam, i.e. the immersion lens positron gun and transport optics. I don't pretend to claim any originality in this topic since all of these principles can be found in books throughout the library on electron optics. It is, however, difficult to find these things all in one place and particularly difficult to find them written specifically for a positron beam person. Electron optics is a topic which is very rich when it comes to a mathematical treatment but often you don't see specific design examples worked out in the book itself. I think Chris Kuyatt's lecture notes[3] from a Fermi summer school are probably the best example of a pedagogically effective attempt at making electron optics easy for someone not acquainted with the theory. What we have tried to do, along his lines, is use the same concepts but suitably modified for positron beams. As you know, most of the first generation positron beam work,[4–6] and much of it today,[7,8] is still done with magnetically guided beams. This is mainly done for reasons of simplicity, although some experiments are less ambiguous if done with magnetic transport. I think we will see in the future that most positron surface and scattering will use electrostatically focused positron beam systems. I hope these notes that we publish will make it a little bit easier. Of course it is rare that things work out that way since it's very hard to document things, especially computer programs that other people can use--but we'll try.

The basic elements of an electrostatic positron beam are illustrated in Fig. 1. You start with a moderator which emits slow positrons which, thanks to Allen Mills,[9] I think now is a very understandable and controllable process. The moderator serves as the "cathode" of an immersion lens gun. Electrostatic transport in most cases would be done by biasing cylindrical lens elements. Deflection, usually 90° to get the beam away from the primary positron source background, and then transport with further stages of acceleration and focusing down to a target completes the system. The transfer matrix method[10] is used in our design procedure here at Brandeis. Basically it is the thick lens equations converted to transfer matrices which take account of the linear relationship between an input ray height r from the axis in cylindrical coordinates and its

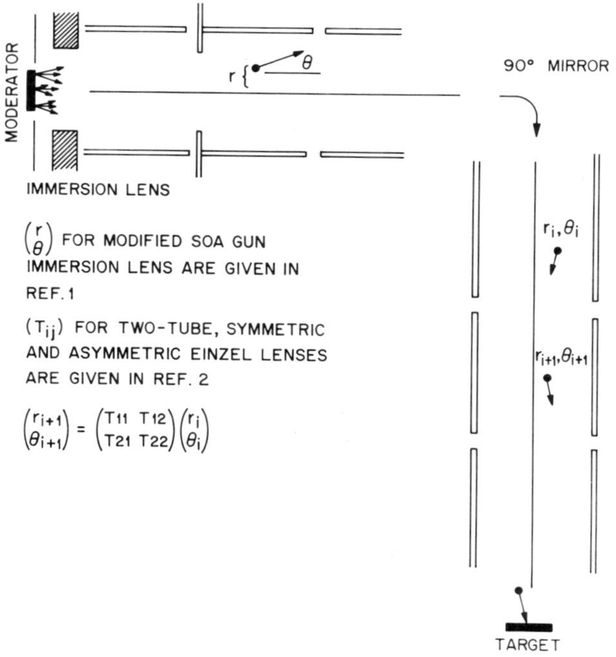

Fig. 1. Fundamental components of an electrostatic positron beam.

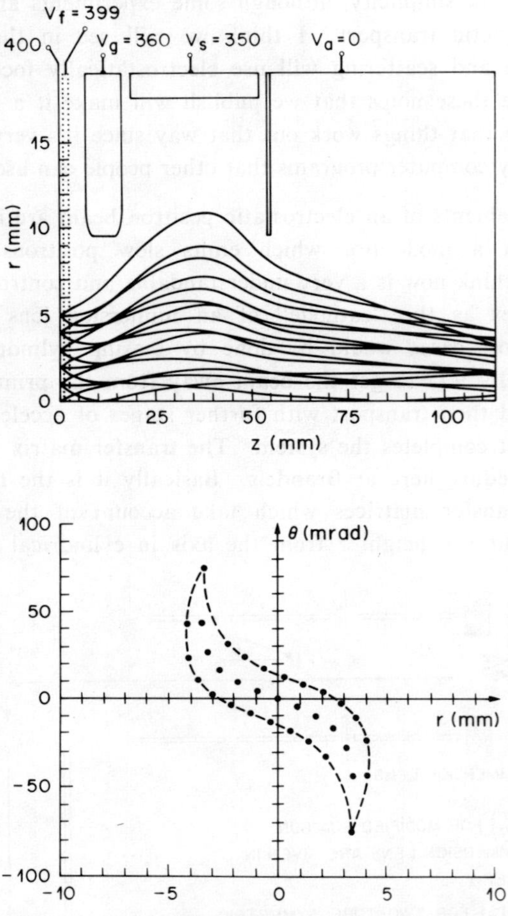

Fig. 2. a) Trajectories and b) Phase space for an electrostatic positron gun.

angle θ with respect to the beam axis. Programs using the thick lens parameters, calculated by Harting and Reid, for two-tube lenses, symmetric and asymmetric Einzel lenses, which are then converted to transfer matrix elements are published in these proceedings.[2]

The trajectories that you need to get out of the gun (which is not a conventional lens element but an immersion system because there is an electric field where the positrons are being emitted) are summarized in these proceedings[1] in a form which is amenable to the transfer matrix method. What is involved in the trajectory calculation is Laplace's equation and Newton's second law. But

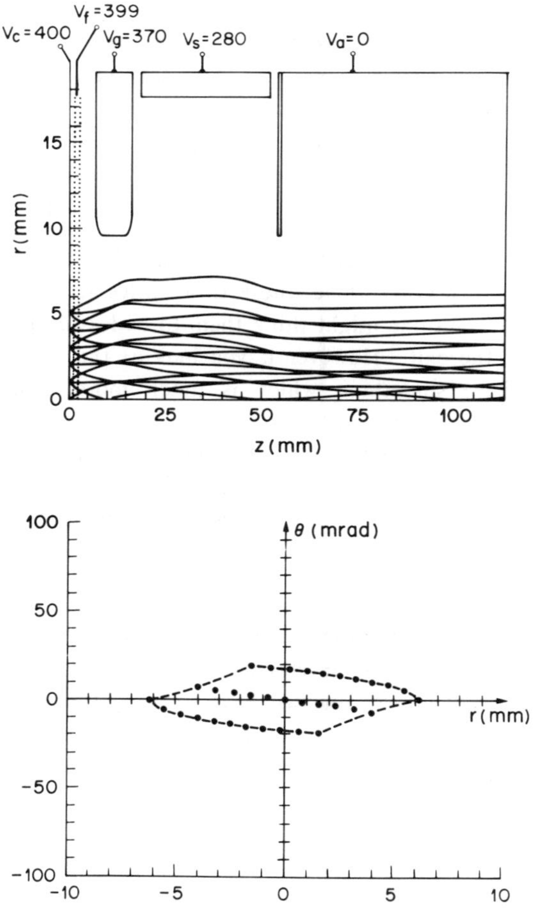

Fig. 3. Plots for a different tuning of the electrostatic gun in Fig. 2.

what is involved in taking the results of these trajectory calculations and putting them in some form that someone else (or even yourself) can use is not so straightforward. For example, in Fig. 2a we show some sample trajectories of positrons launched from a tungsten moderator in the cathode plane of a modified Soa gun.[12] The positrons were launched at 2.5 eV and the angular spread was chosen to be 0°, +10° and -10°. The modified Soa gun consists of four elements: the cathode, Whenelt or grid electrode, and a tube followed by an aperture electrode. What you get back from a computer is a set of exit trajectories leaving the anode. A visual inspection of the exit trajectories is insufficient for ascertaining the optical properties of the gun. What you have to do is plot those exit trajectories in terms of their radius above the axis and the angle they make with respect to axis, as shown in Fig. 2b. You can see the shape outlined by the points in this "phase-space" plot is very complicated; in fact, that's called a highly aberrated image of the cathode. Only a few rays launched from the cathode actually could constitute something that would form an image, virtual or real, i.e. rays which form a vertical straight line in the in the $r-\theta$ plot.

A group of output trajectories for another gun tuning is shown in Fig. 3a. On the basis of visual inspection alone, these trajectories seem to be no better than those in Fig. 2a. In fact one might think them to be less desirable since they look less convergent at the exit. However if you look at the output rays in terms of $r-\theta$ space you have something which is much more amenable to a simple parallelogram. Once you have a parallelogram then you only need two points, two of the corners, to completely describe the entire ensemble of launches from cathode. So the beauty of having low aberrations is not only that it gives you good image properties but the transport design says you only need two rays, i.e. two corners of the parallelogram, to completely define what's going to happen at the target.

In other words, of all of these positrons emitted from this cathode, you only need the two rays which form the corners of the exit space parallelogram. And once you have those two rays, you just have to bring them through the standard lenses with the transfer matrices, so you don't have to follow lots of trajectories to the target.

Present Status

What's typical of a good beam focus (prior to brightness enhancement) is illustrated in Fig. 4. This particular beam was used at Brandeis with Ira Rosenberg and Alex Weiss, to do diffraction[13] and also secondary electron emission studies.[14] At 400 eV, this beam was 6 milimeters diameter and about 10° angular spread; a very broad beam compared to modern electron diffraction studies. So the future of electrostatic experiments with positrons is intimately connected with the ability to get the phase space, that is the angular and radial spread of the positrons, down into smaller areas. This of course is the brightness enhancement goal that has been pursued as a result of Allen Mills' idea of re-moderation.[15] That is, the phase space area of a $r-\theta$ diagram times $E^{1/2}$ is constant so you can decrease the phase space by simply increasing the energy. That's how you get small spots. You then take the high energy positrons in the small spot and impinge them on some material which will thermalize the

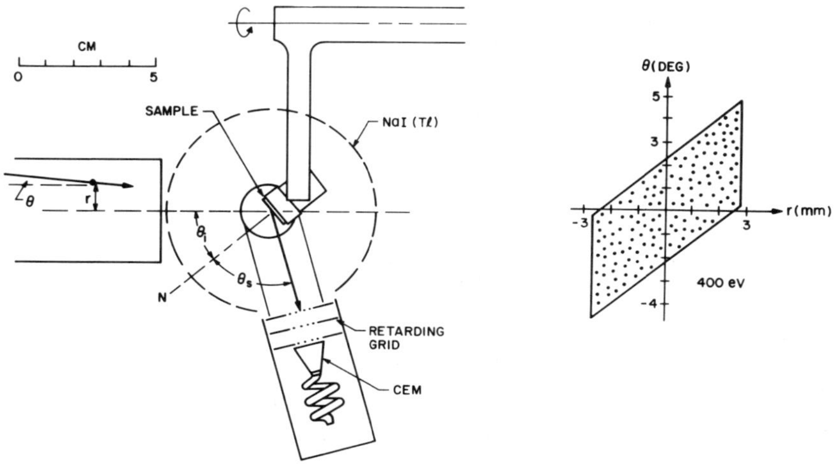

Fig. 4. First generation electrostatic positron beam.

positrons and reemit them; either through a foil in transmission geometry, or near the surface of a bulk solid where they can be reemitted in the reflection mode. In either case the thermalization process doesn't obey Liouville's theorem so you can have positrons reemitted from the surface, again at low energy, but from a much smaller area.

Eventually we will not talk about positron beams in terms of flux alone. We were brought up in an era when we started talking about 10 positrons per second, 100 positrons per second and so forth. We're now starting to talk about fluxes per unit area and per unit solid angle, i.e. the brightness. Experiments where you want to have good angle-resolved small spots, you're going to have to consider both parameters, r and θ. This is discussed elsewhere[16] so I won't spend time on it here.

The reflection mode approach to brightness enhancement is the most obvious. It forgoes the material science problem of making thin moderators, where the positrons come in from one side and slow positrons come out of the other side. The reflection mode done in two stages as illustrated in Fig. 5 turns out to be actually four times more efficient than the transmission mode done in one stage through a foil, and it is very straightforward.[17] We used our transfer matrix, by the way, to calculate the 1 mm diameter spot at the first remoderation gun and this gun is very similar to the Soa gun,[12] so there's actually no new technology or optics involved in this type of remoderation approach. The remoderation scheme that Kelvin showed you,[18] is similar but with different angles and different optics. The Brandeis enhanced beam is designed to use weak focusing in the reflection mode twice and I think there's a very conservative factor of 1000 enhancement in brightness-per-volt to be gained this way.

K. Lynn

Why is reflection mode brightness better by a factor of four? How did you get that?

Canter

Because when you go through a foil half the positrons will come out the same side they went in, so you lose a factor of two there and in the aberration involved in focusing down to a spot strongly there is another factor of two spread in spot size. The trouble with trying to focus down to a small spot at high energies is that you still end up with aberrations. There is a limit on how steep the convergent angles can be, so there's a little cheat in $r\theta\sqrt{E}$ = constant. If you try to make θ big in order to make r small, aberrations take over.

K. Lynn

You do get out of the problem of extraction through the entrance aperture as you pointed out a long time ago.[17] So there are some advantages by going to transmission versus reflection. You just can't say a simple factor of four because you can get larger angles going into a transmission moderator.

Canter

Right, but larger angles give larger aberrations.

Comment

Sure, so you have to be a little more clever in your optics.

Canter

Well I'll mention the problem of aberration, but I think that the aberrations involved with axially symmetric focusing elements become the primary limitation on microbeams. When we start talking about how small can you make a positron beam, (for doing scanning microscopes, microbeams, grazing incidence, etc.), aberrations will be your fundamental limit, not brightness. But in one stage, on a foil, if you try to focus down with a 20° angular pencil angle and that's the type of angle that you're talking about, you've already picked up an enormous θ^3 aberration. Even in the reflection mode you're on the limit of aberrations at 10° focussing.

So, the 'present' of positrons: first, there is brightness enhancement, with some practical results already achieved at Brookhaven, as you saw with those beautiful diffraction spots.[18] Obviously the goal for Brandeis and other people is to keep pushing at higher fluxes by working with a half Curie Co^{58} source or maybe a 50 milliCurie Na^{22} source and a transmission moderator. One will talk about 500,000 positrons per second focused down to a 1 mm spot with a 1° divergence at about 40 eV, that's about the best we're shooting for. The experiments being done at Brandeis include some diffraction and other microbeam applications. The system under construction (Fig. 5) is a collaboration between Allen Mills and Woody Crane at Bell Labs and Phil Lippel, Tom Horsky, and myself here at Brandeis.

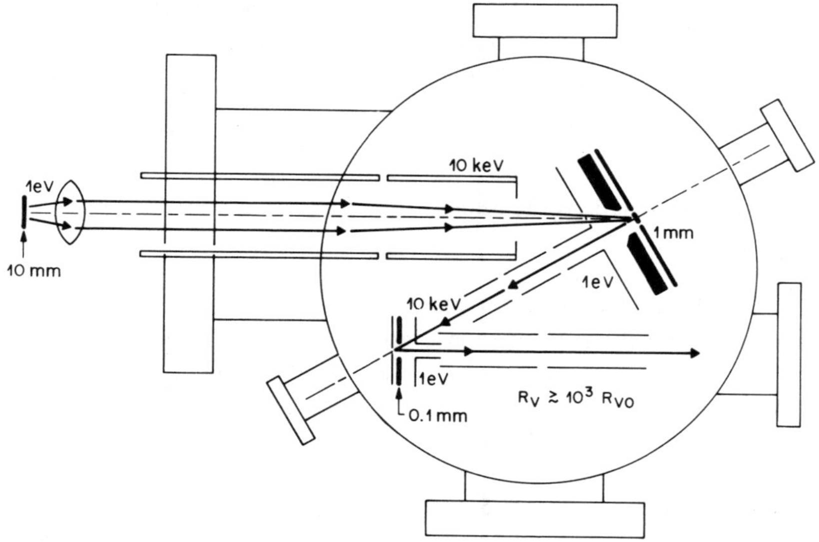

Fig. 5. Brightness enhancement apparatus employing two-stage remoderation.

Future Prospects

The proceeding was a summary of what I learned in about 6 years. It's not very much but that's the trouble with optics; it takes a long time just to find simple things. In fact, the simpler they are, the longer it takes you to accept them, because you think there must be something more complicated behind them. What I'd like to do now is talk about the future. There's a lot of physics that can be done with a positron beam which will be achievable with the first generation two-stage remoderation approach. The bright beam I just showed you in Fig. 5 would be sufficient to do diffraction and quite a few other experiments for many years to come.

Question

What are the beam characteristics coming out, in millimeters and ...

Canter

Well it's a 1 millimeter beam spot with 1° divergence at 40 electron volts energy, and an overall loss of 10. You loose a factor of 10 of all your positrons starting from the primary moderator going through 2 remoderations. If you clean up everything you lose about a factor of ten. So if you talk about 5,000,000 per second from a 1/2 curie Co-58 source and using the best tungsten moderators, you end up with about 500,000 per second in 1 mm, 1°, at 40 eV and that turns out to be roughly the best you'll ever want for diffraction studies, for example.

P. Platzman

How monochromatic is the beam?

Canter

Usually about about 0.2 eV FWHM. If these remoderators are cooled, which is easy to do with the reflection mode geometry then a few meV widths may be possible. It's very easy to introduce even liquid helium temperatures at your moderator stages. Kelvin has said that Nickel reemits the positron with a transverse energy which seems to go down as kT, Allen thinks there might be a lower limit on that, but at least in liquid nitrogen temperatures you could probably get another factor of 5 each stage, and end up with a factor of 25 in brightness per volt. The point I'm making is if you just stay with this technology and play with the moderators and geometry a little bit, cooling and surface preparation, you'll have all the positrons you ever need in a small hole. I don't know if any of you remember that fad in the 60's of trying to shove a piano through a six inch hole. Remember fraternities used to do that in college? Crowding phone booths was a fad of the same era ... they would chop up a piano and stuff it through a hole. Don't ask me where that came from, but obviously that's part of the game of positron beams; how small a hole you can stuff your beam through.

K. Lynn

One thing I think you have to lend a little bit of concern to (or anyway, its been true for the beam at Brookhaven) is that the source configuration in a backscattering mode is not ideal for obtaining a clean phase space. You have this needle, or whatever source you have in there, and that causes phase-space distortions in your brightness end.

Canter

Yes. We gained a lot by having our source as thin as possible on a tungsten ribbon using Allen Mills' very efficient slow positron production method of having a 2 μm layer of copper plated on the tungsten.[9] The source is thus a thin ribbon held in a ring with a grid serving as an equipotential plane. The ribbon itself forms an equipotential plane so electrostatically we have a nice axial system except for shadowing. You're right, looking from the side of the filament it's actually two little beamlets. So all these $r-\theta$ diagrams that you see, they still stay nice parallelograms but there's something missing in the middle. However, we're not making them bigger. Of course, if you're lucky enough, the aberrations will tend to fill those up. So when you're working in the aberration limited spot size, you will start to see a round spot again. But yes, as much care goes into that source configuration and how its placed as all the other parts of the system.

This is now science fiction; these are things that haven't been done with positrons. Some things have been done with electrons which I haven't had another 6 years to understand. Maybe 6 years from now I can talk about this topic with more intelligence. What I'm going to try to do now, in the spirit of looking towards the future, is to talk about *line focusing*.

All the work, so far, that has been done by the slow positron beam community has been done with round beams. We all talk in terms of a circular spot, how small you can make your spot and so forth, but there are a lot of experiments in which the beam does not have to have a circular cross section. There are experiments in which spatial resolution in only one dimension, transverse to the beam axis, is valuable. So, if you take a beam and make it into a rectangular line or ribbon, experiments needing good definition in this direction would become feasible. You would not have to play the game of trying to make spots small in both transverse directions, which is a brightness enhancement phase space problem.

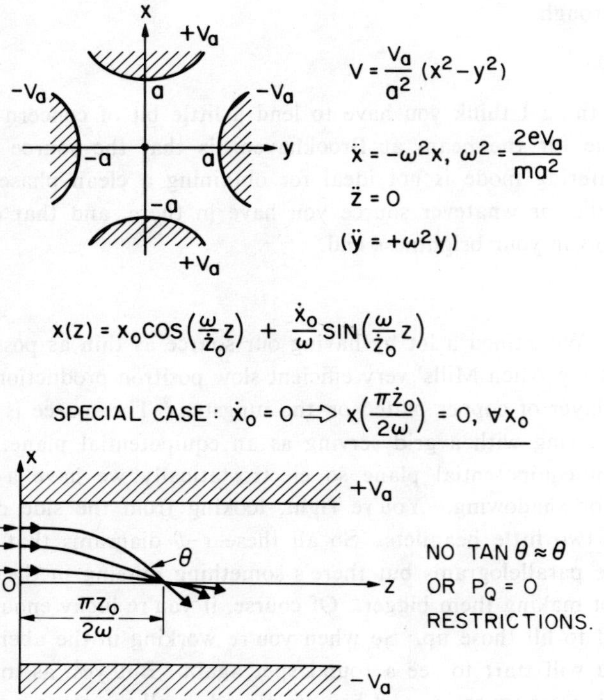

Fig. 6. Line focusing quadrupole lens.

Instead of discussing applications, I'll deal with the principle of line focusing. Figure 6 shows a quadrupole lens. It is an incredibly beautiful system because if you have the right shape electrodes, hyperbolic surfaces, (or any round surfaces at all put far enough away from the center axis) then you get a good hyperbolic potential off axis.[19] It's analytic; you don't need relaxation methods. It's separable; the x part and the y part are separate potentials. Moreover the equations of motion are exact; they look like harmonic oscillators, except there is a central attractive force along the x direction only. Thus positive charges are repelled from the electrodes along the x direction and will be attracted to the electrodes in the y direction. So what happens is that you get compression in the x direction which oscillates sinusoidally. The y motion, of course, diverges like one of these hyperbolic trig functions but just looking at the x direction you can see you get focusing. If you launch all of your particles in the quadrupole field with no vertical velocity component $\dot{x} \neq 0$ (that's infintely bright by the way), you find they will focus to a point. This is not surprising. Lenses should focus to a point if particles are launched perfectly parallel. However, in an axially symmetric lens systems, (tube lenses, and so forth) there's an additional requirement that the angles they make with respect to the axis is small. There is no small-angle approximation in this lens and this focus is perfect regardless of how steeply they come into that point. Also, there's no restriction that this particle be launched within the plane of this drawing. If a particle were coming out of the plane of the drawing in Fig. 6, it would have angular momentum P_ϕ about the z axis. This would prevent it from crossing the axis in the axially symmetric system. There's no such restriction on this cartesian type system. So, in principle, you can focus a beam, (an infinitely bright beam,) down to a *line* "spot" of an infinitessional width; no abberations whatsoever. If you add transverse energies to these positrons, phase space considerations require that you go up to high energies and remoderate to approach infinite brightness, but the nice thing about this is that this lens performs flawlessly. Of course, life isn't that nice because you don't get fields that start perfectly quadratically from zero on one side. There are edge effects. There are off-axis corrections to this hyperbolic potential so these quadrupole lenses are not *that* good but they are usually better than axial tube lenses that we've been using in positron optics, as far as beating aberration limited beam widths.

If you take two quadrupoles and arrange them at 90°, you tend to get some sort of round spot, in fact, a series of these quadrupoles arranged 90° to each other takes a beam this way and that way, in x and y, and as you know that's the alternating gradient focusing of a synchrotron and other transports. Quadrupole lenses are usually relegated to high energies because they can focus to short focal length with very little aberration for high energy as well. Lower energies, in the few keV's, are appropriate for a positron beam microprobe. If you want to probe a surface you don't want to go in too deep, so a few keV is your limit. If you want to focus your beam down to around 1000Å, which is the diffusion limited size of a spot once the positron gets in, those restrictions can be overcome, at least as far as the aberration, by using a quadrupole lens. So I would suspect that if you wanted to hit a small area with a very good resolution grazing incidence or a very small spot, then a line focus would be the best way to go. Given we don't have that many positrons to burn, I think obviously the best thing to do is to go to line sources rather than try to waste so much time getting small points. To get information in the y direction, you could rotate your sample and do a projection by rastering along different directions. If you rastered the beam across the sample and look at the positronium as a function of position and then orienting the sample differently then you'd get some rastered information giving the microstructure of the sample. That's a microprobe being used as a scanning microscope which was proposed by Allen Mills.[15] There is also a microscope proposed by Lester Hulett[20] called an emission microscope which is very exciting. You just focus on a thin foil and make an image of what comes out. That type of microscope would be defect sensitive and not require the rastering capabilities of the type I've mentioned here.

I just leave one other question about the future, as far as this quadropole lens, and it's the following. When you have a potential which is separable in x and y coordinates, i.e. $V(x, y) = V_x(x) + V_y(y)$, we have $(\Gamma_x)_{in} = (\Gamma_x)_{out}$ and $(\Gamma_y)_{in} = (\Gamma_y)_{out}$ assuming (for simplicity) $E_{out} = E_{in}$. Γ_x and Γ_y are the phase space areas for trajectories plotted in $x,\dot{x}$ and $y,\dot{y}$ spaces, respectively. This is directly analogous to our r, θ plots in rotationally symmetric systems with the small angle approximation. Thus you can get the beam skinnier in the x direction but the angles (or $\dot{x}$) also get steeper, so you're not getting any more brightness. The nice thing is that the angles get steeper but there's no aberration, so if you do your foil remoderation along the line you could get brightness enhancement in the Γ_x space. The question I wish to address here is whether there is a potential $V(x,y)$ which mixes the x and y forces such that the general conversation law which conserves the total phase space product $\Gamma_x \Gamma_y$ allows you to decrease the phase space Γ_x for the x,z plane at the expense of increasing the phase space Γ_y in the y,z plane. In other words, not only do you stretch the beam out along y and compress it along x but you also get some extra brightness along x, i.e. a decrease in Γ_x. If you stretch the beam out along y you could perhaps throw more large angular junk into it, thus making it occupy more phase space than it should, then in x you'll get a smaller phase space, for example better collimation. With something like this quadropolar approach, but with a proper cross term, you might be able to mix the two phase spaces, making small spots with good collimation also.

The last thing is the question of microbeams. There's two types of experiments involving small spots, in terms of microscopic structure: the emission microscope of Hulett (there you only have to focus your positrons down to a few micron diameter spot and then blow that up as positrons are reemitted) and the scanning microscope that Allen suggested where you raster a beam and look for some change in signal associated with position. Either experiment requires making the beam itself small, and aberrations eventually become the predominant limits on focussing down to small beams. At energies of a few keV, given everything else we're doing, we would be incredibly lucky if we ever get a 1000Å beam, line or spot, regardless of the brightness. In fact we could have infinite brightness and we still wouldn't get it. Even recently, one of the best line focussing systems for a beam having a large width of 0.1 millimeter was only able to focus down to a 100Å. This is according to computer calculations,[21] and hasn't been done experimentally yet. Assuming an infinite brightness beam, i.e. no transverse energy E_T, Daglish et al. predict that they can get a 0.1 millimeter spot reduced to 100Å. For us, we would have to go through another brightness enhancement stage to begin to approach 100Å because even emitted positrons from a cooled moderator with $E_T = 10$ meV and a diamter of 0.1 mm would have too much phase space. So we wouldn't waste time on 100Å, but 1000Å for positrons might be feasible with low aberration lenses. Figures 7a & 7b show the new approach to line focussing of Daglish et al.[21] that promises to beat quadrupole lenses. The novel feature of this lens is that it uses two compensating cylindrical deflectors. In the figures, imagine the round circles being cylinders coming out of the drawing, and the beam is going between the cylinders. You then put a symmetric plus and minus voltage on the cylinders and the beam is deflected off the axis. The deflection $\theta(y)$ has a leading quadratic term, i.e.

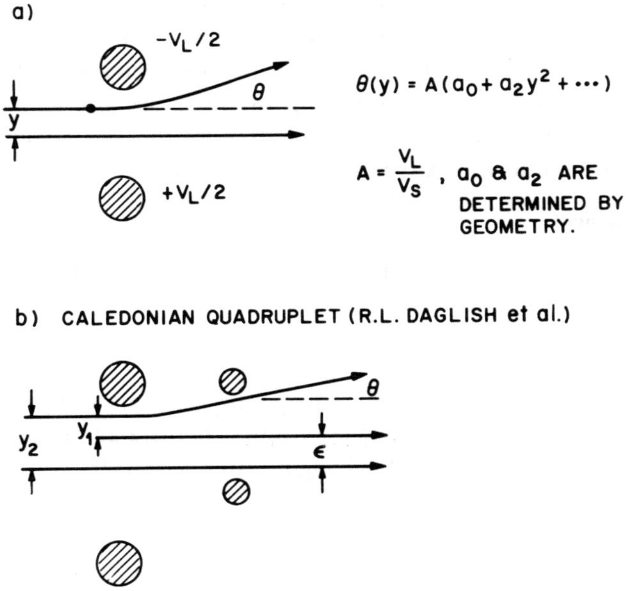

Fig. 7. Normally dispersive cylindrical deflectors a) can be used in pairs b) to form a caledonian quaduplet (R. L. Daglish et al.).

there's no linear term. To a good approximation it is purely quadratic, so this thing will never image. In other words, charges at different distances in along the axis will go off at different angles according to the square of inital diplacement, which is normally useless. It's very good for dispersion, energy analyzers, position sensitive detectors and scanning, but useless for forming a small spot. What Daglish et al. have introduced is the novel concept of taking another pair of rods and putting them off-center by an amount ϵ (ϵ can be large) as shown in Fig. 7b. You get two quadratic deflections. As you go through the first pair you get a quadratic deflection $\theta_1 = A_1\ (a_{01} + a_{21}\ y_1^2)$ and when you go through the second pair you get a second quadratic deflection $\theta_2 = A_2\ (a_{01} + a_{22}\ y_2^2)$.

Because of the offset ϵ, when you introduce $y_2 = y_1 + \epsilon$ you get a cross term for the total deflection $\theta = \theta_1 + \theta_2$ and if you adjust the amplitudes of the deflection powers such that $A_1/A_2 = -a_{22}/a_{21}$ (which can be done with the voltages very easily), you can make this cross term survive and the quadratic term cancel. Thus you can actually get an angle θ which comes out with only a linear dependence on the input height above a certain axis. So, the combination of two quadratic dispersive powers of these two pairs of rods, cleverly tuned, can give you a linear one with incredible aberration coefficients. It's exceptionally good. The linear dependence guarantees imaging. This will take a beam and really focus it to a line. Of course nothing's happening out of the plane, i.e. along the cylinders, so you're not getting any small spots, you're getting a small line again. I think this is the future of positron microbeams because this the only way we'll get down to 1000Å type scanning capability, and O.K. I'll stop.

Acknowledgement

I would like to thank Phil Lippel for bringing the work of Daglish et al. to my attention.

Question

I was wondering if you could make positron mirrors using gazing incidence reflection as is done for neutrons and x-rays? Is there any hope at all for using very low angles of incidence and making focusing mirrors?

Canter

And using diffraction as the collimater?

Question

No, not diffraction, just reflection from the surface potentials.

Canter

No, the only thing I know on grazing incidence would be the positronium formation stuff that Gidley has talked about doing. The problem's the charge, I don't even think electrons can do grazing incidence experiments. It would have to be x-rays or neutrons.

A. Mills

It turns out that LiF has an 80% reflection coefficient for perpendicular energies of about 10 eV, so you could make a grazing incidence mirror.

P. Platzman

What kind of correlation lengths do you think you're going to be able to measure? Suppose I had something with a correlation length of 10,000Å could I measure that with positrons, better than electrons?

Canter

You mean on the surface?

P. Platzman

Yes. On a perfect surface I would see a Bragg spot that would be a delta function, and then there's some resolution limit that comes from the monochromaticity of the beam and the angular spread and all these things; how well can you do with a positrons ? 1000Å, 5000Å?

Canter

About 200Å due to the angular spread and spatial spot; it depends also on the acceptance angle of the detector. We have 20Å for the first diffraction which is really mainly limited by the large angle, and how you focus it. If you have the brightness but still have a large energy spread it still would have been only about 100Å, I guess. It's hard to get below a few hundred.

K. Lynn

Why do you say its so hard Karl, just from aberrations. I mean, the beam that we have now is certainly comparable to or better than, say, a standard Φ

[Physical Electronics] or Varian LEED beam, so if you want to go through another stage of brightness enhancement and you keep your aberrations down, I don't see why you can't go to large transverse coherence length and then measure critical exponents on a surface better than you can with electrons.

Question

What would you be measuring?

K. Lynn

For example, a critical exponent for a two dimensional surface phase transition.

Canter

I think too much depends on your angular acceptance as well. I mean, it's no good to have a 1° beam and have a 10° angular acceptance. That will give you an equally bad coherence length as a 10° beam and 1° acceptance angle on your detector which also goes into the expression.

P. Platzman

No, not if it's a surface. It only diffracts in certain spots.

K. Lynn

I don't see why you can't do it. I agree that with aberrations you have to be somewhat careful. But unlike electron beams, even with LaB_6 or something, here you have the capability to continue to brightness enhance.

Canter

So you think it would actually get better than what you can do with electron beams?

K. Lynn

Well, I think that depends on the aberrations.

Comment

Well certainly people try to measure the critical exponents with electron beams and we're already capable of making a positron beam with better phase space characteristics. Now, if in the next stage of brightness enhancement, the aberration starts to kill you, for some reason, then maybe you can't do it. That's the only limitation, I think.

K. Lynn

I think it's a real nice area that positrons could more than just complement LEED.

Comment

One trouble is getting a surface that warrants going out to such large coherence lengths.

K. Lynn

Yes. But there are predictions and there are people now trying to measure them.

P. Platzman

Take the case of physi-absorbed anything on graphite. The big breakthroughs have come from using x-rays. With correlation lengths of 10,000Å you can do it, but the surface warrants it because you're looking at these commensurate-incommensurate transitions.

Comment

I don't know that they actually look at the surface in the graphite but look at quasi-surfaces in the exfoiliated case.

P. Platzman

Exactly, but with x-rays, you have the resolution.

REFERENCES

1. K. F. Canter, P. H. Lippel, W. S. Crane, and A. P. Mills, Jr., this volume.
2. K. F. Canter, P. H. Lippel and D. T. Nguyen, this volume.
3. C. E. Kuyatt, unpublished lecture notes. A large portion of these notes appear in *Building Scientific Apparatus*, J. H. Moore, C. C. Davis and A. Coplan (Addison-Wesley, Londong, 1983).
4. D. G. Costello, D. E. Groce, D. F. Herring and J. Wm. McGowan, Phys., Rev. **B5**, 1433 (1972).
5. P. G. Coleman, T. C. Griffith and G. R. Heyland, Proc. Roy. Soc. **A331**, 561 (1973).

6. K. F. Canter, A. P. Mills, Jr., and S. Berko, Phys. Rev. Lett. **33**, 7 (1974).

7. A. P. Mills, Jr., Appl. Phys. Lett. **37**, 667 (1980).

8. K. G. Lynn and H. Lutz, Rev. Sci. Instrum. **51** (7) 977 (1980).

9. A. P. Mills, Jr., Appl. Phys. Lett. **35**, 427 (1979).

10. D. DiChio, S. V. Natali, C. E. Kuyatt and A. Galejes, Rev. Sci. Inst. **45**, 566 (1974).

11. Electrostatic Lenses, E. Harting and F. H. Read (Elsiever, Amsterdam, (1976).

12. E. A. Soa, Janaer Jahrbuch **1**, 115 (1959).

13. I. J. Rosenberg, A. H. Weiss and K. F. Canter, Phys. Rev. Lett. **44**, 1139 (1980).

14. Alex H. Weiss and K. F. Canter, *Proceedings of the 6th International Conference on Positron Annihilation*, P. G. Coleman, S. C. Sharma and L. M. Diana, eds., North-Holland, Amsterdam, pp. 162-164 (1982).

15. A. P. Mills, Jr., Appl. Phys. **23**, 189 (1980).

16. K. F. Canter and A. P. Mills, Jr., Can. J. Phys. **60**, 551 (1982).

17. K. F. Canter, Positron Scattering in Gases, J. W. Humberston and M. R. C. McDowell, eds., Plenum Press, New York, pp. 219-225 (1984).

18. W. E. Frieze, D. W. Gidley and K. G. Lynn, Phys. Rev. **B**, to *be published* (1985).

19. O. Klemperer, *Electrostatic Optics*.

20. See for example: J. M. Dale, L. D. Hulett, Jr., and S. Pendyala, Appl. Spect. Rev. **19** (1), 105 (1983).

21. R. L. Daglish, T. Winchester and A. M. Smith, Nuc. Inst. Meth. 218, 7 (1983).

Positron Ejection From Ionic Solids and the Possibility of Moderators for Slow Muons

A. P. Mills, Jr.

AT&T Bell Laboratories
Murray Hill, NJ 07974

I'm really glad so many of us have gotten together to say happy birthday to Steve. I'm going to talk a little bit about positron moderation and about maybe extending the concept of moderation to muons. Slow positrons means positrons with about an eV of energy that are obtained by thermalizing MeV positrons in a solid moderator which then reemits some of them. Slow positrons are very useful for surface experiments and for atomic physics experiments. After much work by lots of people, we're starting to understand how a positron moderator works. In other words, we're beginning to understand the positron surface interaction. A recent experiment suggests that we ought to be able to make slow muons by a similar technique. That would be very for useful for muon measurements on surfaces (muon spin resonance), for making muonium for precision optical experiments, and possibly even for making the $\mu^+ \mu^-$ atom, "muium"; that would push QED to the limit.

The outline of my talk is as follows: I'll just say a few words about the history of slow positrons, a little bit about new results on slow positrons at metal surfaces, something about how positrons are emitted from ionic crystals and then, just very briefly, I'll mention how the fact that we see very energetic positrons coming out of some ionic crystals seems to indicate that maybe we should be able to see muons coming out of the same materials.

This is the history, very briefly, with lots of things left out. In 1958, Cherry, at RCA Labs., first saw slow positrons [1]. To look for slow positrons coming out of a solid was suggested by Madansky and Rasetti in 1950 already [2], and Cherry was the first to succeed. Groce, Costello, McGowan and Herring repeated and confirmed Cherry's result in about 1968 [3] and then lots of people started making positron beams. The big step forward was when Canter, Coleman, Griffith and Heyland in 1972 discovered positrons coming from MgO powders [4]. They suggested that positronium was being made in this magnesium oxide, and was coming to the surface and getting broken up somehow, possibly field ionized. It turns out that that mechanism may be the correct one, at least for some of the ionic solids [5]. Actually magnesium oxide looks like it has a negative workfunction for positrons [6].

Then, in 1974, Pendyala, Bartell, Girouard and McGowan started looking at slow positrons coming from a tungsten moderator and other kinds of materials and they actually measured shapes of the emission spectra [7]. We didn't believe it because they didn't have ultra-high vacuum conditions. But it turns out that the new experimental results that I'm going to show today are very similar to what they got then.

In 1974, Steve Berko, Karl Canter and I showed that there was positronium and excited states of positronium coming from metal surfaces that were being bombarded with low energy positrons [8]. We were using Canter's magnesium oxide moderator which produced enough slow positrons so you could do a lot of experiments. Since that time these same experiments that were done early on have been repeated in ultra-high vacuum. The result is that we're starting to understand the mechanisms for slow positron and positronium formation by surfaces.

I'll now show you Phil Platzman's Feynman diagrams which may be some partly correct approximation of what's going on (see Fig. 1). There are three things that happen when a thermal positron comes to a surface from within: it can either jump out of the solid as a bare particle, it can make positronium, or it can get stuck on the surface. It looks like, at least today, that each of these processes involves an interaction with one electron-hole pair. The one that we're sure about is the emission of positronium. An experiment with Phil Platzman and Loren Pfeifer [9] shows that when the positronium jumps out of a metal crystal it has an energy distribution that has a break at an energy corresponding to the Fermi break. This means that we're leaving a hole behind in the crystal and the positronium emission seems to be proportional to the density of hole states. Presumably the positron surface state is formed in the same way, only the positronium or the positron gets emitted with not enough energy to escape from

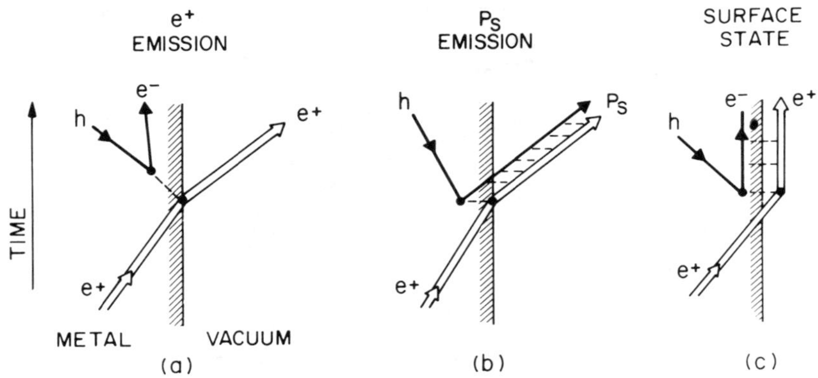

Fig. 1. Likely processes for thermal positrons after diffusing to a surface.

the metal. I'm going to show you, in a few minutes, an experiment I'm doing this week with Ben Brown and Woody Crane. It seems to indicate that when a positron is emitted, it also leaves an electron-hole pair behind. This is a real surprise because for all this time I thought that there's a maximum energy that the positron comes out with (the negative of the positron workfunction). It looked like maybe half the positrons come out of the metal with this maximum energy, which I thought was a delta function distribution. Apparently, with better resolution, it looks like we have a continuum of energies with a cutoff at the maximum.

So, this one electron-hole pair model explains why the probability for these three processes in Fig. 1 are roughly the same: they all occur by roughly the same interaction. The model explains why we see the Fermi break in positronium emission and why the positron yield is roughly proportional to the square root of the negative workfunction [10,11] (that's equal to the density of final states for this positron). I'm wondering if the model can also explain why the positrons are emitted with almost their full energy, why positrons seem to be emitted perpendicular to the surface, as Cherry Murray and I showed several years ago [12], and why the shake-up tail, namely the distribution of energy losses, seems to go like one over the loss energy. I can't answer those questions today but I'll show you the data that prompts the questions.

Here's a typical measurement of positron emission [Fig. 2]. This is a retarding field plot of positron energy in electron volts vs. the number of counts. You see that there must be a distribution of positron energies and that there's a sharp cutoff at the maximum energy. The experiment has been done at three

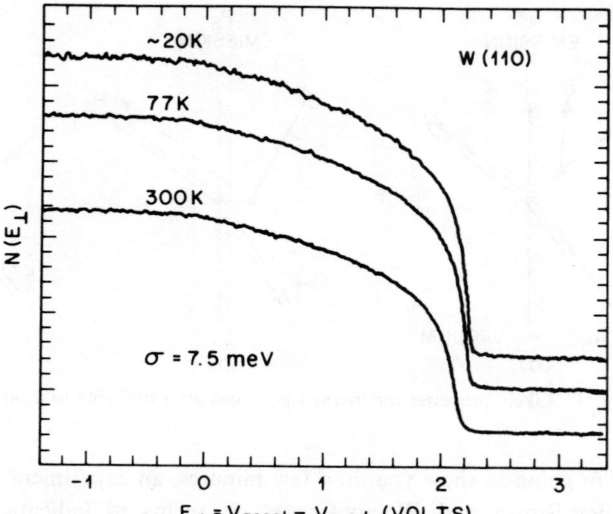

Fig. 2. Typical distribution of $E_\perp$ (component of energy along surface normal) for slow positron emission.

different temperatures; roughly 20°K, 77°K and 300°K. If you look very carefully, (this is reminiscent of Alec Stewart's measurements, because we're looking at a thermal effect on a little corner here) the 300K curve is rounder than the curves taken at low temperatures. Now let's look very carefully with better precision. I'm going to take the derivative of this data so we can examine what it really looks like. Fig. 3 shows what happens if you just take a point by point derivative and fold it with a Gaussian. This is the room temperature curve, and if our resolution were a little bit poorer, we might think that the curve is composed of a delta function peak with a continuum distribution of energy losses. But this experiment has been done with maybe 10 meV resolution and it looks like you can't make a separation between the peak of the curve and low energy part of the curve, it's all the same thing. You can see that easily from Fig. 4. This is the same experiment with better statistics plotted on a log scale. The room temperature peak is shifted from the low temperature data because there's a workfunction shift with temperature that's presumably irrelevant to the present arguments. At room temperature the high energy side of the curve looks exponential. That's exactly what you would get if you have positrons thermalized in the solid, hopping out, gaining an amount of energy equal to their workfunction. The total energy is then the thermal energy plus the workfunction,

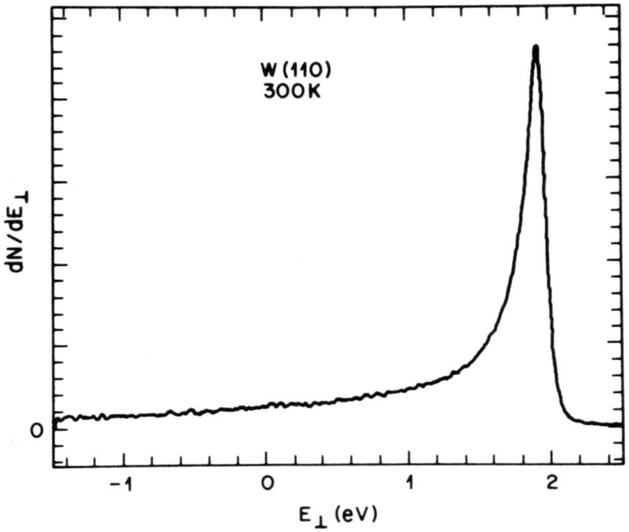

Fig. 3. Differential perpendicular energy spectrum for slow positron emission.

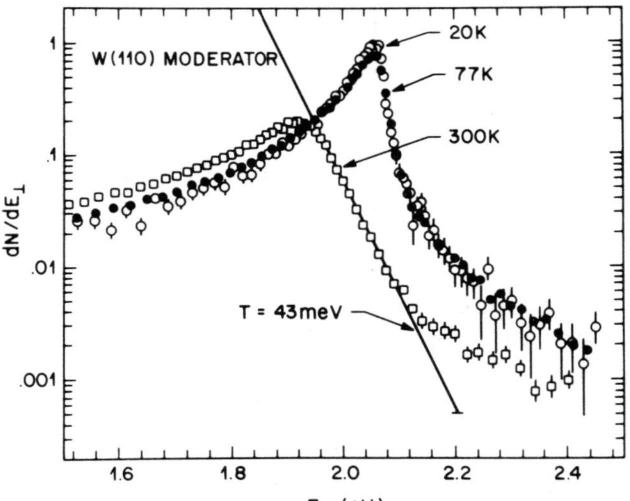

Fig. 4. Differential perpendicular energy spectra for slow positron emission. Log scale.

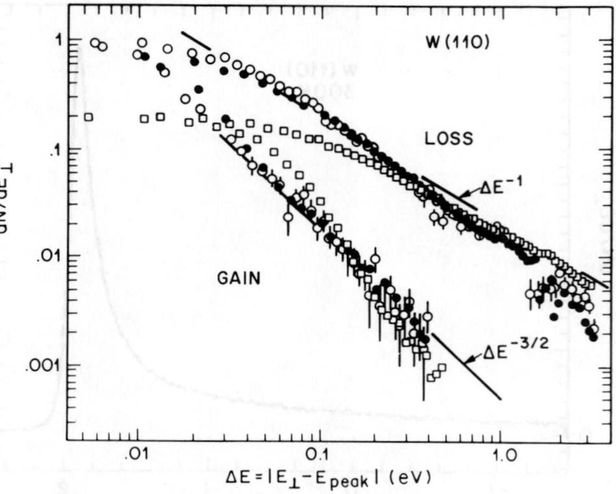

Fig. 5. Data of Fig. 4 re-plotted to illustrate the energy dependence near the emission peak (negative workfunction).

so it should be a Boltzmann distribution [actually a beam Maxwellian] which gives you this exponential tail on the high side. The indication that that's the correct interpretation is, first of all, the slope roughly corresponds to room temperature, 40 milivolts. At low temperatures the slope gets very sharp which seems to indicate that it has a thermal origin.

At the same time, this low energy part of the curve is going up steeply. You can't tell from this what the energy dependence is, but if I show a log-log plot then you can see the energy dependence [Fig. 5]. The energy gain side no longer looks exponential. In fact, the data taken at low temperatures has a power law spectrum which seems to indicate that the positrons are coming out before they thermalize. This experiment actually is being done exactly the way McGowan and Pendyala did their experiment; I'm just looking directly at positrons coming from the moderator, and so that's equivalent to using very high energy positrons, a cheap way to make a high energy positron beam. Even so, the energies aren't high enough to insure that the positrons are all thermalized at liquid helium temperatures, at least the ones that escape. Let's now look at the loss part of the curves. Plotted on the log-log scale, it has roughly a one over energy-loss dependence, indicated by the straight line. This energy-loss tail could possibly be due to phonon emission but phonon emission would give you too much angular spread to the beam and too little energy loss, only milivolts. These energy losses go like the same power out to a volt. [A Lorentzian distribution of angles would give a $1/\Delta E$ distribution of energy losses in the direction normal to the sample surface]. There could be bremsstrahlung as the positron leaves the surface but that gives you a flat energy loss spectrum. There could be dirt, but I don't think so because now that I've looked at this data I looked at my old data and it looks the same. Now that I've seen this data, nothing looks like a delta function anymore. So, we're left with electron-hole pair shakeup which is something like an infrared divergence process as Hopfield pointed out long ago [13]. Why it's $\frac{1}{\Delta E}$, is something that you can probably derive if you make the right approximations [14]. I tried last night but I got $\frac{1}{\Delta E^2}$ so I threw that one out.

Earlier Karl Canter was talking about how bright you can make a positron beam and Ben Brown will tell you something a little bit later about how you can make positronium with a precisely defined energy. That's actually why we did this experiment at low temperatures; to see how narrow in energy the beam would be, and it's about 80 milivolts wide at liquid nitrogen temperature. So even though the Boltzman part of the distribution seems to have gone away, there's still this tail.

K. Lynn

How do you argue the width should go as a function of temperature?

Mills

Well, presumably the narrowest part of the curve is the resolution of the apparatus, the flat top in the 20K data of Fig. 4. If there were no apparatus resolution, [this is a theory problem, or else do-a-better-experiment problem] presumably the curve would keep on going like $\frac{1}{\Delta E}$ but rounded off by the temperature. Actually that's a stupid thing to say because that means at zero temperature you get an an infinite number of positrons, so somebody has to do some theory here. Anyway, the net result is that the width in this picture is mostly due to the energy losses at low temperatures.

Question

How do you distinguish between energy loss and going off-normal directions with a magnetic beam?

Mills

I can't distinguish in this experiment, but if you think about it for a while, if you make up any $\cos^n(\theta)$ distribution, there's no way it will look like $\frac{1}{\Delta E}$, that's my only argument about it. Obviously it would be nice to repeat these experiments, I would like Kelvin to do his REPELS experiment more accurately [15] then you'll know for sure. Apparently then energy loss is not accompanied by a huge momentum spread. If you write down some simple matrix elements it looks like low parallel momentum transfer is highly favorable. [Editors note: These statements are not correct — see E. Gullikson, A. P. Mills, Jr., W. S. Crane, and B. L. Brown, Phys. Rev. B (1985)].

K. Lynn

I don't understand the answer. If you just put a FWHM on your curves how does that go as a function of temperatures? Why can't you measure the width?

P. Platzman

One side is thermal and the other side is something intrinsic.

Mills

Well let's just do it, then you won't have to ask. Here it's about 150 milivolts, at room temperature, at liquid nitrogen temperature it is about 80 milivolts, ok, so it's going as $\sqrt{T}$ but it didn't keep on working because at liquid helium we are limited by something else.

K. Lynn

Just to make a statement, that's inconsistent with our data on the tungsten (110) sample. Even with the non-thermals it looks like 90 meV at room temperature, so there seems to be a difference between the data.

P. Platzman

There's no drastic change in the amount of positrons coming out as you go to down to helium temperatures?

Mills

There's a small increase that might be consistent with the change in the positron diffusion constant.

Comment

But not consistent with shutting off all the positrons [16].

Question

So how does this relate to the Feynman diagram where you have the hole coming in and an electron falling into the hole? Is that equivalent to the shake-up energy loss?

Mills

More or less. I'm not going to talk about that, being totally ignorant.

Comment

But it seems if you have an electron falling in a hole, you would get energy out.

Mills

Yes, and if you're at finite temperature you do get energy out, but at zero temperature you only have losses.

Ok, now on to the next piece of talk. I'll say a little bit about emission from ionic solids which is where the strong beams began long ago with Canter et al. Actually the reason I wanted to look at ionic solids was not just that I had some samples lying around but I wanted to look at some surface resonances. The resolution of the beam wasn't good enough to see these but some other things happened instead. It turns out that when you put a positron into a piece of salt, NaF or something like that, you make positronium inside and you also get positrons, free positrons that didn't happen to catch an electron. These diffuse back to the surface. The positron workfunction of these salts, can either be ±0.5 eV, depending on the dirt on the surface and things like that. But also, the positronium that gets to the surface seems to break up and spit the positron out about half the time. So here's the picture, time going upward [Fig. 6]. The positron comes in, and by an inelastic process it grabs an electron leaving a hole behind. The positronium diffuses to the surface and somehow there's a surface hole, or something, that grabs the electron and Auger emits the positron. The maximum energy you could possibly get would be when the electron in the positronium that's located at the bottom of the conduction band, drops to the bottom of the valence band in a hole whose presence is presently not understood. It shouldn't be there in thermal equilibrium; possibly the positron makes some holes when it goes into the sample and then it comes back and the positronium finds the same hole again. Possibly, as Dupasquier suggested, the positronium makes triplet exitonic positronium which goes to the surface where the exciton recombines giving it's energy to the positron, but that doesn't seem to work out

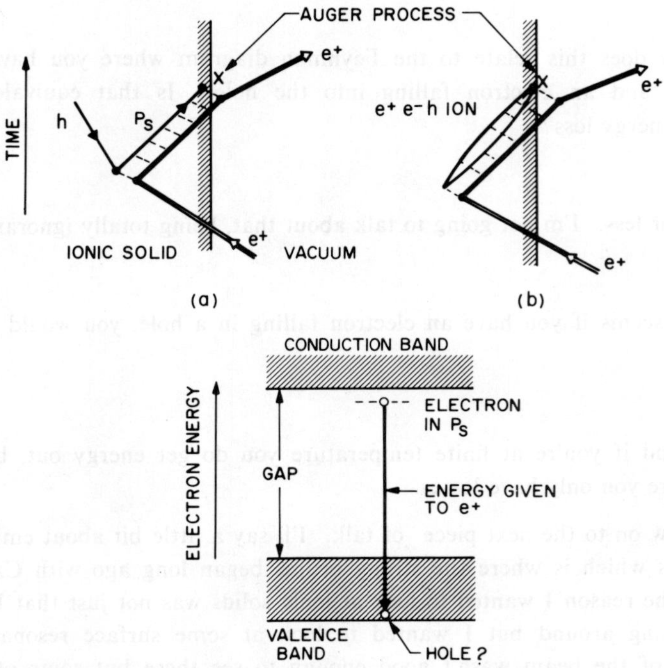

Fig. 6. Probable mechanisms for slow positron emission from ionic solids.

energetically. Figure 7 shows what the measurement looks like. We are plotting the derivative of the positron yield just like the other curves. vs. the perpendicular component of the positron energy. This experiment, which I've done with Woody Crane, has been done on vacuum cleaved and air cleaved lithium fluoride and sodium fluoride and in this particular case the air cleaved samples have a sharp peak at low energy, especially the lithium fluoride. That looks like emission of the positrons that are in the bulk by a negative workfunction mechanism; the rest of the distribution looks like the breakup of the positronium. When you vacuum cleave the surfaces so that they're as clean as you can get them, then you get a big broad energy spread with no sharp component. We've actually measured the positron workfunction by a different method by using positron diffraction to measure the innerpotential [17]. Fig. 8 shows the positron diffraction measurements; you can see a whole bunch of Bragg peaks. If you fit this data to a model where you have positrons inside the crystals in a certain innerpotential and you adjust the lattice constant (that's two parameters for three unknowns) you can get out the positron workfunction which

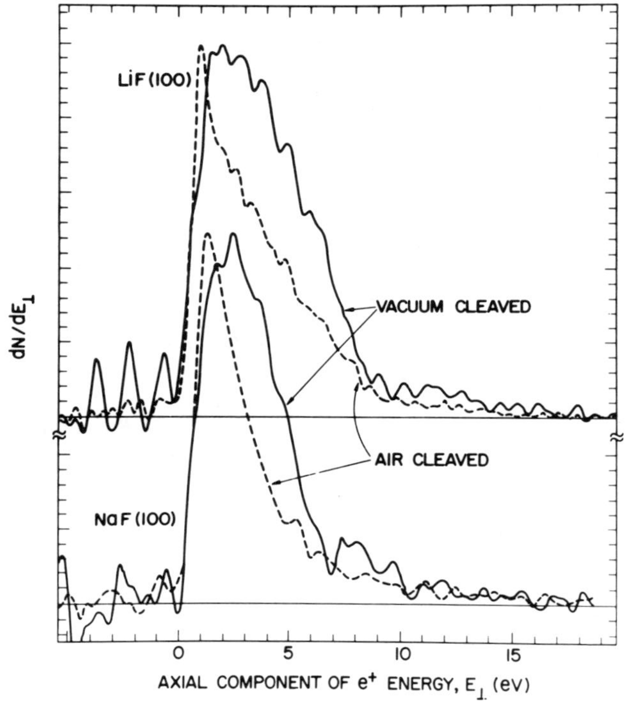

Fig. 7. Differential perpendicular energy distribution of re-emitted positrons from two ionic solids.

is minus the innerpotential. It turns out that the lattice constants are exactly the same, within a percent of the bulk values and the innerpotentials for the vacuum cleaved samples of LiF and NaF are negative. The positron workfunction is therefore positive for the vacuum cleaved surfaces by this measurement. So that means probably it's positronium coming to the surface breaking up to emit the positron.

There's some question as to what is the nature of the surface state that may be causing this. Steve Kevan has taken some photoemission data to see if there's any electron surface states; Fig. 9 shows a narrow peak that is the image of the valance band for sodium fluoride. In the energy gap, which should be from 0 to 10 eV above the peak, there are some electron states, and it's actually peaked up a little bit. So something is going on in the gap that you wouldn't ordinally expect. The NaF sample was air cleaved. When he did it with a vacuum cleaved sample he got the same results. This isn't as surprising as having holes below the

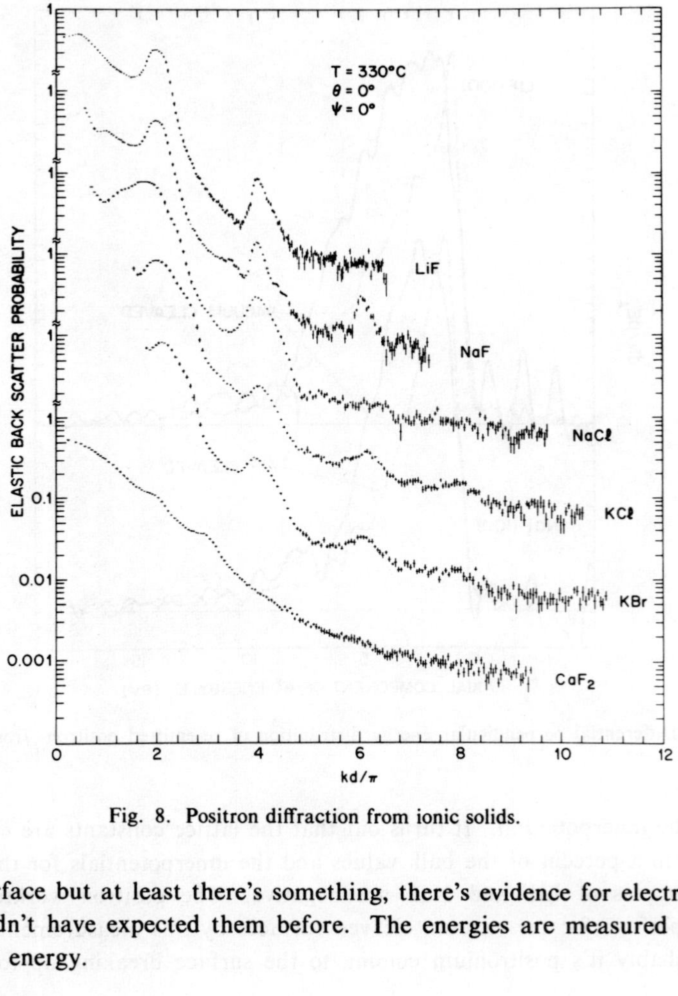

Fig. 8. Positron diffraction from ionic solids.

Fermi surface but at least there's something, there's evidence for electrons where you wouldn't have expected them before. The energies are measured relative to the Fermi energy.

P. Platzman

Different curves are for different what?

Mills

Different angles of incidence. That means that whatever this surface state may be, it is pointing in some direction.

You can also see that there are electrons in the gap, by looking at positronium production. First I'll show you a curve you can understand, Fig. 10. We have

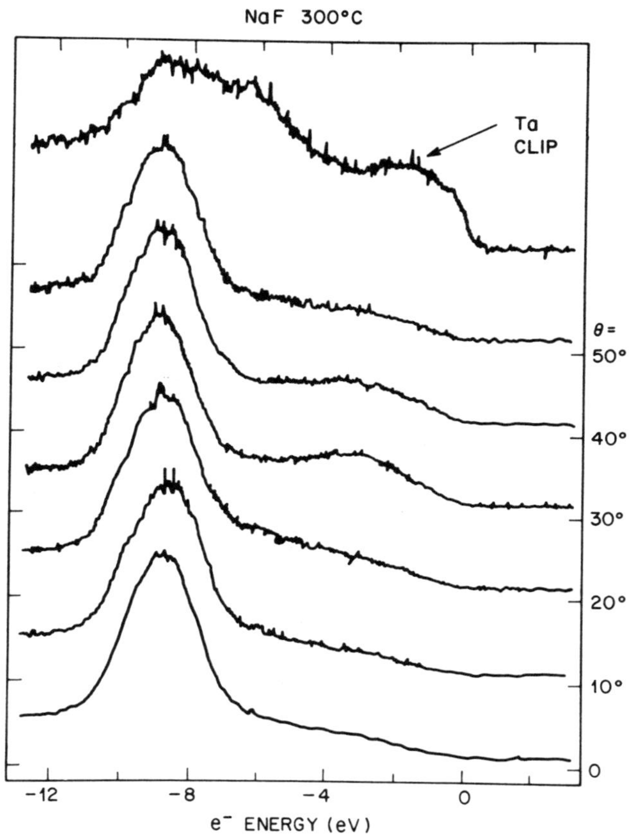

Fig. 9. Photoemission measurements on NaF.

plotted the positron reflection coefficient and the positronium formation probability vs. incident positron energy in electron volts for a NaCℓ sample. The big bump in the reflection coefficient is the Bragg peak. The second Bragg peak is not very well defined because I haven't defined the angles at all, this is the total reflection coefficient.

Question

What's your first peak? Your very first peak going towards zero?

Mills

That probably would go down to zero if the resolution were perfect but when

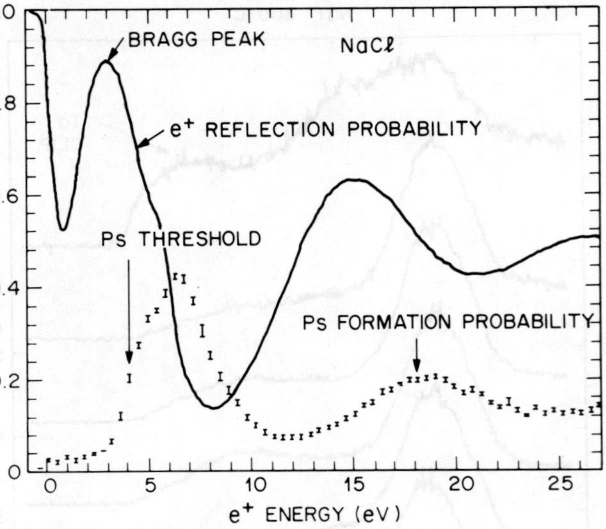

Fig. 10. Surface processes for slow positrons on NaCℓ.

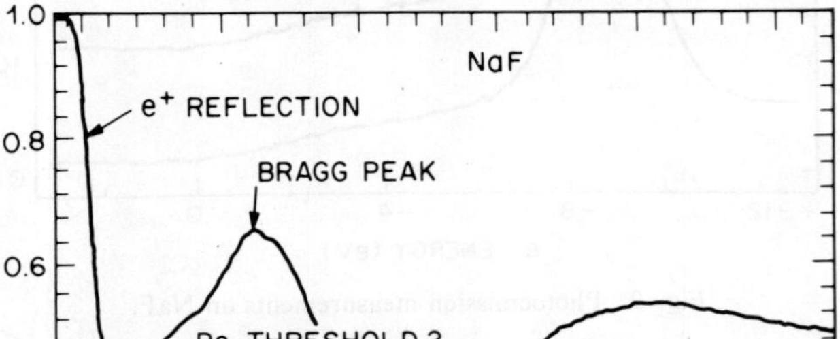

Fig. 11. Surface processes for NaF.

you reverse bias the sample nothing hits it, right? That's not reflection off the sample, that's being retarded because you don't have positrons with enough energy to get to the sample.

K. Lynn

So you think it would come down to zero?

Mills

Well it's a matter of debate, I'm going to do the experiment better.

K. Lynn

I think it's there.

Mills

Good. When you form positronium you have to have enough energy to grab an electron out of the material, so if you come in with too little energy you can't make positronium. The curve of the positronium formation probability in Fig. 10 agrees with that. At the place where the Ore gap should start, ie just below the threshold for making positronium, sure enough, the positronium fraction drops pretty much to zero. Above it, there are wiggles which Kelvin Lynn and also the Michigan group later explained in terms of higher Ore gaps or something [18,19]. So, this curve we understand for sodium chloride. But now lets look at a wide bandgap material, sodium fluoride, (Fig. 11). Where has the Ore gap gone? There should be no positronium below about 7 eV, but apparently there are low energy electrons available for the positron to grab. That's exactly the same thing that Steve Kevan was seeing in his photoemission experiment.

The question of why there are holes, apparently holes below the Fermi energy, isn't answered yet. But we can infer from the fact that we can get positrons out of these materials that we ought to be able to see muons coming out too. It turns out that in lithium fluoride, say, or sodium fluoride, if you implant high energy muons, roughly half of them form muonium and half of them stay as bare muons [20]. Also by analogy with H measurements, muonium should have a sizable diffusion constant in the alkali halides, so that if you implant high energy muons into these materials a good many of them ought to get back to the surface. If a muonium atom gets back to the surface, from the positron experiments, we know that there seems to be roughly 15 volts worth of energy ready to rip it apart. Since the muonium binding energy is only a Rydberg, that means muons may come out of say, lithium fluoride, with a couple of volts of energy, which means we ought to be able to make a muon moderator. That's an exciting prospect for the future, at which point I will end my talk. Thank you.

Comment

Allen, why is that a prediction? It sounds to me like all you have to do is a stick in a piece of lithium fluoride, turn on some muons and look.

Mills

Right, but it's a matter of getting beam time.

Comment

You need a long diffusion length too.

Mills

It is long, it's the same as the positron about 1000Å.

Question

How do you tell you have a muon?

Mills

We've designed an electron optics and muon optics...

Question

Oh, so you're going to make a beam?

Mills

Yes, and suck them away and do time of flight to make sure they're not MeV's and try to turn them off with a low voltage.

S. Berko

On this photoemission experiment, can you spray the surface before you do photoemission with some slow electrons and see whether they get stuck on the surface or not?

Mills

That would be a nice idea. The samples were operated hot to keep them from charging, so I doubt if there are any extra electrons from that, they're conductors. I really don't understand what's going on.

S. Berko

One could always say that you have all these electrons as the positron slows down. You don't get a lot of ionization, you're using slow positrons, right?

Mills

No, but the positron has to slow down, so if it's a kilovolt, that means you've got 50 electron-hole pairs; it may be enough density to explain it.

Question

I just wondered, if you were serious about the muium. You put the numbers in and it seems to be a less feasible thing?

Mills

Well the Japanese are serious about it [21] and they haven't thought about the moderator so this makes it maybe a million times easier.

Question

Why don't you comment on some advantages and differences between that and positronium in QED studies?

Mills

One neat thing about muium is it should very often decay into two neutrinos so the actual rate should give you the number of neutrino flavors. That's a low energy physics experiment to study a very interesting high energy physics question. I won't say any more, I don't really know anything about it. Also, to answer your question, to see if it was ionization, we could put positronium minus ions into the sample and see if you get three times as much emission or something, because then you would have all the ionization in exactly the same spot.

Question

When you say this positronium exists in the alkali halides do you mean as correlated pairs or do you mean real ground state positronium?

Mills

I mean it has a double lifetime [22] and also it's known from other people's experiments that it's magnetically quenched [23].

S. Berko

It's not only that but the angular distribution really looks like a delta function [24]; you get a broad curve and a very narrow peak. It's also known to be delocalized. It's also known to move.

REFERENCES

1. W. Cherry, PhD Dissertation, Princeton University (1958), available from University Microfilms Inc, Ann Arbor Mich, U.S.A.
2. L. Madansky and F. Rasetti, Phys Rev *79*, 397 (1950).
3. D. E. Groce, D. G. Costello, J. W. McGowan and D. F. Herring, Bull Am Phys Soc *13*, 1397 (1968); D. G. Costello, D. E. Groce, D. F. Herring and J. W. McGowan, Phys Rev B*5*, 1433 (1972).
4. K. F. Canter, P. G. Coleman, T. C. Griffith, and G. R. Heyland, J Phys B*5*, L167 (1972).
5. A. P. Mills, Jr., and W. S. Crane, Phys Rev Lett *53*, 2165 (1984).
6. J. Van House and P. W. Zitzewitz, Phys Rev A*29*, 96 (1984).
7. S. Pendyala, D. Bartell, F. E. Girouard, and J. W. McGowan, Phys Rev Lett *33*, 1031 (1974).
8. K. F. Canter, A. P. Mills, Jr., and S. Berko, Phys Rev Lett *33*, 7 (1974); ibid *34*, 177 (1975).
9. A. P. Mills, Jr. L. Pfeiffer, and P. M. Platzman, Phys Rev Lett *51*, 1085 (1983).
10. A. P. Mills, Jr. and C. A. Murray Bull APS *25*, 392 (1980).
11. A. P. Mills, Jr. in *Positron Solid State Physics,* W. Brandt and A. Dupasquier, eds (North Holland, NY 1983) p. 432.
12. C. A. Murray and A. P. Mills, Jr. Solid State Commun. *34*, 789 (1980).
13. See review by J. W. Gadzuk in Photoemission and the Electronic Properties of Surfaces, B. Feverbacher, B. Fitton and R. F. Willis, eds, (Wiley, NY, 1978) p. 111.
14. R. M. Nieminen, to be published.
15. D. A. Fischer, K. G. Lynn and W. E. Frieze, Phys Rev Lett *50*, 1149 (1983).
16. R. M. Nieminen and J. Oliva, Phys Rev B*22*, 2226 (1980); K. G. Lynn, P. J. Schultz, and I. K. MacKenzie, Sol. State Commun *38*, 473 (1981).
17. A. P. Mills, Jr. and W. S. Crane, Phys Rev B*31*, 3988 (1985).

18. M. Eldrup, A. Vehanen, P. J. Schultz, and K. G. Lynn, Phys Rev Lett *51*, 2007 (1983); ibid *53*, 954 (1984).

19. J. Van House, A. Rich and P. W. Zitzewitz, Phys Rev Lett *53*, 953 (1984).

20. R. Kiefl et al., Phys Rev Lett *53*, 90 (1984).

21. K. Nagamine, Invited Talk at IIIrd LAMPF II Workshop (Los Alamos, July 1983); A. Ohsaki, T. Watanabe, K. Nagamine and K. Iguchi, Abstract A62, IXth Int. Conf. Atomic Phys. (Seattle, USA, July 1984); Q. Ma, X. Chang, Z. Liu, Y. Liu and T. Watanabe, *ibid*, Abstract A63.

22. T. Hyodo and A. T. Stewart, Phys Rev B*29*, 4164 (1984).

23. A. Dupasquier, in *Positron Solid State Physics*, op cit, p. 510.

24. T. Hyodo, J. Kasai, and Y. Takeausa, J Phys Soc Jpn *79*, 2248 (1980); K. Fujiwara in *Positron Annihilation*, P. G. Coleman, S. C. Sharma, and L. M. Diana eds. (North Holland 1982) p. 615.

Precision Measurements of the 1S-2S Interval in Positronium

Steven Chu

AT&T Bell Laboratories
Holmdel, NJ

I'm here to talk to you today about some experiments that Allen Mills and I have been doing over the past couple of years measuring the 1S to 2S transition in positronium.[1] For the second version of the experiment, we were joined by John Hall of the National Bureau of Standards.[2] As you will see, this subject is another example of one of the many areas of physics to which Steve Berko has contributed.

Let me start by explaining why it's important to have high precision studies of positronium in addition to the high precision studies of hydrogen. Positronium is a little bit different than hydrogen. It is a leptonic atom for which the Hamiltonian is known to a very high degree of precision. We think we know what the electro-weak Hamiltonian is for a two body system, but unlike hydrogen, which is a one body system, you can't really have something like the Dirac equation that you can solve analytically and then apply QED corrections.[3] The reason is that a two-body system is actually a very complicated system. As you probably learned in first year physics, you can make a simple transformation to center of mass coordinates leaving a one-body system and therefore you can use something beautiful like the Dirac equation. In relativistic quantum mechanics that's not the case. You can see intuitively that the electron and positron circle around one another, and it takes a finite amount of time for them to talk to each other. Time is now a coordinate, not a parameter as in non-relativistic quantum mechanics, so you can't make this coordinate transformation with t_1 and t_2, corresponding to the time coordinates of the electron and positron.

In addition, the QED corrections in positronium are substantially different from those in hydrogen. That is because the e^+ and e^- are antiparticles of one another, so there's a wealth of diagrams that you can have in which the positron and electron come together, annihilate into a virtual photon and then re-emerge as an electron and a positron. The type of Feynman diagrams that you can see are shown in Fig. 1. The basic interaction in hydrogen is indicated schematically as an electron scattering off of a fixed potential. In positronium, we have an analogous electron-positron interaction diagram, but there is an additional annihilation diagram which gives a sizable shift in the energy levels. The first

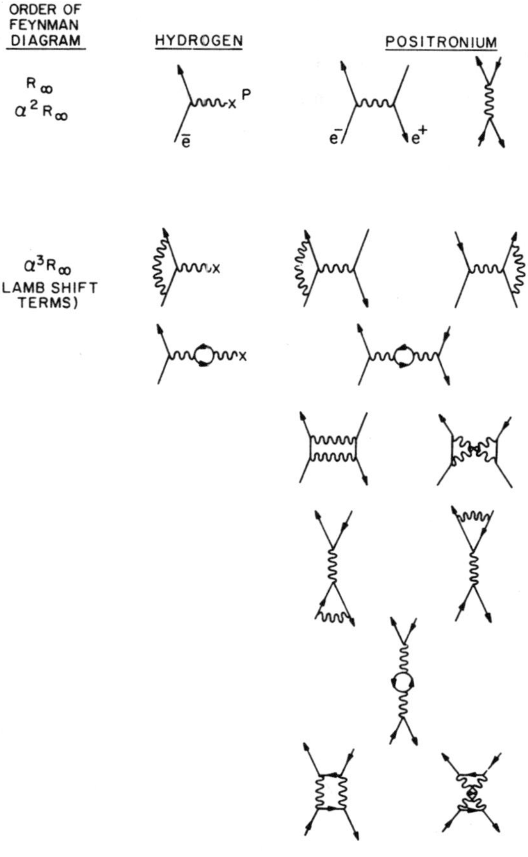

Fig. 1. Feynman Diagrams of hydrogen and positronium showing calculations of order R_∞ and $R_\infty\alpha^2$ and corrections to order $R_\infty\alpha^3$.

order QED corrections in hydrogen are the vacuum polarization and vertex correction. In positronium, first order corrections involve many more diagrams. Thus, positronium is an interesting system in that we think we know the Hamiltonian; we have a relativistically covariant description of an e^+ and e^- system, and yet even approximate solutions to the equation are difficult to obtain. In a certain sense, precision measurements done on positronium would be a test of the basic properties of QED having to do with two body systems.

Now, let me review the history of excited states in positronium. It took a long time for people to actually observe the Lyman-α decay.[4] In fact, it was done at this university in 1975 by a group led by Professor Berko. They formed a slow positron beam that hit a copper target,[5] and looked at coincidence between an annihilation gamma-ray and Lyman-α decay. They obtained the beautiful signal shown in Fig. 2. They used interference filters that were tuned to the predicted Lyman-α decay of positronium, and 15Å or 20Å to either side. The background coincidence rate was about two coincidences per thousand seconds, and the Lyman-α signal is about one coincidence per thousands seconds. That doesn't

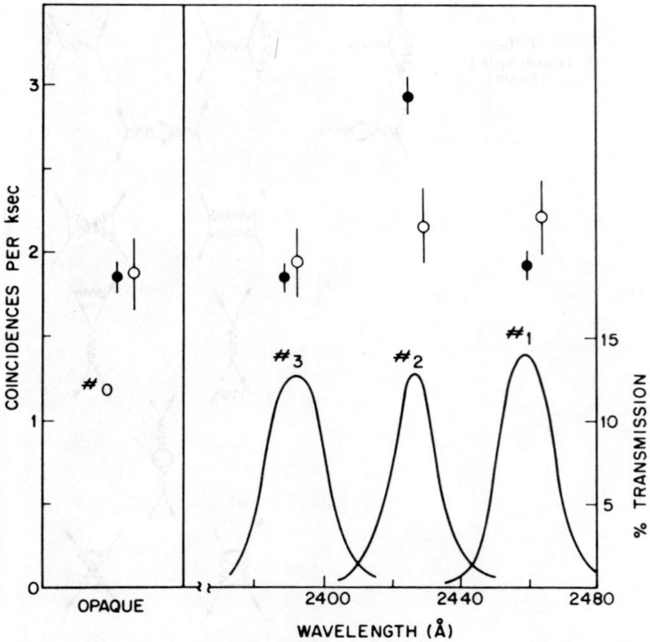

Fig. 2. Observation of Lyman-α fluorescence from positronium by Canter, Mills, and Berko. Interference filter #2 shows an excess counting rate of almost 1 coincidence every 1000 seconds.

really seem like much of a signal, but they used it to make a very nice microwave measurement.[6] They tuned the microwave to the 2S-2P splitting and saw an enhancement of the Lyman-α decay. Once these microwaves are tuned to the proper splitting, a nice resonance appears. (Fig. 3)

If you want to improve the accuracy, one of the troubles is that the inherent line width of this microwave transition is about 44 MHz. It's been known for quite some time that if you really wanted to do precision measurements on any simple atom like hydrogen or positronium, you'd be much better off to do a precision measurement between the 1S state and the 2S state. Those states live a considerable amount of time. In fact, since the triplet ground state lives for 140 nanoseconds, the inherent linewidth of that transition is about 1 MHz.

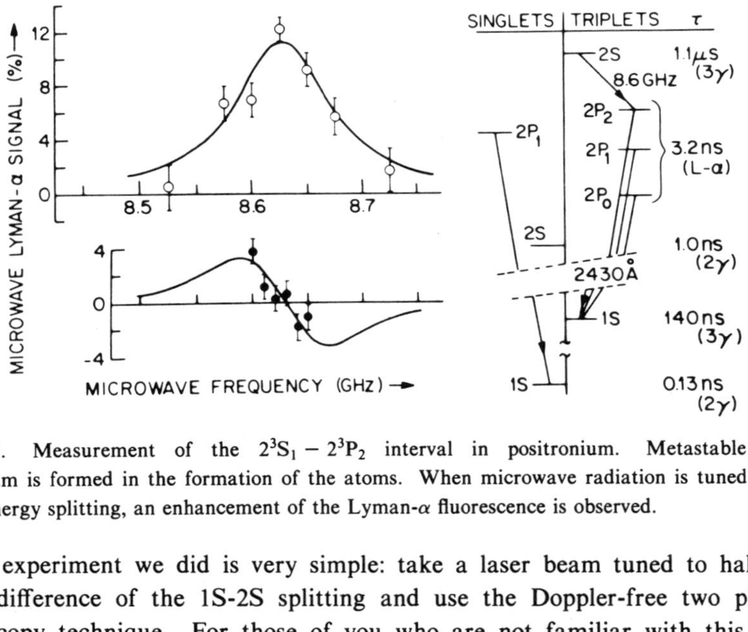

Fig. 3. Measurement of the $2^3S_1 - 2^3P_2$ interval in positronium. Metastable 2^3S_1, positronium is formed in the formation of the atoms. When microwave radiation is tuned to the correct energy splitting, an enhancement of the Lyman-α fluorescence is observed.

The experiment we did is very simple: take a laser beam tuned to half the energy difference of the 1S-2S splitting and use the Doppler-free two photon spectroscopy technique. For those of you who are not familiar with this laser technique, the idea is that we use a transition that requires the absorbtion of two photons. If the photons have equal frequency and are going in opposite directions, a moving atom would generally see one of the photons blue-shifted, but would see the oppositely directed photon red-shifted by the same amount. Thus, all first order Doppler widths are automatically canceled. The beauty of this technique is that as soon as you tune to the correct frequency, all the atoms come in resonance. Otherwise, all the atoms are out of resonance. That's particularly important when you do the positronium experiments where there aren't that many atoms. Our idea is to go, via two photons, from the ground state to the first excited state and then photionize the atom. After that, we collect the e^+ fragment. The general schematic of this experiment is outlined in Fig. 4. We get our positrons from an e^+ emitter. We will use a moderator, much like the ones already talked about at this conference. We store these positrons in a magnetic bottle, bunch them, throw them on another surface and create thermal positronium atoms. Using a laser, we photoexcite the atoms, ionize them, and count the ionization events. The other part of the experiment is to measure the frequency of that laser.

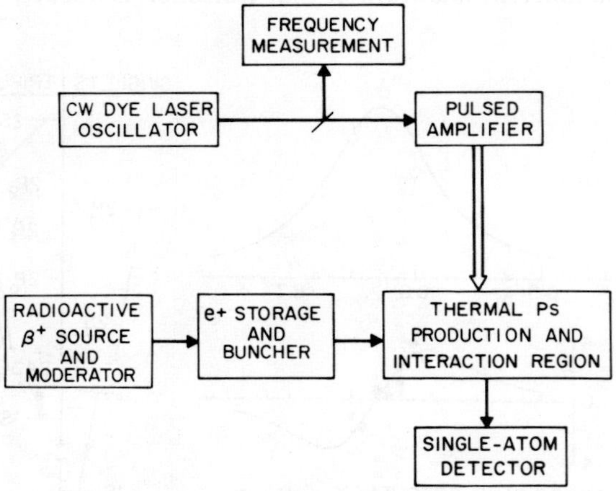

Fig. 4. Block diagram of the experimental apparatus.

Briefly, the schematic of the positron moderator is shown in Fig. 5. The positrons go in, thermalize with the electrons in the metal, and then come to the surface. Approximately several million positrons per second come out of the surface and are extracted from the surface with an average energy of a few eV and a spread of a few hundred meV. After the positrons come out of the surface we put them in a magnetic bottle. The purpose of this magnetic bottle is to match the laser source with the positron source. The laser is a high power, pulsed laser that is on for only a few nanoseconds at a time. The reason is that in order to induce this two photon transition, we need about one megawatt per square centimeter intensity. We want to take the cw positron source and make a bunched positron source. That's done by having a magnetic bottle shown in Fig. 6. The positrons are guided in the magnetic field until they reach an

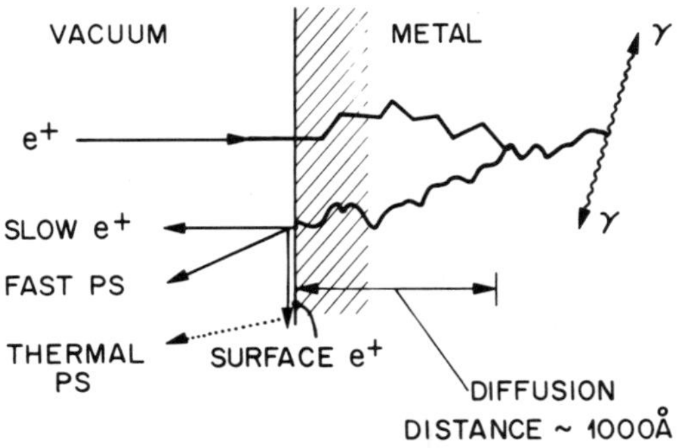

Fig. 5. The Positron Moderator. Positrons incident on a metal surface quickly thermalize with the electrons in the metal. They either annihilate inside the metal or diffuse to the surface where several possible outcomes are shown. We use the slow positrons emitted with energies on the order of a few eV.

electrostatic mirror. Every time they fly past the RF plates, which we tune to the cyclotron resonance of the positrons, the cyclotron orbit increases. Thus, when the positrons come back, they see a pinched magnetic field and adiabatically turn around. What we are doing is localizing the positrons along one direction, but giving them some transverse energy. This process continues for about 50-100 microseconds, at which time the positrons get so much transverse energy, they slip out of the bottle.

Then we put on a high voltage pulse which creates an E field. The field causes the positrons to hit another target. The shape of the potential is simply that of a harmonic oscillator well; it goes as z^2. That's because the longitudinal energy of these positrons is very very small, while the voltage on the plates is typically hundreds of volts to a kilovolt. Thus, the positrons can be regarded as nearly at rest, and will all swing down in ¼ of a harmonic oscillator period and hit the target. The ¼ period is, of course, independent of the amplitude. The

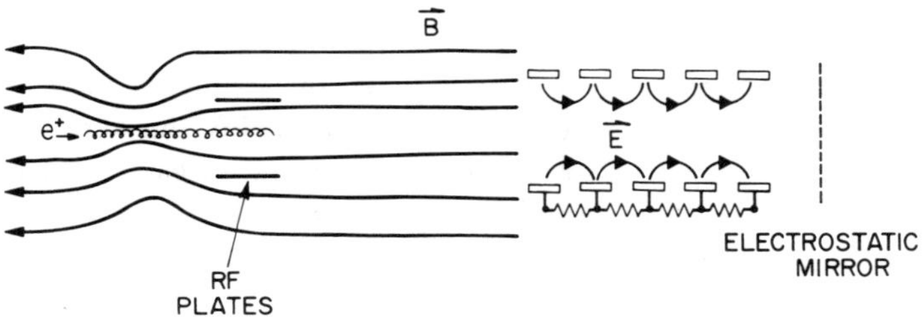

Fig. 6. The Positron Bottle and Buncher.

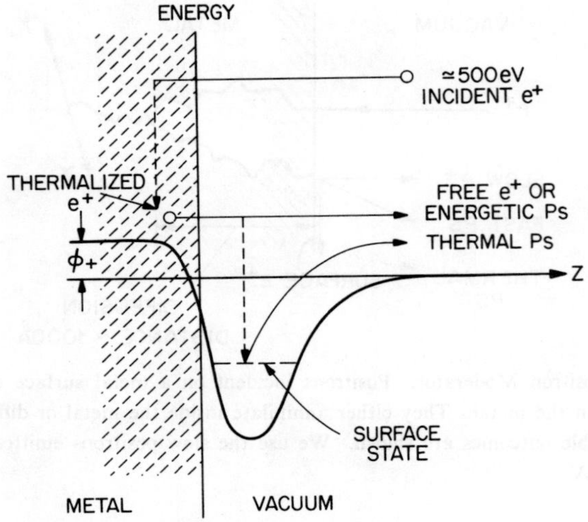

Fig. 7. Formation of thermal positronium in vacuum.

positrons hit the target and the same processes occur as before. (Fig. 7) Energetic positrons could come off the surface. Also, free positrons or positronium with a few eV of energy can come off the surface. The positron can also be trapped in a surface well which is defined by some image charge at large distances from the surface and by a repulsive dipole layer at short distances from the surface. Then, if you heat the surface, positronium can come off, analogous to thermionic emission. After all this work, we now have free positronium atoms in very good vacuum. Also, within about 5-10 nanoseconds we know that the atoms are coming off the surface in an area about 1 square centimeter.

To briefly review what's happened see Fig. 8. Starting with a half curie source, the positrons are moderated, stored in a magnetic bottle, and pulsed onto a metal target. The positronium atoms are then desorbed from the surface of the target. At last count we were able to get as many as 20 thermal atoms per pulse.

The rest of the apparatus is used to make the frequency measurement. We start with a very narrow-band cw dye laser and amplify this very well controlled laser in a series of amplifiers so that we get to several megawatts of pulsed power which we then shine onto the atoms. The basic idea is to make a frequency comparison between the positronium resonance and some well known energy interval. To zeroth order, hydrogen energy levels are very much like positronium energy levels in that the energy levels go as $1/n^2$ and the transition frequency

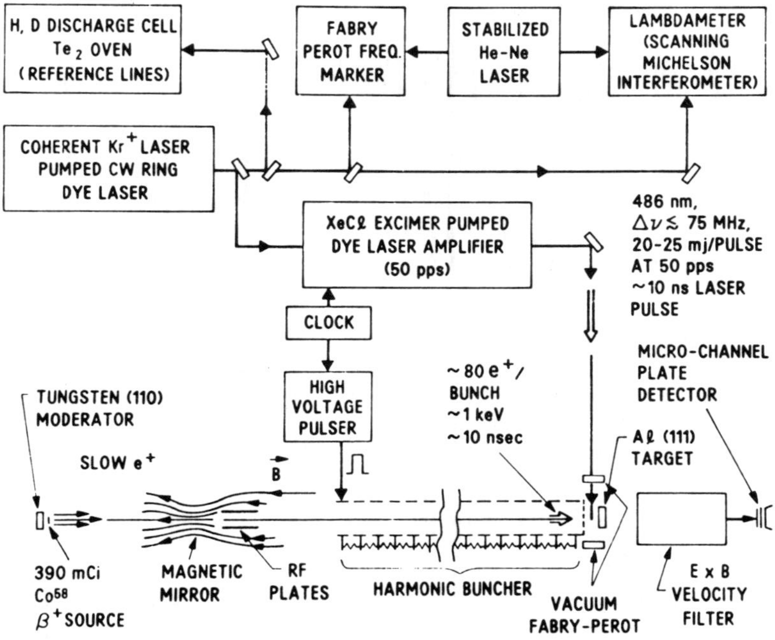

Fig. 8. Schematic of the experimental apparatus.

n = 2→4 in hydrogen is equivalent to half of the frequency of the n = 1→2 in positronium. It would be nice to make the energy comparison between the positronium resonance 1S to 2S and hydrogen reference, n = 2→4.

Unfortunately, hydrogen (actually deuterium) is ~114 GHz away from positronium and although this a near coincidence by optical standards, the cw dye laser was not able to continuously scan between the two lines. Fortunately there are molecular spectra that have a lot of lines. In Fig. 9, we show an example of a very exciting molecular spectra. It is one page out of one volume (there are two volumes) of the Te_2 molecular spectra. To appreciate the density of the molecular lines, we show in Fig. 10 a portion of the Te_2 spectra shown by the arrow taken in our lab with saturation spectroscopy. As you see, the previous coarse spectra didn't really pick up all the lines; we have also superimposed the Doppler-broadened hydrogen fine structure n = 2→4 on this Doppler-free spectra of the Te_2 molecule. The idea is to measure the frequency shift between positronium and one of these molecular lines, and then measure the frequency shift (at your leisure) between one of these molecular lines and the hydrogen

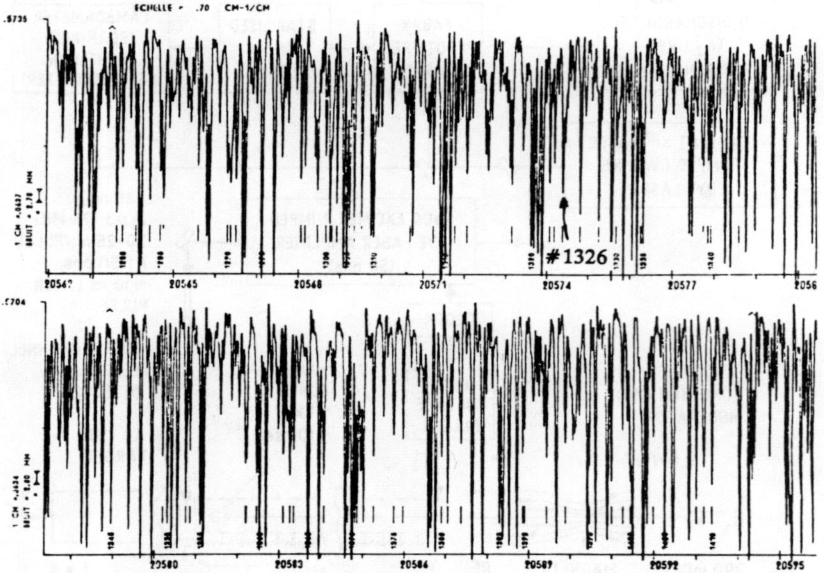

Fig. 9. $^{130}Te_2$ molecular spectra in the spectral range near hydrogen n=2−4 and positronium n=1−2. Taken from *Absorbtion Atlas of Molecular Tellurium*, J. Cariou, P. Luc Laboratorie Aime-Cotton CNRS II, 91405, Orsay, Format (1980).

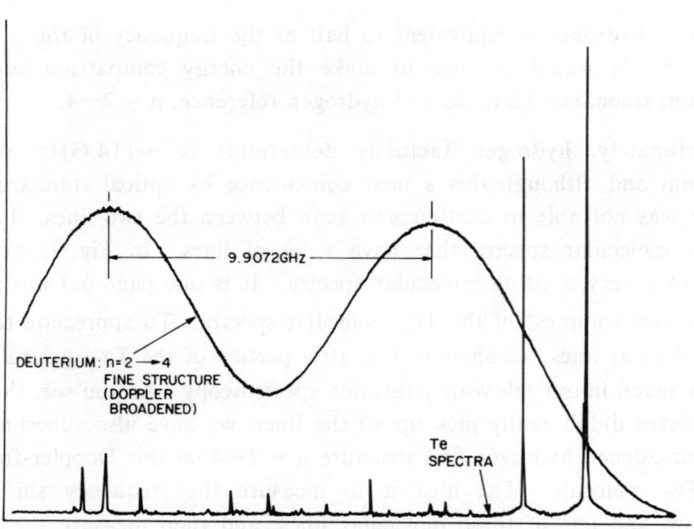

Fig. 10. Simultaneous 20 GHz scan of Te_2 spectra and n=2−4 transition of deterium. In this portion of the Te_2 spectra there is an absorbtion line roughly every 2 GHz.

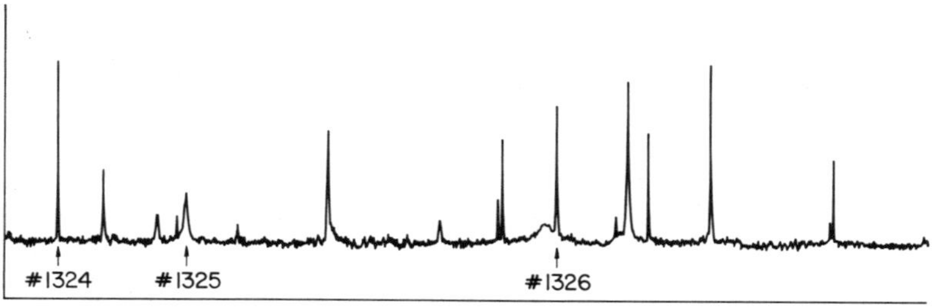

Fig. 11. Doppler free spectra of Te_2 molecular lines in the vicinity of the Ps resonance. The numbers correspond to the C.N.R.S. designations. The absorption cell was run at $T_{body} = 540°C$ and $T_{stem} = 510°C$.

standard. The Te_2 spectra in the vicinity of the Ps resonance is shown in Fig. 11. The actual line of interest, #1326, has a characteristic asymmetric shoulder associated with it that allows us to identify the line among the many other Te_2 lines.

We use normal saturation, Doppler-free spectroscopy with a heated Te_2 cell, and a Wood's discharge cell to make the atomic hydrogen. The frequency difference between the Te lines and the hydrogen lines is made using a long Fabry-Perot interferometer locked to a stablized He-Ne laser. We also use various gadgets that are too complicated to go into at the moment in order to tell us what the wavelength of our laser is on a crude (part per million) frequency scale.

The light comes into the vacuum chamber as shown in Fig. 12. The metal target is hanging from a manipulator in the middle of the chamber. A flat-flat Fabry-Perot interferometer was placed inside the vacuum chamber, and the positronium atoms are made inside this Fabry-Perot. The cavity forms the counter-propagating beams necessary for two-photon Doppler-free spectroscopy and narrows the frequency of the pulsed dye laser to 15 MHz. The sequence is: 1.) pulse the buncher, causing the positrons to hit the target and produce positronium and 2.) shine the laser on it. The Fabry-Perot is a fairly long cumbersome beast hanging in a uhv chamber but it actually works! In Fig. 13, the laser pulse into the Fabry-Perot is shown. The double humped shape is due to the electrical ringing of the eximer laser. What comes out of the Fabry-Perot is also shown in Fig. 13. The little steps are due to the actual round trip time of the photons in the Fabry-Perot as they load up the cavity. The built-up laser light inside the interferometer is very nearly a Fourier-transform limited beam in both frequency and space.

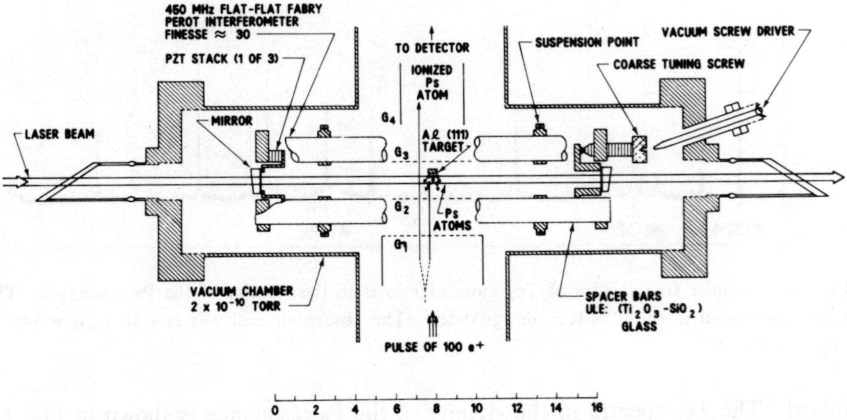

Fig. 12. Detail of the apparatus where the positronium atoms are made. Counter-propagating laser beams are formed by a flat-flat Fabry-Perot interferometer that surrounds the aluminum target.

Fig. 13. Time dependence of the laser pulse before entering the interferometer and after exiting the interferometer.

We introduce a frequency shift when we use the Fabry-Perot interferometer. The metrology is done using a cw laser but the measurement uses a pulsed laser. Even though the interferometer is locked to the frequency of the cw laser, if the pulsed laser enters the Fabry-Perot as a different angle with respect to the cw laser, the pulsed light will not be at the same frequency as the cw light. Figure 14 shows the measurement of the pulsed laser light that exits the interferometer along with the cw light. The measurement was made with a confocal Fabry-Perot interferometer which is insensitive to incident angles. An rf sideband was put on the cw laser with an acousto-optic modulator for a frequency calibration.

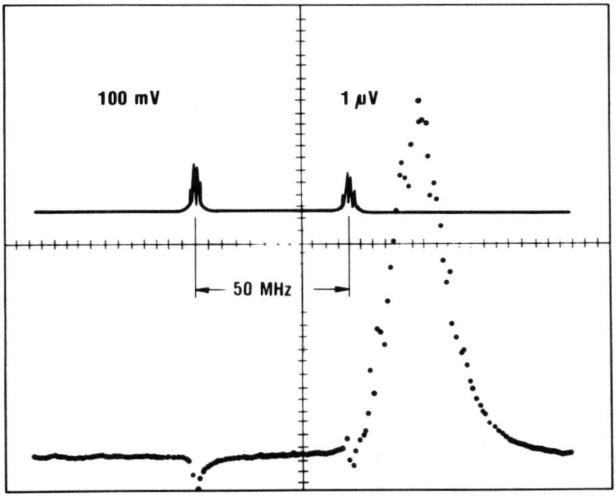

Fig. 14. Frequency off-set measurement of the cw and pulsed light exiting the Flat-Flat Fabry-Perot as measured with a confocal Fabry-Perot interferometer. A 50 MHz frequency sideband is added to the cw laser beam in order to calibrate the frequency scale.

Fig. 15 shows a schematic of the single atom detection scheme we used. Positrons enter from the left (P) and Positronium atoms are made on the target surface. We use two photons to excite to the n=2 state and a third photon to ionize the atom. The e^+ fragment is the collected. Since the whole apparatus is in a magnetic field, the positron is extracted by pulsing the positrons back up the harmonic potential. (There is an electric field between the grid G and target C.) The positrons enter an E×B drift region that deflects them sideways and around the target. They then go through another E×B velocity selector to eliminate false positive ion counts, and finally to a microchannel plate.

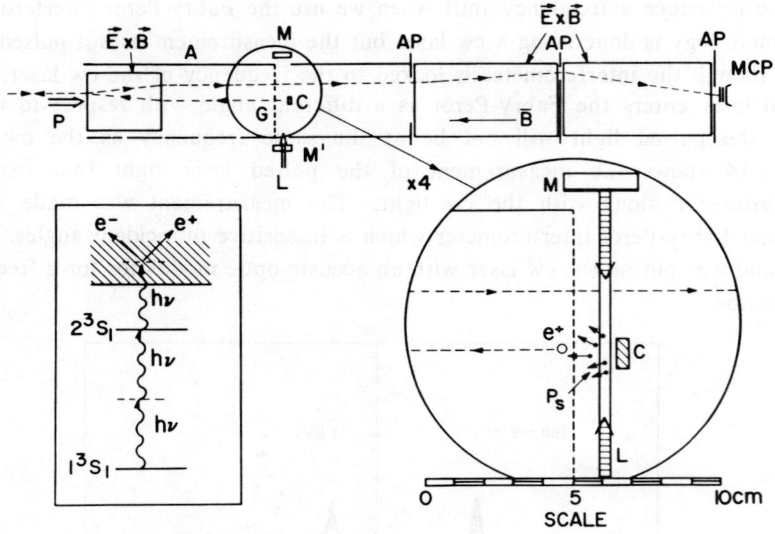

Fig. 15. Schematic of the detection scheme. Positrons denoted by (P→), hit the target C to produce positronium atoms. Counter-propagating light between mirrors M and M′ excite and photoionize the atoms. The ion fragment (e^+) is directed by a series of $\bar{E} \times \bar{B}$ field plates through aperatures AP that separate out heavier positive ions. The positron finally hits a microchannel plate (MCP) and is counted.

You might suspect that this apparatus didn't work the first time it was turned on. In the absence of a signal, we needed a way of lining up the positron optics and detector timing. The way we did it was to use the low energy positrons that are re-emitted from this target (re-moderated). About a third of the incident positrons come off the surface as low energy positrons. If the potential difference between the grid and target is large compared to the 1 or 2 eV low energy positrons, the re-emitted positrons will follow the same tortuous route as the ion fragments. If a phosphor is placed in back of the microchannel plate, one can actually see where the positrons hit. (By the way, positrons are colored green!) By adjusting the electric fields and magnetic field, or perhaps putting some magnetized files in strategic places, the positron can be made to hit in a little 3/8" diameter circle on the back of the micro-channel plate. The e^+ fragment due to the Ps photonization is detected by delaying the pulsed laser for 15 nsec or more after the positrons hit the target and gating the signal from the microchannel plate 15 nsec after the arrival of the re-emitted positrons.

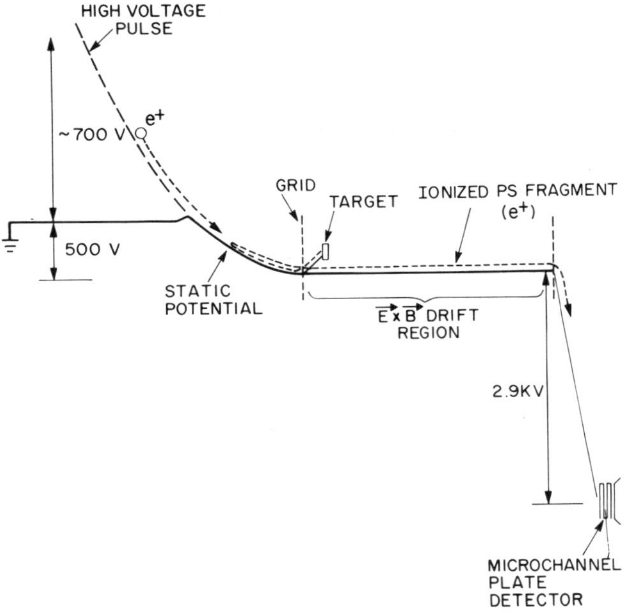

Fig. 16. Electrostatic potential diagram.

Figure 16 shows what the positron sees in terms of the electrostatic potentials. The high voltage pulse produces the harmonic well used to sweep the positrons towards the target. The diagram is self-explanatory.

After all the adjustments have been made, the first signal that appeared is shown in Fig. 17. The first resonance is shown along with a frequency number (Fabry-Perot interferometer with a 75 MHz free spectral range) and a Te_2 spectra. Note the easily identifiable shape of the Te_2 resonance in the center of the diagram: a narrow peak sitting on the high frequency side of a broad low peak. For those of you who don't believe there really is a resonance, I have underlined where it is. Once a signal is seen, the apparatus can be "tweaked" to get a much better signal. Figure 18 is an example of a bigger signal. Again, we show a Fabry-Perot marker, the Te_2 reference line, and background measurement taken when the laser is not on. The signal-to-noise is not bad considering the fact that this resonance is taken with ~10 atoms pulse in a ~1mℓ volume. Notice that the Te_2 reference line now has two comparison resonance lines. These lines are due to a 50 MHz sideband on our laser so that an apparent resonance coincides with the positronium resonance. The taller, central line is a cross-over line.

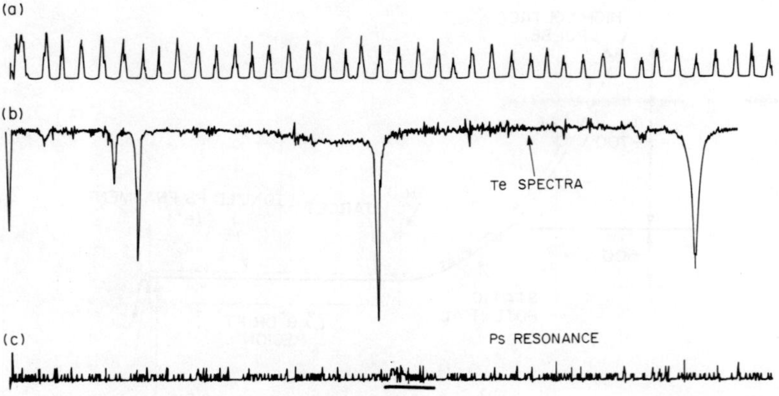

Fig. 17. Experimental scan #7. (a) Fabry Perot frequency marker, (b) Te_2 spectrum, (c) Ps 1S-2S signal.

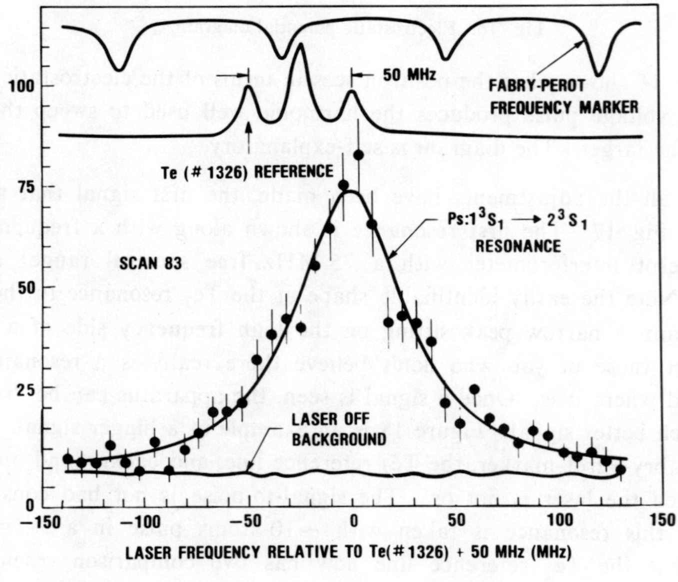

Fig. 18. Experimental scan #83

One of systematic effects we have to account for is the Doppler shift. Although the first-order Doppler shift is eliminated with the two-photon technique, there's a second-order Doppler shift due to relativistic time dilation. A clock riding with the positronium is seen to go slower in the laser frame of reference. We can try to account for this effect by calculating a correction term, but it's much better to measure it. We use a time-of-flight technique to measure the second order Doppler shift. At t=0, we throw the positrons onto the target and immediately a thermal distribution or nearly thermal distribution of positronium atoms come off the target. By delaying the laser beam relative to the time the atoms were made, we can sample either very fast moving atoms or very slow moving atoms. If we do this for all delay times you can actually get a frequency shift as shown in Fig. 19. At very early delay times there are a lot of fast atoms which have a large frequency shift. At longer delay times, there are fewer atoms, but these atoms have less of a shift. We simply extrapolate to infinite delay or zero velocity. This shift due to time dilation is actually quite large because the thermal atoms are very light.

S. Berko

If you do that then the shape cannot be Lorentizian.

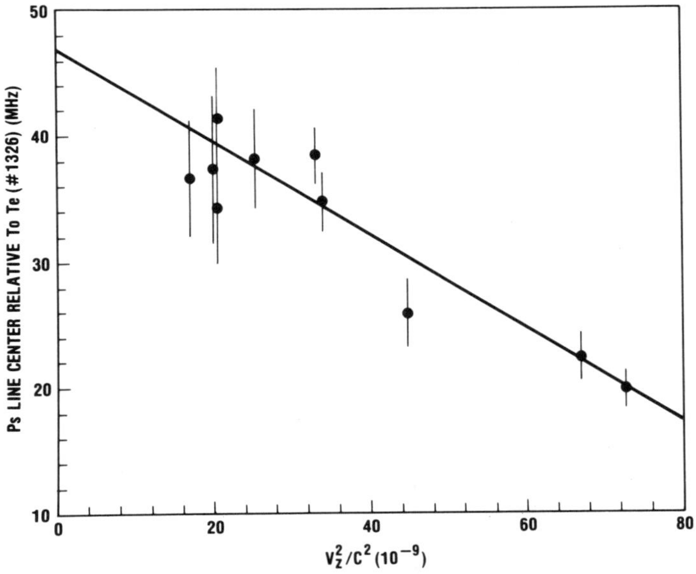

Fig. 19. Second-order Doppler shift (time dilation).

Chu

That's right, and it's especially important if you look very very early in time, but in most of our data we are sampling a spread of velocities and the data was not good enough to warrant looking at deviations from a Lorentzian.

There are other systematic effects as well. The most serious effect is the ac Stark shift. This is a very high intensity experiment because we want to excite

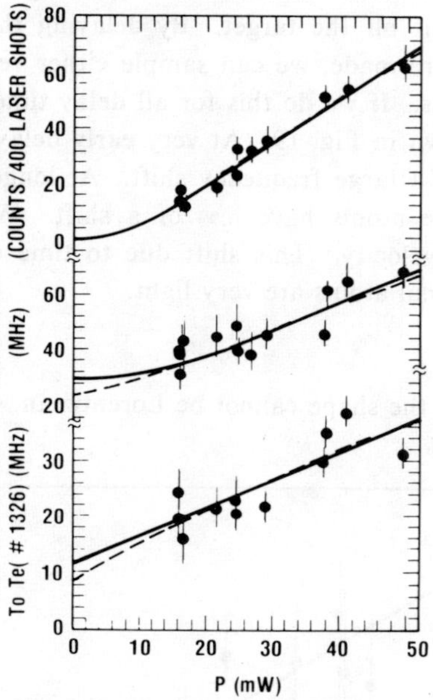

Fig. 20. AC Stark shift, line width, and amplitude vs. laser power (relative).

all the atoms; i.e., we have to saturate the two photon transition. At saturation intensities, large ac Stark shifts are expected, and in Fig. 20, we plot several measured quantities as a function of the power of the laser, which we take to be average power coming out of the Fabry-Perot. The line center, relative to the Te line, shifts roughly linearly with intensity (quadratic in electric field) as expected. The solid lines are obtained using perturbation theory, and the dotted lines are an attempt to try to get a better fit by accounting for saturation.

Table I contains our preliminary data. We call it preliminary because we intend to go back and measure the frequency a little bit better. The deuterium frequency is based on theory. Hydrogen energy levels are fairly well-known

TABLE I. Measurements and calculations used to obtain the Ps $1^3S_1 - 2^3S_1$ interval. Frequencies in megahertz.

$\frac{1}{2}\nu_{Ps} - \nu_{Te_2}$ $(v_z^2/c^2 \to 0)$	+45.1 ± 4.2	+46.7 ± 3.3
ac Stark shift	−15.2 ± 7.3	−21.6 ± 5.0
$\frac{1}{2}\nu_{Ps} - \nu_{Te_2}$ $(v_z^2/c^2 \to 0, P \to 0)$	+26.7 ± 4.9	
$\nu_{Te_2} - \nu_D$	+114410.4 ± 1.8	
dc Stark shift	+17.0 ± 5.0	
$\nu_D - \frac{3}{16} R_\infty c$ (theory)	−156226.041	
$\frac{1}{2}\nu_{Ps} - \frac{3}{16} R_\infty c$	−41771.9 ± 4.9 ± 5.3	
Motional Stark and Zeeman	−1.14	
dc Stark Shift	−0.07	
First-order Doppler shift	0.0 ± 1.9	
ν_{Ps} $(1^3S_1 - 2^3S_1)_{exp}$	$\frac{3}{8} R_\infty c$ − 83546.3 ± 10.5 ± 10.6	
ν_{Ps} $(1^3S_1 - 2^3S_1)_{th}$	$\frac{3}{8} R_\infty c$ − 83533.06(2) ± (10)	

because of the radio-frequency Lamb shift measurements. These measurements agree with theory to a precision of ~10 kHz for the 2S-2P splitting. Using the theoretical deuterium frequency as the beginning absolute frequency, we have measured the Te_2 to deuterium frequency difference. A major systematic in this part of the experiment, is the dc Stark shift in the deuterium discharge cell. The atoms in an ugly plasma give a large frequency shift and a frequency uncertainty because we are uncertain about the exact nature the plasma.

In the measurement of the $\frac{1}{2}\nu_{Ps} - \frac{3}{16} R_\infty c$ frequency in Table I, we have divided the error into two parts: a 4.9 MHz error associated with the $Ps - Te_2$ measurement and a 5.3 MHz error due to the $Te_2 - D$ measurement. The motional Stark and Zeeman shift of the Ps atoms in the 150 Gauss magnetic field, the dc Stark shift of the Ps atoms, and the residual first-order Doppler shifts are also shown. Using the calculation of Fulton[7] and the 1981 value of the Rydberg[8], experiment and theory differ by 13 ± 10.5 ± 10.6 MHz. The next higher order terms $\alpha^4 R_\infty$ have not been calculated and could lead to a modification of the theoretical value by ~10 MHz. Unfortunately, we agree with theory. So far, we have about a 10^{-8} measurement of the 1S-2S, and that corresponds to about ½ percent measurement of the QED corrections.

The corollary that comes out of this measurement is that if one believes in the QED calculation, we can turn around and say what we've measured is the Rydberg constant. However, the Rydberg constant is not the same constant as measured using the hydrogen atom. It is a mix of the e^- and the e^+ Rydberg constant. Since the two constants agree to ~12 parts per billion, we can deduce that $\frac{m_{e^+}}{m_{e^-}} = 1$ to the order of 24 parts billion. This is the best measurement of the e^-/e^+ mass ratio to date.[9] Now, what about the future? How good a measurement can we make if we try very hard?

S. Berko

Can I ask you something while you are still there? There is also a shift of the ground state by QED, which you have to include in fitting the data. At what power of alpha do you see that shift?

Chu

We are testing the Fulton and Martin calculations that can be expressed as

$$E(2S) = E_{RYD}\,[-\frac{1}{8} + C_1\alpha^2 + C_2\,\alpha^3/n\alpha + C_3\,\alpha^3 + ...]$$

Crudely speaking, the factor of 1/8 is the hydrogen approximation, the factor α^2 is the Pauli fine-structure term, and the $\alpha^3/n\,\alpha$ and α^3 terms are the first order radiative corrections calculated by Fulton.

How can one improve this measurement? One of the things we can do is get a better positron source. The free electron laser beam dump that Phil talked about[10] will certainly give us about four orders of magnitude more positronium atoms. Ideally, we would also like to measure the frequency difference between hydrogen and positronium and eliminate the tellurium cell.

Our present level of accuracy is about an order of magnitude away from the uncertainty in the Rydberg constant. We can probably improve the measurement by two orders of magnitude. Since we trust the hydrogen values to ~10 kHz (for the 2S states), a comparison of the frequency ratios of hydrogen to positronium factors out the Rydberg and will reveal any anomalies in the positronium calculations at the level of our expected experimental uncertainties (~ 100 kHz). One possibility is to take a laser, put a 114 GHz frequency sideband on it. We use this sideband to make a Doppler-free measurement of the 2S-4P deuterium transition in an atomic beam, and use the unshifted frequency to excite positronium atoms. In a good build-up cavity intensities of several kilowatts/cm^2 of cw power should be possible. The major line broadening will be due to the transit time of the Ps atoms across a submillimeter diameter laser beam and a source of cryogenic Ps atoms would also be helpful.

The expected linewidth of this positronium resonance would be about 1-5 MHz and the linewidth of the hydrogen 2-4 if we can go 2S to 4S and with a little admixture of P would also be a few MHz. Thus, we can maybe expect an uncertainty in the 100 kHz range. I think after we get to the 100 kHz range, the experiment will begin to get difficult, and more clever people will have to take over.

REFERENCES

1. S. Chu and A. P. Mills, Jr., Phys. Rev. Lett., **48**, 1333 (1982), A. P. Mills, Jr., and S. Chu, in *Atomic Physics* **8**, ed. A. Rosen and S. Svanberg (Plenum, New York, 1982), p. 83. The relevant reference on the production of thermal positronium in vacuum are contained in these early references.
2. S. Chu, A. P. Mills, Jr., and J. L. Hall, in Laser Spectroscopy VI, ed. H. P. Weber and W. Lüthy, (Springer-Verlag, Heidelberg, 1983), p. 28; S. Chu, A. P. Mills, Jr., and J. L. Hall, Phys. Rev. Lett., **52**, 1689, (1984).
3. See, for example, H. A. Bethe and E. E. Salpeter, in "Quantum Mechanics of One and Two-Electron Systems," *Handbuch der Physic*, ed. S. Flugge (Springer-Verlag, Berlin, 1957), p. 281.
4. K. F. Canter, A. P. Mills, Jr., and S. Berko, Phys. Rev. Lett., **34**, 187 (1975).
5. K. F. Canter, A. P. Mills, Jr. and S. Berko, Phys. Rev. Lett., **33**, 7 (1974).
6. A. P. Mills, Jr., S. Berko, and K. F. Canter, Phys. Rev. Lett., **34**, 1541 (1975).
7. T. Fulton, Phys. Rev. A**26**, 1794 (1982).
8. S. R. Amin, C. D. Caldwell, and W. Lichten, Phys. Rev. Lett. 47, 1234, (1981).
9. See P. B. Schwinberg, R. S. Van Dyck, Jr. and H. G. Dehmelt, Phys. Lett. **81A**, 119, (1981) for the previous best measurement of m_+/m_-.
10. P. M. Platzman, this volume.

Positrons in a Low Density Gas: Understanding the Galactic Center Annihilation Line, and Forming Monoenergetic Beams of Positronium

B. L. Brown

AT&T Bell Laboratories
Murray Hill, NJ 07974

I am very pleased and honored to be able to talk to you today. I haven't known Steve Berko very long but we are going to be working together soon and I'm very excited about that. We will be at Brookhaven National Labs with a consortium working on an intense positron beam and some of the ideas that I present here on monoenergetic positronium may be incorporated into this work.

The two areas that I am going to talk about today are somewhat different in their application. The first idea is a novel technique for creating monoenergetic positronium in a gas and I will discuss some of the properties that this monoenergetic positronium should have. I will also include a preliminary observation of the annihilation spectrum of monoenergetic positronium in helium.

The second subject that I will present is related to the gamma ray observations of a 511 keV positron annihilation line from the direction of the galatic center. This line was thought to be very narrow and on this basis neutral hydrogen was ruled out as an annihilation medium. I have found that a correctly constructed annihilation spectrum for neutral hydrogen is consistent with the galactic lineshape. Included in some of the comments I'll make will be some experimental results that I've obtained working with Marv Leventhal, Allen Mills and Dave Gidley at AT&T Bell Laboratories.

To start off I would like to summarize the positron-gas atom scattering channels. The five principle interactions of positrons with gas atoms are: elastic scattering, excitation (where you leave the atom in an excited state), ionization, positronium formation (where positronium is formed leaving an ion), and direct annihilation of the positron with an atomic electron. The latter two will be the principle processes that I'll be discussing this afternoon. The positronium formation threshold is the same as the lower limit of the Ore gap and is defined as $E_I - 6.8$ eV, where E_I is the atomic or molecular ionization energy and 6.8 eV is the positronium binding energy.

A monoenergetic positronium beam could be a useful tool for doing a number of different experiments. The positronium-atom collision cross-sections would be interesting to measure and it would be possible to look for positronium molecular bound states. We could also look for resonances that Dick Drachman has predicted for collisons between positronium atoms and hydrogen atoms, and Dick has a number of different ideas along that line. We can also use variable energy positronium for investigating surface structure and Karl Canter has proposed some ideas in that direction.

I have drawn a simple picture of positronium formation in a gas (Fig. 1). A positron comes in, collides with an atom, helium for example, and departs in the positronium state leaving an ion. The approximate energy relationship here is the energy of positronium coming off, E_{ps}, is equal to the incoming positron energy, E_0, minus the ionization energy, E_I, plus the positronium binding energy 6.8 eV minus the recoil energy of the atom, E_r. The atom is taken to be at rest. The incoming positron energy is spread by the energy resolution of the positron moderator, ΔE, which typically is about 0.2 eV. This is the main limiting factor in the positronium energy spread. The recoil energy E_r also causes some spread

$$E_0 - I + 6.8 = E_r + E_{Ps} \qquad \text{ENERGY}$$

$$\vec{P}_0 = \vec{P}_r + \vec{P}_{Ps} \qquad \text{MOMENTUM}$$

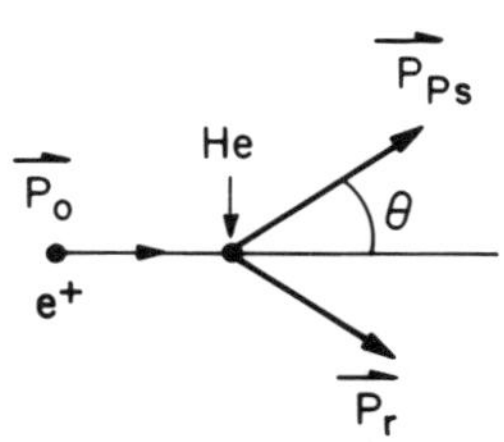

Fig. 1. A simple schematic of the positronium formation shown for helium. The approximate energy relationship is also given (the motion of the gas atom is neglected).

in the positronium energy. With conservation of energy and conservation of momentum we can derive an approximate expression for the recoil energy E_r of the atom:

$$E_r \approx \frac{m_e}{M} [3E_o - 2Q - 2 \cos \theta \sqrt{2E_o (E_o - Q)}]$$

where $Q = E_I - 6.8$ eV, m_e is the electron mass, M is the gas atom/molecule mass, and θ is shown in Fig. 1. In Fig. 2 the angular dependence of the recoil energy E_r is shown. Again the energy spread of E_r is smaller than that of the typical moderator. Right now we're limited by the moderator and not by the dynamics of the collision. Using a cold moderator with a reverse bias, a less efficient but narrower positron beam can be obtained. Woody Crane, Allen Mills and I are presently testing a cooled moderator at Bell Labs (See, A. P. Mills, this volume).

Let's look for a moment at an Ore diagram for He (Fig. 3). In the energy range labeled A, monoenergetic positronium can be formed with the first inelastic collsion. If an ionization or excitation collision is made in this range the positron drops down below the Ore gap and it is impossible to form positronium. Above this, in the energy range labeled B, it's possible to get a monoenergetic

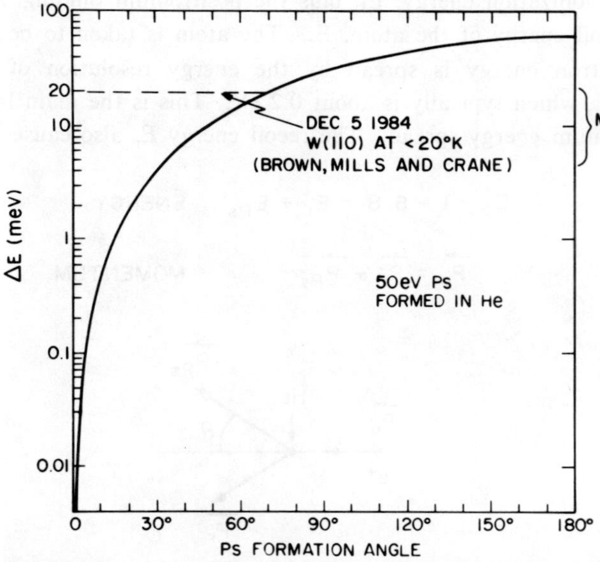

Fig. 2. The recoil energy of the gas atom as a function of positronium angle. The incident positron is assumed to come in at $\theta = 180^\circ$ and be at an energy of exactly E_o (no energy spread).

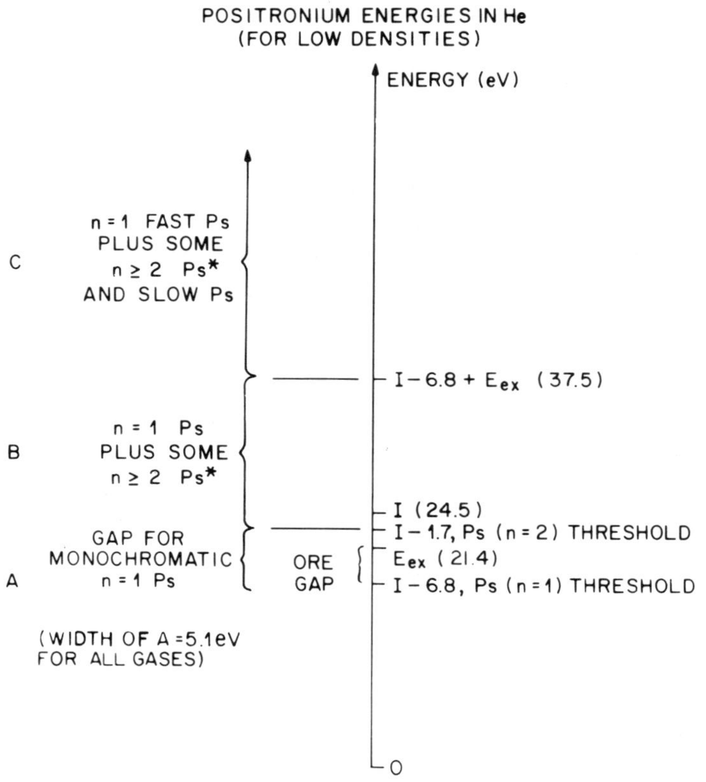

Fig. 3. An Ore diagram for Ps formation in He. All the noble gases have a similar pattern.

component followed by some positronium at lower energies that would be separated by at least the minimum excitation energy (about 20.4 eV for an He atom). A usable monoenergetic component would be present with some lower energy positronium.

A. Stewart

Why are you talking about helium only?

Brown

Helium is just a specific example of the very general principles of Ps beam formation in gases that I am describing here. I am using He because there are reliable calculations done in He and it is a simple system to think about. We could use almost any gas atom or molecule. Helium is interesting because it has a high binding energy. It can produce higher energy pure monoenergetic positronium, and there is a larger gap between the monochromatic positronium formed at higher energies and the positronium formed after first making an inelastic collision.

The differential positronium formation cross sections in He have been calculated with several models.[1] The Born approximation and all other models show essentially the same angular dependence, so the model we choose is not important. The differential cross sections are shown with a linear scale in Fig. 4 and their form is nearly Gaussian. An appreciable amount of the positronium comes out in a narrow angle so it's very collimated. That's why this can be considered a beam of positronium atoms. This curve is for positrons with an incident energy of about 60 eV forming positronium at 42 eV in helium. All the curves along this energy range have basically the same functional form but below 10 eV, they start to spread out quite a bit and the positronium becomes more isotropic.

P. Platzman

That's for helium? For scattering from helium?

Brown

Yes, these calculations were done for helium.

P. Platzman

And for hydrogen, it's narrower?

Brown

For hydrogen, I don't know.

P. Platzman

Well hydrogen is less tightly bound so you would think it would be narrower. I mean what produces the peak in the forward direction is the fact that the binding energy of the target provides the angular spread. It must get narrower.

Brown

It might get narrower in hydrogen, but you would be bothered more by contamination if you are interested in just getting a monoenergetic beam. In hydrogen the minimum excitation energy is much less than that of helium, so you would get more contamination from lower energy positronium. For example

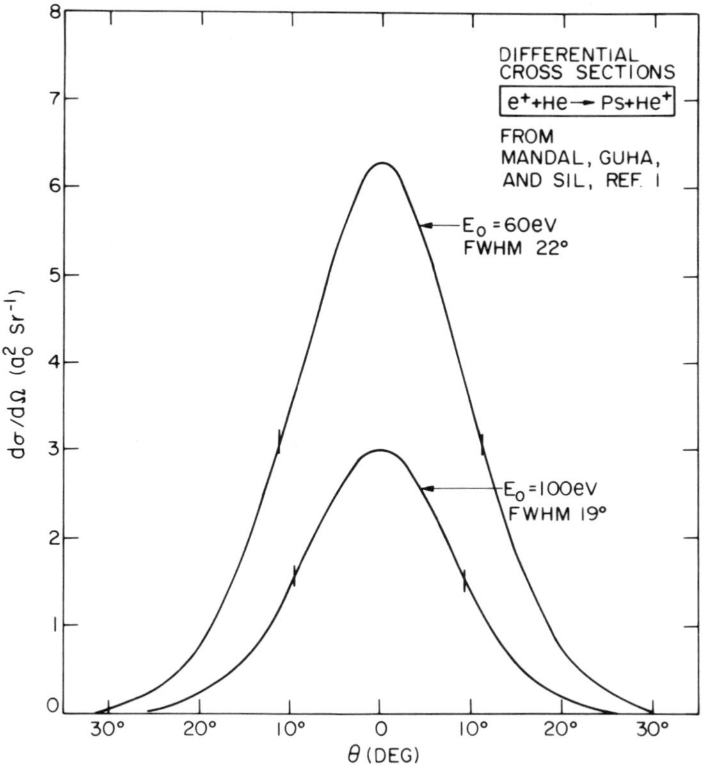

Fig. 4. The solid lines show the calculated differential cross sections ($d\sigma/d\omega$) for Ps formation in He from Ref. 1. at two incoming positron energies.

50 eV positronium formed in hydrogen has some contamination starting at 40 eV and below. Other noble gases such as Ar or some molecules with very low binding energy could also be used. Again I am using He as an example and of course other gases could have advantages over He in a given situation. For example there are many gas atoms with higher Ps cross sections than He which may produce Ps more efficiently. Molecular gases with very low binding energies could produce a more highly colimated Ps beam. I am not saying He is the only gas to use.

In Fig. 5 the apparatus used for a preliminary observation of monoenergetic positronium is shown. Briefly, a ^{58}Co positron source is used in conjunction with a tungsten moderator to produce slow positrons. The positrons are magnetically

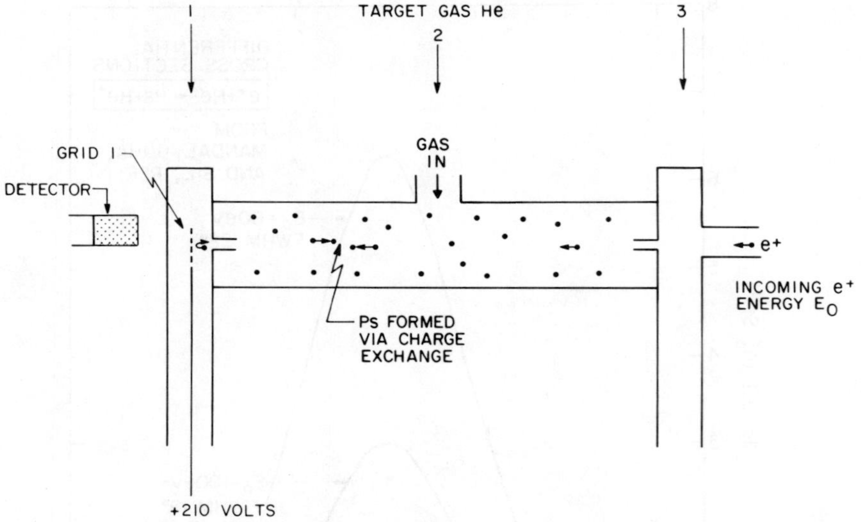

Fig. 5. The experimental apparatus used to demonstrate in a preliminary way the annihilation of monoenergetic positronium. The apparatus was originally used to simulate the galactic medium for positrons. (See Ref. 2).

confined and travel through a vacuum chamber and into the gas region where the collisions occur between positrons and gas atoms.

The preliminary observation of monoenergetic positronium was carried out with positrons making one round trip through the gas cell. The grid in region 1 is set at a reflecting potential. The singlet positronium formed in a collision with a gas atom will produce a blue-shifted 511 keV annihilation line, if it's formed going toward the Ge(Li) detector. It will be red-shifted if it's formed going away from the detector. This gas is of sufficiently low density so that the contributions going each way, red and blue-shifted monoenergetic positronium, are approximately the same. The experimental spectrum is shown in Fig. 6. The side peaks extend out to plus and minus 55 eV and are due to the red and blue-shifted positronium singlet annihilation. The central peak is due to two major sources: contamination from positrons that make an excitation or an ionization collision and then form positronium, and triplet positronium that annihilates on the walls of the apparatus.

This concludes my remarks on monoenergetic positronium beams formed in a gas. I would now like to discuss the galactic positron annihilation line and I'm

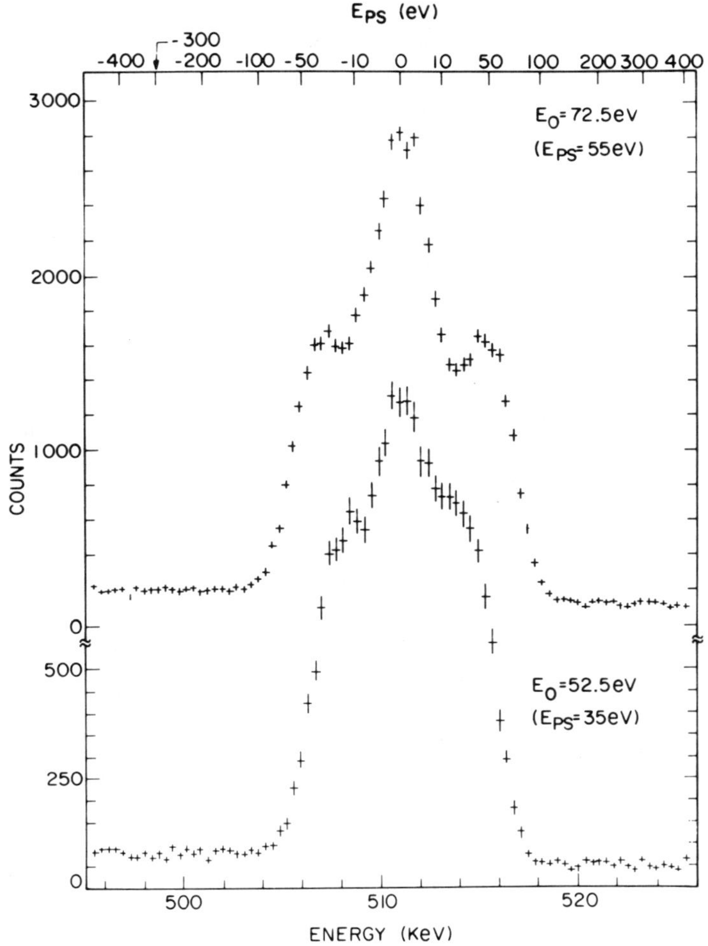

Fig. 6. The first observation of a monoenergetic Ps beam formed in He. The Ps singlet annihilation line is Doppler shifted both above and below the 511 keV annihilation line indicating that a colimated monoenergetic Ps beam is produced.

going to address the question: "Can the annihilation medium be cold neutral atomic hydrogen?" Previous investigators had considered a medium where you would have the direct annihilation line and the positronium annihilation line together. They ruled out atomic hydrogen because the observed line was so narrow. I believe that their interpretation is incorrect and that a neutral hydrogen medium cannot be ruled out based on the data that we presently have. There are two components of the annihilation line spectrum that we expect from a neutral medium. One is the direct annihilation of positrons with atomic electrons, which Dick Drachman has derived from theoretical results by Humbertson, with a width of 1.3 keV at 511 keV. This spread is principally due to the distribution of atomic electron momentum. The other contribution to this narrow line could be the singlet annihilation line which has a Doppler broadened width of 6.4 keV. Leventhal, Mills, Gidley and I have measured this width[2] with the apparatus shown in Fig. 5. Bussard, Ramaty and Drachman calculated the width with a Monte Carlo method.[3] We actually got 6.4 ±.1 keV, and they got 6.5 keV with their calculations so we are in good agreement. If the fraction of the original positrons that survive to below the positronium threshold and directly annihilate is X then the relative singlet positronium contribution would be $(1 - X)/4$. The statistical ratio of positronium formation requires that 25% go into the singlet channel and 75% go into the triplet channel. In what follows I will refer to our measurement,[2] X = 9% and the theoretical value,[3] X = 5.5%.

Now I will mention some of the pitfalls in analyzing the galactic center data that many people, myself included, fell into. One problem was the experimental error limits that were quoted in various papers. Both the experimentalists and others interpreting the results took the one-sigma error limits to be three-sigma error limits.[4,5] This was confusing because some theoretical ideas about what was happening in the galactic center were based on these limits. Confusion was also created when the calculation of the narrower line was first quoted to have a 2.6 keV linewidth. This was subsequently corrected to be 1.3 keV FWHM,[6] but it left some confusion. It was adequately corrected but not everyone knew about the correction. Another fundamental problem is that people who have reported the galactic center linewidth limits have used a Gaussian lineshape model. If the lineshape is really a wide and narrow combination of lineshapes then it's not a Gaussian, so error limits on the full width half maximum of the curve are not too meaningful. It's difficult to compare the width of this strange looking Gaussian-like thing to something thats been fit with a Gaussian, and confusion has resulted. A further difficulty is that the models weren't fit directly to the data, which was actually the reasonable thing to do. So, with all those problems in mind we will look at the picture obtained by correctly constructing a lineshape and then fitting it directly to the data. In doing this I have found that we can't rule out cold atomic hydrogen as an annihilation site as had been previously reported. At the same time other proposed annihilation media cannot be ruled out such as annihilation in an ionized plasma,[3] or annihilation in dust grains.[7] In

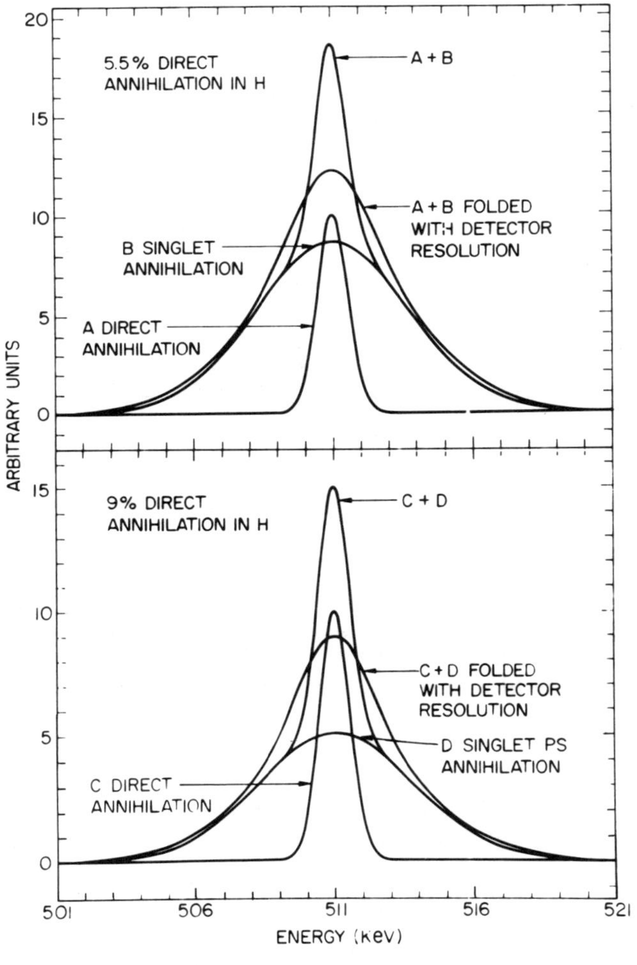

Fig. 7. Simulated positron annihilation energy spectra for neutral hydrogen with two different survival fractions, 5.5% and 9%. Folding with the HEAO-3 detector resolution is shown. For simplicity the spectra are shown without the triplet contribution. The triplet is included with the spectrum when fit to the data in Fig. 7.

Fig. 7 the complete spectrum is constructed with the direct annihilation 1.3 keV width added to the singlet annihilation 6.5 keV width. The sum, A+B, is a very peaked function. When you take the detector resolution and fold it at every point through this curve, you get a function which is still more peaked than a Gaussian of the same area.

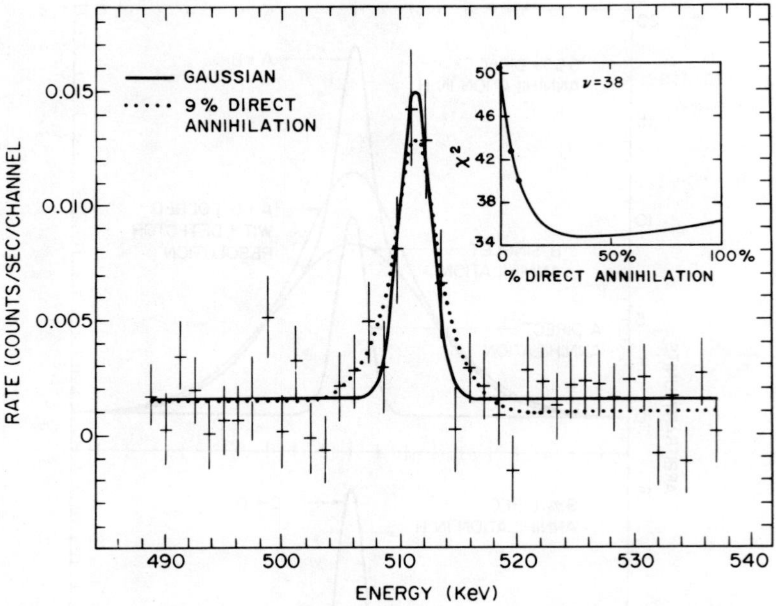

Fig. 8. HEAO-3 cosmic gamma ray spectrometer data from the direction of the galactic center. (The data was obtained from Ref. 5). A gaussian fit is shown with a 3.25 keV FWHM. My two-component model is shown in dots for the 9% direct annihilation fraction. The inset shows χ^2 for other fractions.

I'd like to talk just for a moment about why it's wrong to compare linewidths when the curves are different shapes. The HEAO-3 satellite data, is shown in Fig. 8. Using a Gaussian function to fit the data, a width of 3.14 ± .57 keV (one-sigma limit) is obtained with the detector resolution included. When unfolded the limits become $1.6 \begin{cases} +0.9 \\ -1.6 \end{cases}$ keV. If these limits are compared directly with our lineshapes, agreement is obtained with the unfolded lineshapes, both for 9% direct annihilation fraction and a 5.5% direction annihilation fraction. However, with the folded picture the 5.5% fraction gives rather severe disagreement while the 9% fraction is acceptable. Clearly comparing Gaussian fits to non-Gaussian linewidths is not the correct thing to do. The model must be fit directly. This was done for the survival fraction of 9% that was derived from the experimental measurements. The fitting result for the 9% fraction is shown in Fig. 8. The HEAO-3 satellite data · are used since they are the best, statistically. Marv Leventhal, MacCallum, and Stang also made a very good observation with a high resolution gamma-ray detector and these conclusions also hold for that observation.[4] In the Fig. 8 insert, chi-squared versus the direct annihilation fraction is shown with a dot at X = 5.5% and 9% fractions. The number of degrees of freedom here is 38. A chi-squared per degree of freedom this large would be expected 25% of the time if the 5.5% model is correct. If the 9% model is correct, a chi squared per degree of freedom this large is expected about 40% of the time. So based on a chi-square test these numbers are both reasonable.

One additional point here is that the statistics are poor on this line. Hopefully we can expect more data in the future. I just obtained the latest on the galactic center from Phil Platzman during lunch. Marv Leventhal and his colleagues are in Australia and have flown their most recent Germanium γ − ray detector. The 511 keV line is apparently still "off", but this is a very preliminary result, obviously, since it's less than a week old. [This is still true. See

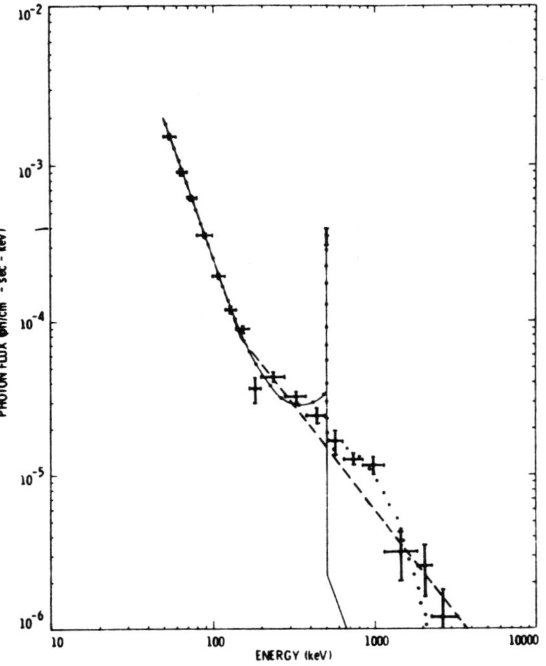

Fig. 9. HEAO-3 data reproduced from Ref. 9. The solid line shows a fit using are power law plus a triplet continuum. The dashed line is a fit using two power laws. The dotted line is a fit using power law, a triplet contribution and a black body spectrum.

Leventhal, MacCullum, Huters, and Stang in Ap. J. **302** (1986)]. This line, some of you may not know, has varied in intensity and it has turned off over the period of a year. However while it was on, it was confirmed by several observers. Dick Drachman, a couple of years ago, proposed a time-dependent model where the annihilation mechanism was a combination of positronium formation and direct annihilation. He reasoned that since the direct annihilation takes longer, one would see a positronium annihilation line that would be wide and then after a period of time the line would become narrow.

Question

How do you know that the line is coming from the galactic center?

Brown

The angular resolution is 35° for the HEAO-3 telescope[5] and they let the telescope drift across the galactic center which is Saggitarius A West. The Bell/Sandia telescope[4] has an angular resolution of 15° and they point their telescope toward Sag A West and then away. The difference spectrum shows the 511 keV line. We are not absolutely sure it is there, but it seems likely.

Now for some astrophysical speculations. In Fig. 9 we see the broad energy spectrum on either side of the 511 keV line obtained by HEAO-3. Several attempts have been made by the JPL group to model this broad part of the spectrum.[9] They obtained marginal fits both to a double power law plus positronium and to a single power law plus positronium plus a black body spectrum. In my fit to their published data I feel that a black body spectrum or something similar does give a better fit to the spectrum above 511 keV. I have tried to fit the results including a Monte Carlo generated spectrum of a hot relativistic low density plasma.[10] I get a higher positronium fraction than Riegler et al., quite consistently.[11] What this might be telling us is that the radiation may be produced in two distinct areas, one in the diffuse interstellar medium due to positrons annihilating. This could be ionized, dusty, neutral or some combination of these. The second area could be a more dense and hot region where pair creation and annihilation might be occuring with gamma-gamma scattering. This might give rise to the distribution seen above 511 keV in the HEAO-3 data. The point is that it is possible to have a significant positronium contribution by fitting with some possible high energy spectra. The intensities of the narrow line and the high energy component could vary somewhat independently since they would presumably be produced in different areas.

Drachman

I will start the discussion by responding to some of these points. The trouble that Ben was referring to is partly my fault, so I'd like to explain what I did wrong. (Theorists are allowed to do that sometimes; you know about factors of two that show up in calculations.) If the width of an angular distribution is calculated to be a certain number of milliradians then you can convert that into an energy spectrum by turning it sideways. Instead of looking at the transverse component you look at the longitudinal component. In the course of making that transformation in our first paper I forgot that the mass of a positron and electron is twice the mass of an electron, so there is a factor of two in velocity. The velocity gives you the Doppler shift which is what gives you the energy spread, whereas the ratio of the momenta gives you the angular correlation. The momentum ratio doesn't care what the mass is. So apparently I did that a couple of times and finally discovered, as Ben did, that it was wrong. However in the original paper it's 2.6 keV, which is exactly twice what I think it should be now. That's based on a calculation by Humbertson of positrons and hydrogen atoms. That's not quite the same as what ought to be in an H_2 experiment because the H_2 width in our paper is also up by a factor of two, but that comes from the experiment of Alec Stewart, suitably jiggled up. He did some angular correlation in molecular liquids, and then we adjusted it a little bit because we said it was so dense that maybe its changed a little bit; a 10% correction. But all of those were overestimated by a factor of two.

I want to mention the remark about the time dependence. There may just be nothing at the galactic center. Stirling Colgate, who is a well-known radical astrophysicist, thinks that occasionally a supernova goes off in that vicinity. It is not visible because there's so much obscuration in the direction of the galactic center. He thinks that we were just fooled by one or two of them going off in a several year period and there won't be any more for a while. So we may not see anything until another supernova goes off, in which case the whole thing is less interesting than we had hoped it would be. We'll see. One more thing, I want to say something about the spirit of astrophysics since I'm a converted physicist. I used to believe in chi-squared tests and precision in measurements, and so on, till I became an astrophysicist and I discovered that it's extremely hard to get any data at all and the Astrophysicists publish as gospel anything they manage to get back. I think Ben is right to say that the data is noisy and we're not sure what the background is, you don't really know what's going on and you ought to be careful. But that's not the right way to do it if you are an astrophysicist. There are for example some time dependent, gamma ray bursts that have been seen, which have a little wiggle on the side of them, and those have been forcefully interpreted as red-shifted positron annihilation lines. They're at 400 keV instead of 511 keV and they've already got models of great sophistication about red-shifted annihilation on the surfaces on neutron stars. I mentioned before the theory that they might be gamma-ray annihilation lasers, that's another possibility. It turns out that a gamma-ray annihilation laser, according to one theory, always has its frequency pulled away from 511 kilovolts, almost exactly to where these bumps show up on the burster. All that is in the literature in black and white, although it has absolutely nothing to do with what's really going on out in space. But it's a lot of fun to talk about it, and if you were always careful you would never have anything to publish. If you read the history of astronomy you discover that every two or three years of the last 2000 years, people have changed their theories about what the universe is like. There's quite a nice book about the spirit of astrophysics. On the scale of the universe, it goes through history from about 1800 to the present about how big our galaxy is and whether the other spiral nebulae are in fact galaxies. Every few years a new piece of information is considered which is mostly erroneous. Various pieces of data have been introduced and whole books have been written about them. First the galaxy was large then the galaxy was small, then it was large again then it got twice as big and then it got smaller. So it's a lot of fun, that's the point I'm going to leave you with.

Question

You only consider direct annihilation and singlet positronium annihilation. Is there any contribution from the triplet?

Brown

Yes there is, and in the fit in Fig. 8 the triplet contribution is included. But this is just for the data close to the line and if you start worrying about the component above the annihilation peak then you get into worrying about the power law spectrum. It is very hard to know what's going on out in the high energy range. I'd like to mention that Gunther Reigler, one of leaders in the HEAO-3 group, has stated that there is a limit (at 90% confidence) of 70% maximum positronium formation. This would limit our positron direct annihilation fraction to greater than 30%. The 9% that we obtained experimentally and the 5.5% that Dick Drachman got, would be ruled out by his data if his model is correct. In my own analysis of the data shown in Fig. 9 I find I can obtain a much higher Ps fraction then stated by Riegler.[11] Also the kind of limit by Riegler et al. is possibly falling into the pitfall of minimizing chi-squared and then asking, what the 1σ, 2σ, 3σ error limits are. They assume a normal variable distribution (giving a parabolic χ^2 distribution). I believe that this is a risky assumption.

R. Drachman

I'd like to comment on the time dependent scenario that Ben was talking about. I tried that out just for the purpose of reducing the amount of positronium because it's in Reigler's data.[11] In fact if you found by a chi-squared test that you would need 40% for X then you could have that with this wonderfully flexible model that I have. You would simply say that a certain amount of time has elapsed after the injection of positrons at which this amount of direct annihilation is left. You can get anywhere from zero to 100% depending on the time and of course the only way you could actually distinguish that is if you had a satellite with a good resolution gamma-ray detector watching the galactic center. Then you could see time variation and linewidths, and everything you wanted and that would be very nice.

Question

I had a question on the first part of the talk. Do you see excited states of positronium?

Brown

The cross-sections for the n=2 excited state formation of positronium are down by more than one order of magnitude from the n=1, from calculations, so I don't believe that I would see it. It might be a factor but it would be small. I might also point out that the overall efficiency of this positronium beam is good. About 1% of the positrons that you put in will come out within a 20° cone at, let's say 50 eV, so it's an efficient and collimated, more or less, source of positronium.

Comment

You get about 1% that form positronium.

Brown

Yes, that's a conservative estimate. It's really slightly higher but the positronium can suffer some gas collisions and break up so I would say that it is something on the order of one percent efficiency.

REFERENCES

1. P. Mandel, S. Guha and N. C. Sil Phys. Rev. A312 2913 (1979). V. Secalso, P. Khan and A. S. Ghosh Phys Rev. A**28** 2181 (1983).
2. B. L. Brown, M. Leventhal, A. P. Mills, Jr., and D. W. Gidley, Phys. Rev. Lett. **53**, 2347 (1984).
3. R. W. Bussard, R. Ramaty, and R. J. Drachman, Astrophys, J. **228**, 928 (1979).
4. M. Leventhal, C. J. MacCallum, and P. D. Stang Astrophys. J. Lett. **225**, 11 (1978). A linewidth limit is given here of $\lesssim 3.2$ keV which is actually the 1.4σ limit. The 2.6σ limit (99% confidence) is 5 keV for a simple gaussian fit with the detector resolution taken out.
5. G. R. Riegler, J. L. Ling, W. A. Mahoncy, W. A. Wheaton, J. B. Willet, A. S. Jacobson and T. A. Prince, Astrophys. J. Lett. **248**, 13 (1981) quote a limit of $1.6 \begin{cases} +.9 \\ -1.6 \end{cases}$. This has been quoted as "<2.5 keV" but this is only the one σ limit. The 2.6σ limit (99% confidence) is <3.9 keV using a simple gaussian fit.
6. R. J. Drachman in *"Positron Electron Pairs in Astrophysics,"* ed. M. L. Burns, A. K. Harding and R. Ramaty (New York: AIP) 1983 p. 242.
7. W. M. Zurek, Astrophys. J. **289**, 603 (1985).
8. For further details see B. L. Brown, Astrophys. J. Lett. **292**, L67 (1985).
9. G. R. Riegler, J. L. Ling, W. A. Mahoney, W. A. Wheaton, and A. S. Jacobsen, in Ref. 6, p. 230. Also a more recent fit is given by these authors in Astrophys. J. Lett. **294**, L13 (1985).
10. P. D. Noerdlinger in Ref. 6, p. 379.
11. I recently determined that Riegler et. al in Ref. 9 were using a different definition for the Ps fraction than others had previously used (for example Ref. 4). This has led to additional confusion in interpeting the galactic Ps fraction since Riegler et. al compared their results directly to the results in Ref. 4. A clarification of the Ref. 9 results will soon be published in the literature. The most recent limits for the HEAO-3 Fall data are now $71\% \begin{cases} +11 \\ -23 \end{cases}$ for the fraction of incident positrons decaying via the positronium channel. These results now agree with my analysis of their data in the sense that a high Ps fraction is possible.

Polarized Positron Measurements of the Electron Helicity in Chiral Molecules and the Origin of Optical Activity

Arthur Rich

The University of Michigan
Department of Physics

A. Introduction

1. Outline and Preliminary Comments

As the last speaker of the session I would like to thank Allen Mills and Karl Canter, and I'm sure I'm not alone in this, for organizing a really wonderful symposium. As I sat here today from 9:00 a.m. until now, I was truly impressed with the breadth of the field that we positron researchers have. It ranges from condensed matter physics right up to the galactic center. I want to take the liberty to applaud our hosts, and I invite you to join me.

Now let me turn my attention to Steve Berko and his birthday. I think that the talk I'm going to give today is probably very appropriate for both Steve Berko and for a birthday. There are two reasons why. One is that Steve certainly was one of the first, if perhaps not the first person, to use polarized positrons. In those early days he investigated ferromagnetic materials with polarized positrons. Our current work is somewhat different but still in the same spirit. The second reason is that we address the question of the origins of biological optical activity and this ultimately relates to the origins of life itself. It is, in a sense, a definition of what life is, what living things are, and that's certainly a birthday. Probably you should do it on the first birthday rather than the 60th, but in any case let me go on to an idea of what I will talk about.

We're going to stay away from positrons in the introduction. I have to tell you a little bit about handed molecules, something I really only learned about myself a few years ago. I will discuss the handed structure of amino acids and then our research into the question of why they have the handedness that they do. That's what we are investigating in a certain sense, in the work that we are doing at Michigan. There will be a theoretical analysis of this, where positrons will come in finally. And then I'll be talking about our method of getting at this problem using polarized positron beams. In the spirit of this symposium (Allen wrote a letter and asked us to speculate and give new data) I will be presenting only new data and, indeed, doing some speculating.

Before getting into the details of our research let me assure you that, to say the least, I'm not doing this work alone. I have a number of collaborators, and I would like to give them credit at the beginning. Figure 1 shows our entire positron group at Michigan and the experiments we are doing. The people who have worked or are working on the optical activity experiment are David Gidley (David, whose name has been mentioned here many times in other contexts, is not currently directly involved with the experiment, but he has been in the past),

UNIVERSITY OF MICHIGAN – EXPERIMENTAL

Ralph Conti – τ (1^3 S_1), CP Test
William Frieze – e^+ Surface Physics
David Gidley – τ (1^3 S_1), Surface Magnetism
Mohammed Hatamian – CP Test
Mark Skalsey – Weak Interactions

Jim Van House – e^- Helicity in Chiral Molecules, Polarized e^+ Develpment
Chris Westbrook – τ (1^3 S_1)
Paul Zitzewitz (U/M Dearborn) – τ (1^3 S_1, e^- Helicity in Chiral Molecules, Polarized e^+ Devel.

UNIVERSITY OF MICHIGAN – THEORETICAL

Bill Ford
Gordon Kane
Bob Lewis
Len Sander
Yukio Tomozawa
Ed Yao

U. of Toronto	**Princeton U.**	**Wake Forest U.**	**GM Res Lab**
D. Paul		R. Hegstrom	W. Capehart
D. Holdsworth	F. Calaprice	e^- Helicity	Surface Magnetism
Weak Interactions			

Fig. 1. Collaborators.

Jim Van House (who I would say is the central figure in this research at this time, got his Ph.D. working on it, and is continuing), and my colleague Paul Zitzewitz (a colleague of many years at Michigan who has been with us from the beginning on this). We are lucky to have some very good theoretical support at Michigan. The list of people that I've given here support us in general in our positron and positronium research. In particular, Bill Ford and Bob Lewis have given us a lot of theoretical support in the optical activity work. Finally, I would like to acknowledge our very strong collaboration with Professor Roger Hegstrom who is a theoretical chemist at Wake Forest University, and has done major theoretical work in this area. With that let me get on to the experimental work.

2. The Handed Structure of Amino Acids

The first thing I would like to do is to discuss amino acids. In a general sense we are interested in all optically active molecules, but since the technical

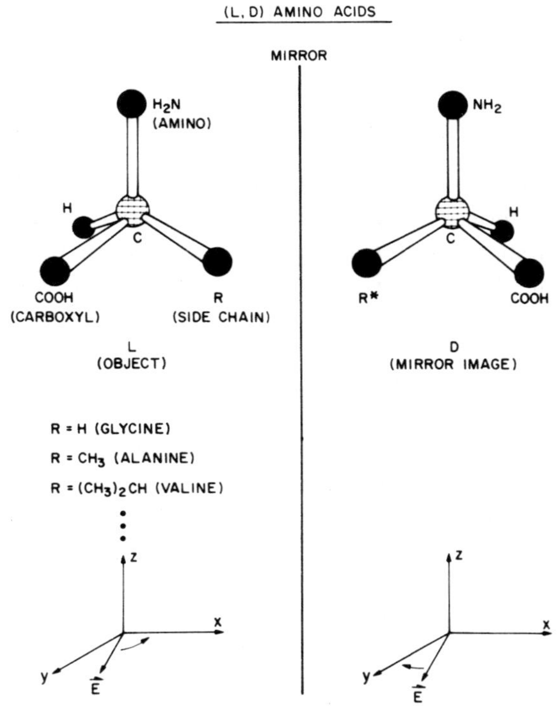

Fig. 2. L and D amino acids.

illustration office made this nice transparency (Fig. 2) of amino acids for a different talk, I thought I'd use it. The main point that I want to make is that the structure of all 26 amino acids is basically the same. You just take the ball labeled R (the side chain) and you change it. For example, if R is hydrogen then you've got glycine. If it's CH_3, you've got alanine, which is something we actually use, and you go right on down through the list of amino acids.

The mirror image of an amino acid is also shown here. An interesting thing is that in nature you don't find this mirror image. In other words, essentially all amino acids found in biological systems have the same (left) handedness. By left handed it is meant that if you start from the H_2N and proceed in the order COOH-R-H you proceed counterclockwise. This is arbitrarily called in L amino acid. There are historical reasons for the label L, but we will not discuss them here. If you went in the order NH_2-COOH-R-H and proceeded clockwise you would have a right handed or D amino acid. The property of handedness implies optical activity--that is, as shown in Fig. 2, handed substances rotate the electric vector of polarized light (say D clockwise and L counter-clockwise or vice versa--the sign of rotation is not absolute but for a given optically active material depends on wavelength).

If you make amino acids chemically, you would make equal amounts of L and D. The general question that we are exploring is not why one handedness exists in nature. That is understood. We should have optically pure substances because stronger bonds are formed if you only combine materials with the same helical structure. The question is, is L determined causally, or was the choice caused by random fluctuation. That's the question we would say we we're investigating.

3. The Vester-Ulbricht (V-U) Hypothesis

Are the weak interactions the cause of the prevalence of L Amino Acids in biological organisms? If L was determined over D in a causal manner you have to look for some interaction that selects a certain handedness. Well, the weak interactions are certainly handed. That is, due to parity non-conservation, the positrons emitted in beta decay are right handed. Their spin is parallel to their momentum. Very shortly after beta decay was shown to violate parity some biochemists at Yale picked up on the phenomenon. They made a hypothesis (now called, after the originators, the Vester-Ulbricht hypothesis) which stated that the selective destruction of D versus L in DL mixtures by polarized electrons caused the present biological state (pure L). That is, they said that life was created racemic (equal amounts of L and D) in chemical reactions and that somehow the D was destroyed selectively by polarized electrons present due to radioactive decay. But they didn't say how this might occur in any detailed way. They said here we have a handed probe, (that means an electron, they were thinking of electrons not positrons), that selectively destroyed the D. Somehow. No numbers were put in. Publication of that thought initiated something like several hundred experiments all over the world in which people, more or less, took bags of amino acids, exposed them to radioactive sources, and looked to see if there was some selective destruction going on. And they usually saw something. In other words, they would put a bag of equal amounts of L and D into a beam of electrons or reactor or whatever it was. It would come out and they would analyze the products, and find more L than D. It was generally at their limit of sensitivity which was always about one part in 1000 and the L greater than D results were never reproducible.

I got involved as follows. There were experiments that were performed first by groups at the Central Research Institute for Physics and the Biophysics Research Institute in Hungary which I thought were rather good experiments when I first read about them. They looked for asymmetric triplet positronium (o-Ps) formation in L versus D amino acids. There was no quantitative theory which predicted the size of a selective effect if you used polarized positrons, but such preferential o-Ps formation is not ruled out by symmetry considerations, and so could exist. I studied this a little bit and I showed that in principle they had made some mistakes in their analysis, and that the experiments should not have given an effect, although they did give an effect. I then got hooked on the whole thing myself, which may or may not have been a mistake. I realized that there was a correct way to do the experiment (which I will describe in more detail below), and we started out working on it. We have made some real progress, and that's what I want to tell you about. It's now a relatively solved problem at least qualitatively.

B. Theoretical Analysis

1. Molecular Electron Helicity - A New Property of Chiral Molecules.

I'll now tell you a little bit about the physics of what we have here, and then I'll go right to 1984 and we'll discuss the results that we have obtained. The first question that we have to deal with is, specifically, how can you get a differential effect? How can irradiation of these amino acids or any chiral systems, (in fact, we are no longer dealing only with amino acids in our work--there are other chiral systems which should show larger effects) by handed particles, by positrons with helicity, give rise to a differential effect? Let me show you. The key resides in a new property of optically acitve molecules: the bound electrons have a net helicity density (helicity per unit volume) associated with them. Here I should give credit to Peter Hrasko (Central Research Institute for Physics in Hungary) who I think really, in a semi-quantitative way, first put this idea forward. What we have discovered is that if you consider a molecule or an atom you can show that there is a helicity density, as defined in Fig. 3. You're taking the spin dotted into the momentum, the normal definition of helicity that we all use, and sandwiching it between the wavefunctions for that system. This expectation value

I. Helicity Density ($\mathscr{H}(\vec{x})$)

$1e^-$ state: $(\Psi_s(\vec{x}),\ s = \pm 1)$

$$\mathscr{H}(\vec{x}) = \mathrm{Re}\{\Psi_s^*(\vec{x})\vec{s}\cdot\hat{p}\Psi_s(\vec{x})\},$$

$$h = \int \mathscr{H}(\vec{x})\,d\vec{x}$$

$2e^-$ (Singlet) State: (+ spin up ($\uparrow$), - spin down ($\downarrow$))

$$\Psi(1,2) = 1/\sqrt{2}\left[\Psi_+(1)\Psi_-(2) - \Psi_-(1)\Psi_+(2)\right]$$

$$\mathscr{H}(\vec{x}) = \mathrm{Re}\left[\left[\left(\int d\vec{x}_1\Psi^*(\vec{s}_1\cdot\hat{p}_1)\Psi|_{x_2=x} + \left(\int dx_2\Psi^*\vec{s}_2\cdot\hat{p}_2\Psi\right)\right|_{x_1=x}\right]\right.$$

$$= \frac{1}{2}\,\mathrm{Re}\left\{\left[\Psi_+^*(1)\vec{s}_1\cdot p_1\Psi(1)\right]_{x_1=x} + \left[\Psi_-^*(1)\vec{s}_1\cdot p_1\Psi_-(1)\right]_{x_1=x}\right.$$

$$+\left[\quad(2)\quad\right]_{x_2=x} + \left[\quad(2)\quad\right]_{x_2=x}$$

II. Properties of $\mathscr{H}(\vec{x})$

A. $\mathscr{H}(\vec{x}) = 0$ for molecules with a center of inversion symmetry, or mirror plane or if H(spin-orbit)=0.

B. $[\mathscr{H}(\vec{x})]_{average} = 0$ when the average is taken over an ensemble of randomly oriented molecules with superposable mirror images.

C. $\mathscr{H}(\vec{x}) \sim (\alpha Z)^2/a_0^3$ for a chiral molecule or when averaged over an ensemble of randomly oriented chiral molecules ($\alpha = e^2/\hbar c \sim 10^{-2}$, Z = atomic number and a_0 = Bohr radius).

D. $\mathscr{H}_L(\vec{x}) = -\mathscr{H}_D(-\vec{x})$.

Fig. 3. The bound helical electron model.

is non-zero, under certain circumstances, for a single bound electron. That's a very important point. You could also do this for two-electron singlet states as shown in the figure. You can go on and do it for anything.

What I'd like to do now is get to the properties of the helicity density, which is, in a sense, the bottom line. This, by the way, has all appeared in print in some fairly extensive articles in Nature.[1,2] The helicity density is zero for molecules which have a center of inversion symmetry or a mirror plane. Point B (Fig. 3.) is that its average value is zero when you take the average over an ensemble of randomly oriented molecules with superimposed mirror images. Chiral molecules do not have mirror images which can be superimposed on themselves, so that would not be true for a chiral molecule. The important thing here is point C which states that the helicity density is of the order of $(\alpha Z)^2$, characteristic of spin-orbit coupling. The presence of a_0^3 (a_o is the Bohr radius) is because we are dealing with a density. So the helicity density is non-zero for a chiral molecule, or when averaged over an ensemble of such chiral molecules. There is no external orientation present in these chiral molecules, strictly a gas that we're dealing with, or a liquid, or a powder. So this gives you a feeling for the size of the thing, and it's interesting. For low Z (say $Z = 6$) the size of the helicity density is parts in 10,000. But we don't have to stay with low Z if we decide we just want to demonstrate the effect and forget about the amino acid problem for awhile, and that's our immediate goal. Later I'll discuss our work in high Z chiral molecules.

Finally, an interesting property is shown in D and one that you could guess on symmetry grounds: If you look at a particular point in an L and then a D molecule, that is, an object and its mirror image, and look for their helicity density, you'll find that they are the negative of each other. That is what you would expect, if you think about it a little bit. Actually, none of this is too obvious. At the time we first thought about it, the question was whether a helicity density could even exist on symmetry grounds. We talked about it with some pretty good people who wondered whether it would even be there, and not be zero, but it's there all right.

Let's now say that we've calculated the helicity density. I haven't even given you a hand waving argument as to why it exists. I'm just saying that it is allowed by symmetry, and that it has been calculated by Roger Hegstrom (our theoretical collaborator) although not too accurately yet because of uncertainty associated with the wavefunction of chiral molecules. What we've got is chirality, or a helicity, associated with the molecule, and we're attacking it with a particle which has helicity. Now you can immediately believe, as those who have thought about this a little bit in other contexts have, that you will get differential effects.

Let me first tell you what happens when you do this with electrons. In Fig. 4 I've given you an answer. What we're going to do here is to ask what happens when we send in an electron. It's energy is E_e and it has helicity h_e which we can take to be essentially one for electrons in beta decay. There is a target molecule with its helicity density that I just discussed. Notice that we don't know the sign of the helicity density. As I said earlier the calculation is in a rather primitive state. The quanitity η is a molecular asymmetry factor which basically depends on how chiral the system is. If it's "flat", η is zero. If it's spiral, then η gets to be a little bigger, so η characterizes the chiral state of the molecule.

What happens if you look for ionization of this molecule by longitudinally polarized electrons, incident first on L molecules and then on D molecules. You calculate the cross-section for inonization for L and D molecules and compare

INCIDENT ELECTRON — Energy-E_e, Helicity-h_e

TARGET MOLECULE — Helicity Density, $\mathscr{H}_L = \mathscr{H}_D \sim \pm\, (\alpha Z)^2/a_0^3$

Molecular Asymmetry — $\eta \sim (10^{-1} - 10^{-3})$

$$A_{ion} = \frac{\sigma_{ion}\,(L, h_e) - \sigma_{ion}\,(D, h_e)}{\sigma_{ion}\,(L, h_e) + \sigma_{ion}\,(D, h_e)}$$

$$E_e \sim 10^5 \text{ volts} \; - \; A_{ion} \sim \frac{h_e\,(a_0^3\,\mathscr{H}_L)\,\eta}{E_e}$$

$$A_{ion} \sim (10^{-10} - 10^{-12})\, h_e Z^2 \underset{Z=6}{\sim} (10^{-8} - 10^{-10})\, h_e$$

Fig. 4. Asymmetric ionization of L vs. D molecules by electrons of helicity h_e.

them via the asymmetry A_{ion}. The result for A_{ion} (Fig. 4) is the first quantitative expresseion of the Vester-Ulbricht hypothesis. If you send the same longitudinally polarized electrons onto L or D molecules there is an asymmetry in ionization. The expression is a product of the projectile helicity, the molecular electron helicity density times a_o^3 (essentially this is the helicity of the electron in the molecule), and η; all things that you believe ought to be there. Then there's something that you might not guess: it is divided by E_e, the projectile energy. That's not unreasonable, after all. It turns out that this a spin-spin type effect, an exchange effect, and it goes to zero at high energy, as do all spin effects, because at high energy the scattering cross-section is less spin-dependent. I can't really convince you that it's E_e; maybe it should be E_e^2. In fact it turns out to be more complicated than E_e, but E_e is the basic dependence and I just siphoned off the rest in the interest of simple formulas.

P. Platzman

This is coming from an exhange effect between the electron that's coming in and the electrons in the molecules?

Rich

Primarily.

P. Platzman

If it were positrons it would be different.

Rich

Yes, that's right.

Let's discuss the magnitude of the effect for electrons. It's 10^{-10} to 10^{-12} when you put all the numbers in, putting in the electron energy directly and so on, times the helicity of the incident electron times Z^2. For a typical biological molecule, this gives you 10^{-8} to 10^{-10} as the asymmetry in the ionization. The best experiments that were ever done were at the level of a part in a thousand, so they were 5-10 orders of magnitude low compared to what they needed in sensitivity. The experimenters were looking for something that's there, but they were a little bit off in their sensitivity.

2. Asymmetric Weak Interaction Effects in Chiral Molecules

Let's leave this now; I just wanted to show what this would be for electrons. Now let's switch to positrons (Fig. 5) and now I promise that this is the end of my discussion of anything that has not got to do with positrons. Figure 5 looks a little complex and I'll go through it slowly. What I'm going to do now is to discuss the same problem, but now I'm going to send positrons into a chiral molecule. I'm going to try to motivate why there should be an effect.

Let's take the positron helicity to be +1, just to simplify everything. I want to make another vast simplification. I'm going to take a so-called L molecule, and I'm going to forget about the $(\alpha Z)^2$. I'm going to make its helicity equal to 1 and now let's see what happens. So I'm sending a positron of a certain helicity in, on a molecule whose electron spins are all either aligned (L) or anti-aligned (D) with respect to their velocity. Now it's very late in the day - and especially after nine talks - to go through something this complex, but let's try.

So here's a positron of a certain velocity and parallel spin attacking an L molecule. The molecular electron velocity could be parallel to the velocity of the positron. Let's just consider this possibility. If so, its spin is parallel to the spin of the positron, and you form triplet positronium. On the other hand, the positron might, with equal probability, see a molecule whose electron has a velocity antiparallel to itself since the molecules are not aligned. In that case, the electron spin, which is aligned with its velocity, will be anti-parallel to the

I. Heuristic Analysis

Projectile — Molecule(M)

$(h_p = +1)$ — $L(\mathcal{H}_L = +1/a_0^3)$

$\vec{v}\Uparrow\vec{s}\uparrow \rightarrow \Uparrow\uparrow$ $(\Psi_T - \text{Triplet})$

$\vec{v}\Downarrow\vec{s}\downarrow \rightarrow e^+ + M\ (\sigma_{Ps} \sim 0)$

$(\vec{v}_p \Uparrow \vec{s}_p \Uparrow)$ $\quad |\vec{v}_p| \simeq |\vec{v}|$ $\quad D(\mathcal{H}_D = -1/a_0^3)$

$\vec{v}\Uparrow\vec{s}\downarrow \rightarrow \Uparrow\downarrow = \frac{1}{\sqrt{2}}(\Psi_T + \Psi_S)$

$\vec{v}\Downarrow\vec{s}\uparrow \rightarrow e^+ + M\ (\sigma_{Ps} \sim 0)$

$$A_{Ps}\ (\text{Triplet}) = \frac{\sigma^T_{Ps}(L) - \sigma^T_{Ps}(D)}{\sigma^T_{Ps}(L) + \sigma^T_{Ps}(D)} = \frac{1 - \frac{1}{2}}{1 + \frac{1}{2}} = \frac{1}{3}$$

II. Quantitative Analysis

$$\mathcal{H}_L = -\mathcal{H}_D \sim \pm(\alpha Z)^2/a_0^3,\ \alpha = e^2/\hbar c \sim 10^{-2},\ h_p < 1$$

$$A_{Ps} \simeq h_p(\alpha Z)^2\, \eta_{Ps} \left[\eta_{Ps} \sim .1\text{–}.001\right] (10^{-5} - 10^{-7}) h_p Z^2$$

$$A_{ION} \sim \left[\left(\frac{h_e^-}{h_p}\right) \times \frac{1}{E_e(eV)} \times A_{Ps}\right] \sim \left[(10) \times \frac{1}{10^5} \times A_{Ps}\right] = 10^{-4} A_{Ps}$$

Fig. 5. Asymmetric triplet positronium formation in L vs. D molecules by polarized positrons.

positron spin. In addition, just to make the effect clear, we simplify further and work with one special positron velocity. That is, we take the velocity of the incident positron equal to the velocity of the electron in the molecule. Then for electron and positron velocities anti-parallel you don't form positronium. Of course I'm really saying you get somewhat lower on the formation curve than for relative velocities parallel, so Ps formation is less probable. In the viewgraph, for simplicity, I take the Ps formation cross section to be zero. In fact as we know the cross section depends on the $e^+ - e^-$ relative energy and that's good enough to see the effect. For velocities parallel we form all triplet.

Now I do the same exact thing with the D molecule where the relative velocity and spin of the electron are anti-parallel rather than parallel. As before we get positronium formation for the velocity of the positron parallel to the velocity of the electron. The spins are then anti-parallel, so I form a linear combination of triplet and singlet. In the case where the velocities are anti-parallel for the D molecule I have a different cross-section. I called it zero for the sake of simplicity. So what happens is that (now I give you, once again, the answer), when I form positronium for an L molecule I get more triplet formation than for a D molecule. Of course, the whole thing is smeared out. Now look at the quantity which I might define as the asymmetry in triplet positronium formation. It's 1/3 in this highly oversimplified case. I hope that what I've got across is that there should be a difference in the formation of triplet positronium when you go from L to D. However, let me also point out that if I stick with L, but flip the spin of the positron, that is, flip the helicity of the positron, I would get the same effect. I think you can believe that.

If you do this quantitatively (part II of Fig. 5) and carry out all the calculations (it's all in the literature), you get an answer that looks very much like the answer we got for the electron case: $A_{Ps} \simeq h_p\, (\alpha Z)^2\, \eta_{Ps}$. Now where does that come from? That comes from the helicity density of the electron, whose real value is not 1, but is of order $(\alpha Z)^2$, the helicity of the positron h_p and η_{Ps} which is analgous to η_{ion}. What we don't have in this case that we had for the electron is division by the energy, so A_{Ps} is bigger than A_{ion} by about 10,000 times. That's why we use positrons, otherwise there would be no real reason for it. You could try to do electron scattering experiments. In fact those would yield approximately the same value of asymmetry at Ps formation energies, but can you do a high precision scattering experiment at 10 eV?

The Ps effect can be measured. If you go to high enough Z the predicted asymmetries may be observable. That's an experiment that I'm going to tell you about in a minute. The asymmetry in ionization, that is if you revert to thinking about electrons again, is the thing that we really want to get at if we're interested in the biology. In fact, our purpose is both to study the chiral molecule-positron interaction and also the biological implications. There is a way to relate the asymmetry of ionization by electrons to the positronium asymmetry which we can really measure in the laboratory. We can go through a formula to get that, as shown at the bottom of Fig. 5. There I have just rewritten the formula for A_{ion} given in Fig. 4 in terms of A_{Ps} as given in Fig. 5. What it comes out to, of course, is that the ionization asymmetry is a part in 10,000 of the positronium asymmetry. And it's just due to the need to divide by E_e.

C. Experimental Investigations of Asymmetric Weak Interaction Effects in Chiral Molecules

The next step is to tell you about the experiment in which we're going to try to make this measurement.

This is the positron beam that we have at Michigan. (Figure 6) I'm not going to go through the details of it here. I'll just say that it was basically built

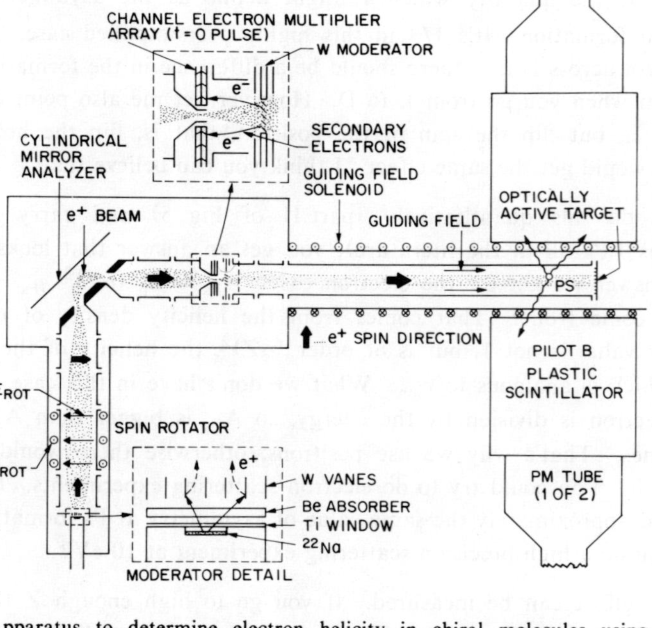

Fig. 6. Apparatus to determine electron helicity in chiral molecules using low energy polarized positrons (1983-4).

by Jim Van House and Paul Zitzewitz. We use a 30 millicurie source which yields 240,000 slow positrons per second. The advantage here is we do have a polarization of 50%. It was measured in the typical way we measure positron polarization which I'll also not go into. You'll have to trust that we know how to do it by now. We have a spin rotator here which is nice because it allows us to change the spin of the positron with respect to the velocity. That's crucial in this experiment. We don't have to do something as crude as changing an L to a D. You can leave the L alone and then just flip the spin. That's changing the helicity of the projectile. After finishing a spin-flip experiment on L, however, we do change to D and then to DL (equal amounts of D and L) as systematic checks on the L results.

After the spin rotater, the beam is steered 90° away from the source. We are down to 150,000 positrons per second by this point, primarily due to losses in the cylindrical mirror analyzer. The positron strikes a vane (shown in enlargement) and the secondary electrons come back to give a t = 0 pulse, separated from the target region, which is very nice. In other words the remoderated positrons are now tagged in time. This is the first use of remoderation to both form and time tag a slow position beam.

Fig. 7.

Fig. 8.

We have described the technique in the literature in a journal of a more biological nature.[3] Anyway it works very nicely, and although it's only about 15% efficient, now every positron is tagged, which is very important in our experiment. The positrons are then magnetically transported to the target region. I should comment on the magnetic transport. We've been using electrostatic beam transport for almost all of our slow positron research, but eventually we did go to a small solenoid here just to get a better beam down at the end. We used an 80 gauss field. It doesn't affect the experiment particularly. Finally, we form positronium on a plate with various optically active targets. So this is the basic experiment.

I now have a couple of photographs of the apparatus, the first showing Jim Van House and myself. Well this is the room (Fig. 7) and you don't see a lot in this. Everything is always buried under lead bricks. This gives you an idea of the scale, which is like the scale of most of the other experiments discussed today. Here's another picture of the whole thing (Fig. 8). This shows the room a little bit better without the personnel in it. And that's what we call a biology lab.

D. Conclusions and Future Research

Now I'll give you the data (Fig. 9), which is unpublished at this time. We had gotten, at one time, a result in leucine.[1] It was published with all kinds of

EXPERIMENTAL LIMITS FOR POSITRONIUM FORMATION ASYMMETRIES IN CHIRAL MOLECULES (1984)

Target	Z	A_{Ps} (DL) × 10^4	A_{Ps} (L) × 10^4	A_{Ps} (D) × 10^4
Leucine	6	−0.5 ± 1.5	−1.2 ± 1.5	−2.7 ± 1.5
Selenocystine	34	+0.1 ± 4.1	−9.0 ± 4.6	+1.4 ± 4.6
Thyroxine	53	−5.9 ± 4.5	+5.0 ± 5.1	+1.3 ± 5.0

ASYMMETRIES IN POSITRONIUM FORMATION (A_{Ps}) AND ELECTRON IONIZATION (A_{ion}) IN CHIRAL MOLECULES (1984)

Target	Z	A_{Ps} (Expt'l Limit)	A_{Ps} (Theor Range)	A_{ion} (Z=6) From A_{Ps} Expt'l Lim	A_{ion} Expt'l Limit
Leucine	6	$+(0.7 \pm 1.0) \times 10^{-4}$	$10^{-5} - 10^{-7}$	10^{-7}	10^{-3}
Selenocystine	34	$-(5.2 \pm 3.3) \times 10^{-4}$	$10^{-4} - 10^{-6}$	10^{-8}	
Thyroxine	53	$+(1.9 \pm 3.6) \times 10^{-4}$	$10^{-3} - 10^{-5}$	10^{-9}	

Fig. 9.

provisos concerning systematics and statistics (it was a three sigma result), and eventually we did find that there was a systematic associated with that result. This data (Fig. 9) is much more sensitive and it shows no result in leucine. However, as I said, it's better to go to higher Z material if we just want to see the effect. In other words, our goal is to investigate what we consider this new property of optially active materials, the fact that there's a real helicity of the electron in those materials. We just want to try to see that. So what we're trying to do is optimize the material in the beginning and drop the biological question entirely. That would come later. So we go to something called selenocystine which is an amino acid but with Se in the cystine molecule ($Z = 34$). The effect goes like Z^2, so we're up a factor of 36 in the realm of sensitivity here. This shows some data that we took when we used what we call

DL; that is, equal amounts of L and D, which should show no effect. Then we looked at just L and then just D. The bottom line of this work in these higher Z materials (thyroxine also is a high Z optically active material), is no effect. So we claim no effect at the level of a few parts in 10,000.

What should we have seen, and how does it compare with what the theory predicts? The experimental limits are given on the bottom of Fig. 9. They are obtained from an average of the results shown at the top of the figure. So that's the experimental limit, and you can satisfy yourself that it's at the level of a few parts in 10^4; so, as yet, no asymmetry. The theoretical range is very uncertain, primarily because the wavefunction for these molecules is not known. If the wavefunction were known, we could do the calculation to within a factor of two or so. So this says that we have entered, basically in only one case (throxine), the regime where we might have seen an effect, were it there. What does this result predict for a limit on the asymmetry in ionization? The limit is very small. Our thyroxine results already tell you that the asymmetry in ionization is below a part in 10^9. Remember that the experimental limit is 10^{-3}. That was done using 100 keV electrons in scattering. So we told people searching at that level that they should not expect to see anything. I think this turned off a number of experiments here and abroad.

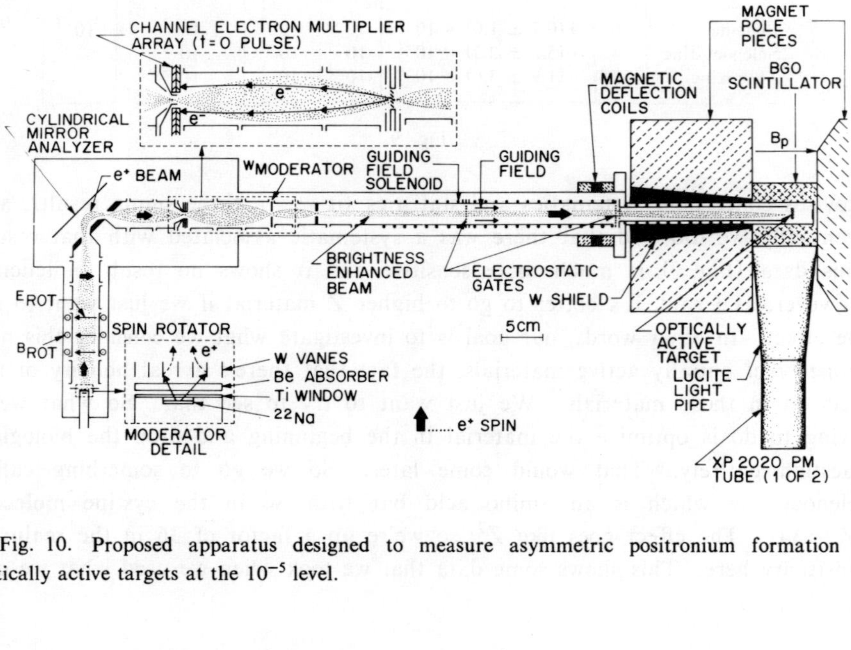

Fig. 10. Proposed apparatus designed to measure asymmetric positronium formation in optically active targets at the 10^{-5} level.

This ends the data we've taken up to this time. What about our future plans? Well we think that in the future we can do much better. We think we can go to asymmetries of a part in 100,000. Let me show you the apparatus which we plan to build (Fig. 10). It incorporates some interesting new features. Using a time tagged, brightness-enhanced beam, we feel we can get another factor of 10-20 out of this apparatus.

That's the first point. The next point is: "what conclusion can we come to concerning the origin of optical activity?" Now I will return to biology. There are people who would say that we have already proven that the so-called Vester-Ulbricht hypothesis is incorrect. The limits we set on the asymmetry are so low that if you believe our work you would say that there's no way that that asymmetry could ever give complete L dominance. We feel that's premature, but I should point out that really it's the biologists and biochemists who have said this. I should say that we feel it's incorrect because you can have amplification of these small effects as long as the small effects are large enough to dominate the statistical fluctuations which would occur in a given pool of amino acids. This is now the highly speculative part of the talk that Allen wanted. In other words, if you have a small asymmetry, yet in an amino acid bath sufficiently large so that the asymmetry overweighs statistical fluctuations, then some biochemical process could cause amplification of the small asymmetry. That's possible. The question is: "is this asymmetry so small that we could never, under any reasonable circumstances, hope to overcome statistical fluctuations?" The results that I'm about to show you will be published in *Nature*, in 1985.[5]

There are some β^- sources shown in the top half of Fig. 11 which are intense enough to give you asymmetries that will overcome statistical fluctuations (see Fig. 11 - lower half). They are around. Now I am not saying that this is how things happened. I'm just saying that these sources existed. You can't find them in the ocean. This is the amount of β–ray material thought to be in the ocean 3 billion years ago. It's not enough. We have already pretty well ruled that out, basically since under reasonable conditions of temperature you need too much alanine to overcome statistical fluctuations, as you can infer from the bottom of figure 11. There are things called uranium detrital deposits. We have probably ruled those out. What we haven't ruled out are meteorites, where you do find ^{26}Al, which is a positron emitter; a very interesting thing. At low temperatures, in reasonable amounts of material, you can actually overcome statistical fluctuations. Now is this a very hare-brained thing, to think that life may have come from meteorites? A number of people, including, for example, Fred Hoyle and his colleagues, think that life may not have originated on planets. In any

SOURCE STRENGTHS AND IONIZATION TIME CONTANTS FOR POSSIBLE PREBIOTIC BETA RAY SOURCES

Beta Ray Source	Source Strength $S(s^{-1}\ cm^{-3})$	Ionization Time Constant $(\epsilon S)^{-1}$ (yr)
^{40}K, ^{14}C in Ocean	0.2	6×10^{10}
Total Crustal Average	3.9	2×10^{10}
Uranium Detrital Deposit	6×10^{2}	8×10^{7}
^{26}Al (Meteorites)	1×10^{4}	3×10^{6}
^{235}U Natural Reactor (Oklo-10 Kilowatts)	1.4×10^{9}	60

MASS OF ALANINE REQUIRED FOR THE WEAK ITERACTION INDUCED CHIRAL POLARIZATION TO EXCEED STATISTICAL FLUCTUATIONS

Beta Ray Source	Temp (°C)	Maximum Chiral Polarization χ (max) $(A = 10^{-11})$ $\chi = \frac{(n_L - n_D)}{(n_L + n_D)}$	N(kG-Alanine) $N \sim \frac{1}{\lvert\chi\ (max)\rvert^2}$	Time Required (Years) $\chi = \chi$ (max)
^{40}K, ^{14}C in Ocean	−20	2×10^{-14}	10^{3}	2×10^{8}
	20	7×10^{-18}		9×10^{4}
	100	1×10^{-22}		2
Uranium Detrital Deposits	−20	3×10^{-12}	0.1	8×10^{7}
	20	5×10^{-15}	10^{5}	9×10^{4}
	100	1×10^{-19}		2
^{26}Al (Meteorites)	−20	3×10^{-12}		3×10^{6}
	20	1×10^{-13}	1	9×10^{4}
	100	3×10^{-18}		2
^{235}U Natural Reactor (Oklo- 10 Kilowatts)	−20	3×10^{-12}	.1	60
	20	3×10^{-12}		60
	100	1×10^{-13}		2

Fig. 11.

case there are amino acids in meteorites and there is enough radioactivity. Finally, and what I think is rather amusing, and a point which I'll end with, is this ^{235}U natural reactor that was discovered in French West Africa (Gabon);

the Oklo reactor. Here you have more than enough radioactivity. In a period of 60 years, typically, and with very small amounts of alanine, (a tenth of a kilogram, which is not unreasonable in a sort of space like that), you would overcome statistical fluctuations. Then, of course, you would have a tiny asymmetry which would have to be amplified, but I'll leave you with the thought that it's possible that the life we now know originated in a nuclear reactor.

REFERENCES

1. D. W. Gidley, A. Rich, J. Van House, and P. W. Zitzewitz, Nature **297**, 639 (1982).
2. R. A. Hegstrom, Nature **297**, 643 (1982).
3. J. Van House, A. Rich, and P. W. Zitzewitz, Origins of Life **14**, 413 (1984).
4. J. Van House, A. Rich, and P. W. Zitzewitz, To be published "Origins of Life" 1986.
5. R. A. Hegstrom, A. Rich, and J. Van House, Nature **313**, 391 (1985).

RELATED PAPERS

Modified Soa Immersion Lens Positron Gun*

K. F. Canter and P. H. Lippel

Physics Department, Brandeis University
Waltham, MA 02254

W. S. Crane and A. P. Mills, Jr.

AT&T Bell Laboratories
Murray Hill, NJ 07974

In this paper we present, in a particularly compact form, the positron trajectories over a wide range of tunings for the positron gun illustrated in Fig. 1. Without the tube element between the Wehnelt and Anode electrodes, this would essentially be the immersion lens introduced by Soa[1] and used as a prototype electron "gun" by Kuyatt and Simpson.[2] Out of consideration mainly of space limitations the intermediary "Soa tube" was introduced into this lens system for the early Brandeis LEPD work.[3] This tube prevented punch-through of fields due to nearby grounded surfaces. Because of the large separation between the Wehnelt and Anode planes, the diameter of the gun would have had to be $\approx$15 cm if a tube had not been used. In addition, this tube allowed one more degree of freedom in controlling exit trajectories.

As discussed in these proceedings, the ensemble of positrons launched from a moderator forms a parallelogram in $r-\theta$ space if the positrons are transported by low aberration electrostatic optics. We have found such parallelograms in a field-free region 60 mm downstream from the anode of the modified Soa gun (113 mm from the cathode). The positrons were traced through the gun optics and into the field-free region using the SLAC Electron Trajectory Program

* Work supported in part by NSF DMR-8217753

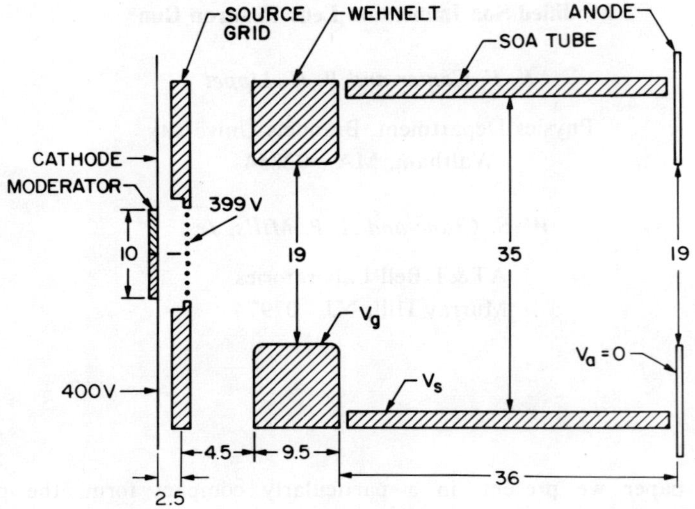

Fig. 1. Modified Soa gun. Dimensions are given in mm. The Wehnelt electrode has 1.5 mm radius rounded edges. A Co-58 source is mounted against the source grid and faces the moderator.

(SLAC Internal Document 226) with space charge effects neglected. Twenty-eight launches were traced for each gun tuning to allow aberrations to be identified; for the low-aberration tunings presented here, knowing only the r,θ values for two adjacent corners of the parallelogram completely specifies the remaining positron trajectories. Equivalently, these two corner trajectories specify the location z_o of the image (virtual or real) of the moderator, the linear magnification M of the image, the beam angle θ_B, and the pencil half-angle θ_p. The relationship between these beam parameters and the parallelogram corners is illustrated in Fig. 2.

For a beam accelerated to energy E at the gun exit, Liouville's theorem simply relates θ_p to M as follows:

$$\theta_p = \frac{1}{M} \sqrt{E_T/E}, \tag{1}$$

where E_T is the transverse energy component intrinsic to the moderator. Thus the three parameters z_o, M, and θ_B are sufficient to specify the exit trajectories corresponding to a positron moderator of a particular diameter D and transverse energy E_T, with the proviso that neither D nor E_T be so large as to cause significantly aberrated exit trajectories (i.e., poor parallelograms). The results

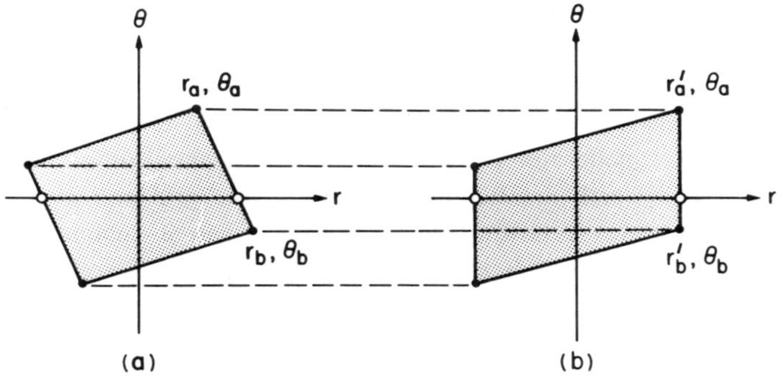

Fig. 2a. Idealized r,θ diagram at gun exit, for z = 113 mm from cathode (moderator).

Fig. 2b. The same ensemble of trajectories field-free extrapolated to the image position z_o with respect to the cathode. Note that the $\theta = 0$ trajectories do not change in their r values, and that θ_a and θ_b are unaffected by the field free extrapolation. The values of r_a, θ_a and r_b, θ_b are used in calculate the trajectories of the beam through the system following the gun.

tabulated here are in this form. Launches were made at 0 and $\pm 10^\circ$ with an intrinsic energy of 2.5 eV, the negative work function of a W(110) moderator. This corresponds to $E_T = 2.5$ eV $\sin^2 10^\circ = 0.075$ eV.[4] If E_T is taken to be 0.17 eV for W(110), corresponding to extremal launch angles of $\pm 15^\circ$, the gun parameters given here still apply with θ_p given by Eq. (1). The acceleration energy E chosen for the trajectory calculations was 400 eV. Over a range of several keV down to $\approx$100 eV, the corresponding trajectory parameters z_o, M, and θ_b can be obtained from the 400 eV results by simply scaling the Wehnelt and Soa tube voltages (V_g and V_s, respectively) by the ratio of the desired acceleration energy to 400 eV. The corresponding θ_p can be obtained from Eq. (1). The lower limit of 100 eV for scaling is a rough estimate of when the intrinsic 2.5 eV launch energy assumed for these calculations begins to differ significantly from a zero launch energy. The critical potential difference in this gun is the voltage between the cathode plane and the Wehnelt electrode. When this difference is less than $\approx$20 times the value of E_T, the scaling of these tuning voltages becomes less reliable.

One other subtlety in scaling the gun to below 400 eV is the role played by the offset voltage between the source grid and the moderator. This gun is used with a Co-58 source on a 0.20 mm thick by 1.5 mm wide W foil which is pressed against a grid and faces the moderator. The moderated positrons pass back through the grid, albeit shadowed by the 1.5 mm wide source ribbon. For all practical purposes, this source grid can be regarded as a planar cathode if the voltage difference between the grid and the moderator is small. The trajectory calculations were carried out with the source grid 1 volt lower than the moderator potential. As a check, we also looked at source filament voltages as low as 5 volts below the 400 volt moderator potential and saw only small differences in the exit trajectories. It is worth mentioning that the source grid can also serve as the retarding element in a retarding field analyzer, when its potential exceeds the moderator potential by $-\phi_+ + \phi_c$, where ϕ_+ is the positron work function of the moderator and ϕ_c is the contact potential between the moderator and the source grid assembly. This allows a quick check on the longitudinal energy spread (and hence E_T) of the moderator as a convenient beam diagnostic. (The overall beam transmission through an electrostatic system is sensitive to E_T.)

The gun trajectories are also suitable for use with a thin film moderator[5] in the transmission mode (i.e., the primary source is behind the cathode plane) by ignoring the source grid altogether.

The calculation presented here was carried out with the moderator's emitting diameter D = 10 mm. The exit trajectory r,θ diagrams can easily be scaled for moderators of different diameters, although a larger diameter (in relation to the gun dimensions in Fig. 1) is not recommended since this would lead to excessive spherical aberrations. Without incurring aberrations one could use a larger moderator with a larger gun, or a more compact gun could be used with D < 10 mm.

The gun described in this paper is presently in use in two electrostatic beam systems at Brandeis; the beam for the original LEPD studies[3] (now used for Ps experiments) and a high brightness beam being used in a Brandeis — Bell Labs collaboration to carry out high resolution LEPD and microbeam experiments.[6] Extensive beam mappings have been carried out in the high brightness beam apparatus and we find good agreement between measured beam diameters and the values calculated from the gun trajectories presented here.

How to use z_o, M, θ_{B10} Tables to Obtain r_a, θ_a and r_b, θ_b for a 400 eV acceleration energy.

1. Specify moderator diameter D (mm) and transverse energy E_T (eV).
2. Scale θ_B using

$$\theta_B = \theta_{B10} * \frac{D}{10} . \tag{2}$$

3. Obtain θ_p from eq. (1) using E = 400 eV.
4. $\theta_a = \theta_B + \theta_p$ (3)

   $\theta_b = \theta_B - \theta_p$ (4)
5. At a distance z (mm) from the cathode (moderator), we have

$$r_a = \frac{M}{2} \cdot D - (z_o - z) \cdot \theta_a \tag{5}$$

$$r_b = \frac{M}{2} \cdot D - (z_o - z) \cdot \theta_b \tag{6}$$

6. Note that we use the convention that M is positive for real (i.e., inverted) images and negative for virtual images.

TABLE OF EXIT TRAJECTORY PARAMETERS FOR MODIFIED SOA GUN

V_g = Wehnelt voltage (volts)

V_s = Soa tube voltage (volts)

10 mm diameter moderator, 400V acceleration voltage.

V_g = 360

V_s	z_o (mm)	M	θ_{B10} (mrad).
0	−50.5	−1.18	33.8
25	−45.7	−1.08	28.7
50	−44.0	−1.02	23.8
100	−38.2	−0.891	13.1
150	−34.2	−0.794	2.22
200	−33.6	−0.814	−6.47
250	−44.6	−0.765	−12.1
280	−58.5	−0.843	−10.9
300	−77.1	−0.966	−6.83
325	−134.	−1.37	3.37

V_g = 370

V_s	z_o (mm)	M	θ_{B10} (mrad.)
0	−81.7	−1.77	55.4
25	−71.7	−1.57	50.6
50	−68.4	−1.46	46.1
100	−56.1	−1.21	35.7
150	−47.7	−1.03	24.7
200	−49.7	−0.953	15.1
250	−58.8	−0.945	7.64
280	−73.7	−1.03	6.83
300	−96.8	−1.19	8.82
325	−176.	−1.77	16.2

$V_g = 380$

V_s	z_o (mm)	M	θ_{B10} (mrad.)
0	−216.	−4.33	79.1
25	−169.	−3.428	74.9
50	−138.	−2.81	70.4
100	−110.	−2.19	61.5
150	−99.2	−1.73	50.8
200	−81.5	−1.47	40.6
250	−90.9	−1.40	31.4
280	−111.	−1.50	28.2
300	−154.	−1.81	28.3
325	−310.	−3.08	31.9

$V_g = 385$

V_s	z_o (mm)	M	θ_{B10} (mrad.)
25	−559.	−10.9	87.9
50	−340.	−6.64	83.7
100	−192.	−3.74	74.8
150	−137.	−2.60	64.9
200	−118.	−2.09	54.5
250	−127.	−1.95	44.6
280	−162.	−2.16	40.6
300	−226.	−2.68	39.4
325	−631.	−6.25	41.2

$V_g = 390$

V_s	z_o (mm)	M	θ_{B10} (mrad.)
0	293.	5.42	105.
25	401.	7.46	102.
50	645.	12.0	97.7
150	−595.	−10.6	80.7
200	−327.	−5.46	70.6
250	−297.	−4.38	60.3
280	−363.	−4.79	54.5
300	−635.	−7.51	51.7
325	1820.	18.1	51.1
350	278.	2.19	56.0
385	111.	0.591	85.0

REFERENCES

1. E. A. Soa, Janaer Jahrbuch **1**, 115 (1959).
2. J. E. Simpson and K. E. Kuyatt, Rev. Sci. Inst. **34**, 265 (1963).
3. I. J. Rosenberg, A. H. Weiss and K. F. Canter, Phys. Rev. Lett. **44**, 1139 (1980).
4. D. A. Fischer, Ph. D. thesis, State University of New York at Stony Brook, 1984.
5. K. G. Lynn, B. Nielson and J. Quateman, Appl. Phys. Lett. **47** (3), 239 (1985).
6. K. F. Canter, T. Horsky, P. H. Lippel, W. S. Crane, and A. P. Mills, Jr., "Proceedings of the Third International Workshop on Positron (Electron) Scattering in Gases," (World Scientific, Singapore, 1986).

Transfer Matrix Subroutines for Common Tube Lenses*

K. F. Canter, P. H. Lippel and D. T. Nguyen

Physics Department, Brandeis University,
Waltham, MA 02254

This paper presents, and documents, Fortran language computer programs (in the form of separate subroutines) for the most commonly used tube lenses, i.e. the two-tube, symmetric einzel and asymmetric einzel lenses. For purpose of illustration, the transfer matrix method is described in detail with the two-tube lens serving as a specific example.

The most commonly used electrostatic lens is formed by two tubes separated by a small gap. A more complex lens system, such as the "zoom lens" that was designed for the early Brandeis LEPD experiments,[1] can be easily built up using a succession of sufficiently separated two-tube lenses.

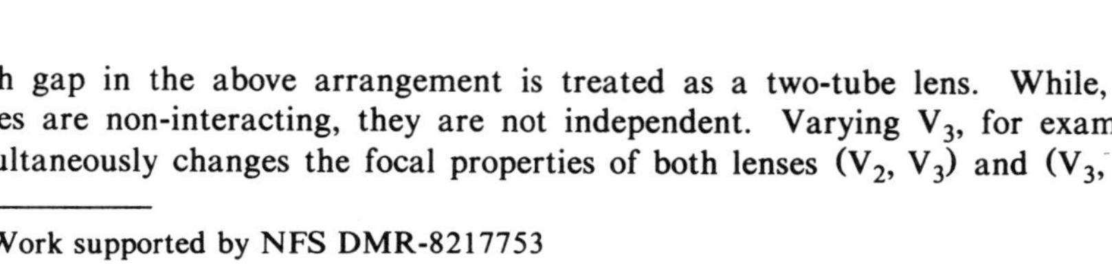

Each gap in the above arrangement is treated as a two-tube lens. While, the lenses are non-interacting, they are not independent. Varying V_3, for example, simultaneously changes the focal properties of both lenses (V_2, V_3) and (V_3, V_4)

* Work supported by NFS DMR-8217753

in the above figure. For the LEPD work, a separation of three times the tube diameter D was considered sufficient to allow the successive lenses to be regarded as non-interacting. This approximation is good since the voltage inside a tube (relative to the other tube) is within 0.995% of the tube voltage at a distance of 1.5D away from the 0.1D wide gap.[2] If space limitations were severe then a 2D lens separation would probably be workable. The point is to keep the lenses far enough apart to allow the use of published two-tube focal properties in one's beam transport program.

Transfer Matrix Method

The method of electrostatic beam design used for the electrostatic beams at Brandeis is based on calculating two positron trajectories launched from an off-axis point on the slow positron moderator surface. This calculation entails first a trajectory calculation from the moderator, through a modified Soa immersion lens or gun,[3] into a field free region downstream from the gun. From this point one can then follow the positrons on the rest of their journey through the system without having to carry out the elaborate trajectory calculations that are needed within the Soa lens system. This is due to the fact that the rest of the system is comprised of standard lenses (two-tube and symmetric einzel lenses) for which the focussing properties have already been tabulated. The transfer matrix method is used to calculate the trajectories through a series of such conventional tube lenses. It should be noted that the transfer matrix method employed in the transfer matrix subroutines given here was obtained directly from Chris Kuyatt's lecture notes on electron optics,[4] with the same sign convention and notation wherever possible. Thus the optical axis is denoted by "x" instead of the more usual "z". Simply stated, the transfer matrix method is a convenient embodiment of the thick lens equation which does not require the use of object-image relationships for trajectory analysis.

For a positron entering a two-tube lens gap from the left, its trajectory at some point past the gap to the right can be expressed as

$$\begin{pmatrix} r_2 \\ \theta_2 \end{pmatrix} = (T) \begin{pmatrix} r_1 \\ \theta_2 \end{pmatrix},$$

where r and θ denote the trajectory height and angle with respect to the beam axis, as shown in the figure below.

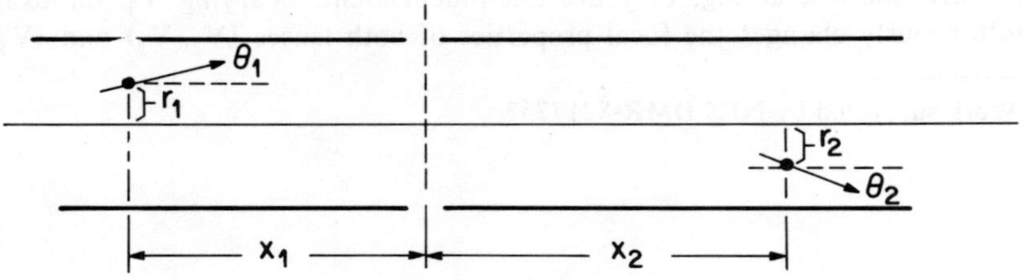

Note the sign convention that x_1 is positive to the left and x_2 is positive to the right of the gap. Also in the above figure r_1 and θ_1 are positive and r_2 and θ_2 are negative. In terms of the focal lengths f_1, f_2 and the mid-focal lengths F_1, F_2 the transfer matrix elements for $x_1 + x_2$ are expressed as

$$T_{11} = \frac{F_2 - x_2}{f_2},$$

$$T_{12} = f_1 - \frac{(x_1 - F_1)(x_2 - F_2)}{f_2},$$

$$T_{21} = -\frac{1}{f_2},$$

$$T_{22} = \frac{F_1 - x_1}{f_2}.$$

The transfer matrix subroutines use the focal lengths calculated by Harting and Read (H-R) to obtain the transfer matrix elements from the above expressions.[2] Fortunately H-R have also provided analytical fits to their focal and mid-focal lengths. For the two-tube lenses, the fits are functions of $V = V_1/V_2$ over the range $2 < V < 90$. Since the zero of potential is always chosen such that $eV_i = mv_i^2/2$ for the ith electrode (v_i is the velocity of the positron if it were to strike the electrode), $V > 1$ corresponds to an "accelerating" two-tube lens. Subroutine TTL incorporates the H-R analytical fits for two-tube lenses and also covers the more complete range $1/90 < V < 90$. The extension of the range is discussed below.

For the "retarding" mode, i.e. $V < 1$, the transfer matrix elements $T(V)$ can be obtained from the matrix elements for the corresponding acceleration mode $T(1/V)$, by a simple time reversal argument. The elements of the matrix $T(V)$ for $V < 1$ are expressed in terms of the elements of $T(1/V)$. The range $1 < V < 2$ is referred to as the "weak lens" range by H-R. In this range the divergence of the focal and mid-focal lengths as $V \rightarrow 1$ prevents a single analytical fit to these lengths from encompassing the stronger lens range as well. It is easy to handle the analyticity problem in the weak lens range by fitting the transfer matrix elements as a function of V instead of fitting the divergent focal lengths. As $V \rightarrow 1$, the transfer matrix approaches the simple "field-free" transfer matrix, i.e.

$$(T) = \begin{pmatrix} 1 & x_1 + x_2 \\ 0 & 1 \end{pmatrix}$$

A simple quadratic fit to T vs. V for $1 < V < 3$ is adequate to obtain trajectory heights and slopes accurate to 1% at a distance of 3D from the lens gap. By now it should be apparent that one must specify x_1 and x_2 in order to evaluate the transfer matrix elements. The beauty of the transfer matrix method is that this is no loss of generality.

To illustrate how the transfer matrix for a given x_1 and x_2 can be used for any other two positions, say $x_0 > x_1$ and $x_3 > x_2$ as shown below

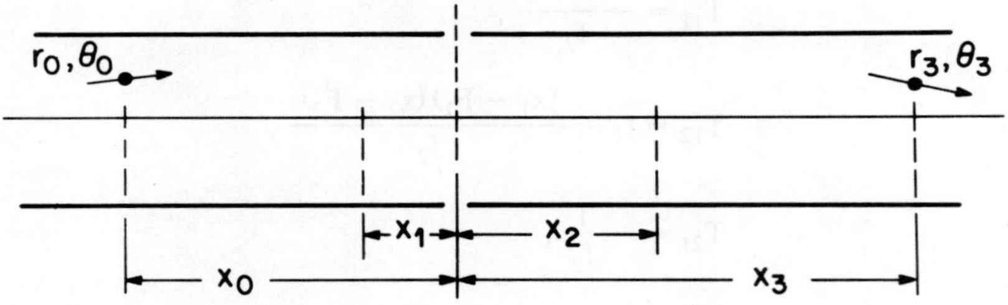

one could easily prove the following:

$$\begin{pmatrix} r_3 \\ \theta_3 \end{pmatrix} = (T)_{0,3} \begin{pmatrix} r_0 \\ \theta_0 \end{pmatrix},$$

$$(T)_{0,3} = (T)_{2,3}\ (T)_{1,2}\ (T)_{0,1},$$

where

$$(T)_{0,1} = \begin{pmatrix} 1 & x_0 - x_1 \\ 0 & 1 \end{pmatrix},$$

and

$$(T)_{2,3} = \begin{pmatrix} 1 & x_3 - x_2 \\ 0 & 1 \end{pmatrix}.$$

In other words, we can use field-free transport for $x_0 \rightarrow x_1$ and $x_2 \rightarrow x_3$, even if these positions are not actually in field-free regions. For all of our transformation matrix subroutines, x_1 and x_2 were chosen to be D/2 = DZ. Although the trajectories calculated this close to the gap will not be too accurate (but are good enough to estimate lens filling and hence spherical aberration),[5] accurate r,θ values are nevertheless obtained by straight-line (i.e. field-free) extrapolation into regions $\gtrsim$2D from the gap.

How to use Subroutine TTL

1. Select a starting point x_0 to the left of the gap. Choose x_0 such that r_0, θ_0 is known. This usually means x_0 is in a field-free region so that r_0, θ_0 is easily obtained from considering upstream apertures, optical elements, guns, etc. A compilation of exit r,θ values for the Brandeis Soa gun is in these proceedings.[3]

2. Field-free transfer to $x_1 = D/2$, i.e.

$$x_1 = x_0 + \theta_0(x_0 - x_1) \ ,$$

$$\theta_1 = \theta_0 \ .$$

3. Specify V_1 and V_2. Remember, these are usually not the voltages on the elements relative to ground in the laboratory. An example is shown below.

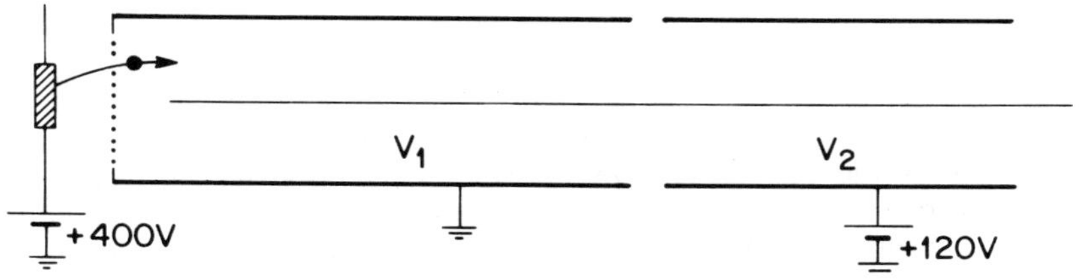

In this case, $V_1 = 400V$ and $V_2 = 280V$, ignoring the positron negative work function ϕ_+. The above arrangement is not recommended as a good positron gun.

4. Call subroutine TTL and operate on r_1, θ_1 with the matrix elements returned to get r_2, θ_2 at $x_2 = D/2$.

5. Field-free transfer away from x_2.

Symmetric Einzel Lens Subroutine SEL

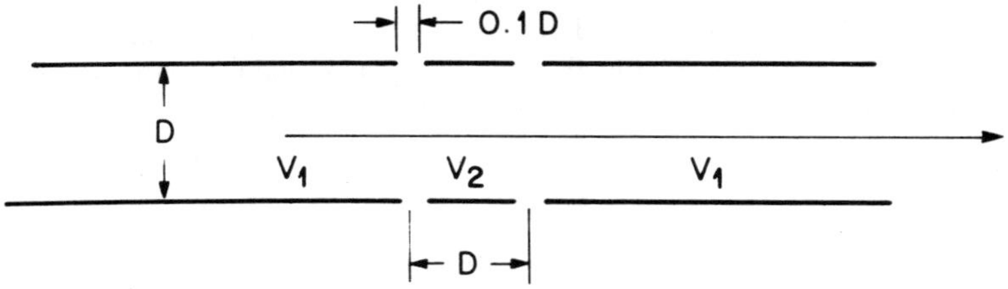

The symmetric einzel lens provides beam focussing without changing the beam energy. For compatibility with the previous subroutine, SEL also returns a transfer matrix which operates between two points D/2 to either side of the reference midplane of the lens. Although this puts the entrance and exit planes in the lens gaps, there is no loss in accuracy by field-free extrapolating the trajectories to asymptotically field-free regions $\gtrsim 2D$ away from the gaps. This allows one to call SEL or TTL interchangeably from one's main program.

Asymmetric Einzel Lens Subroutine AEL

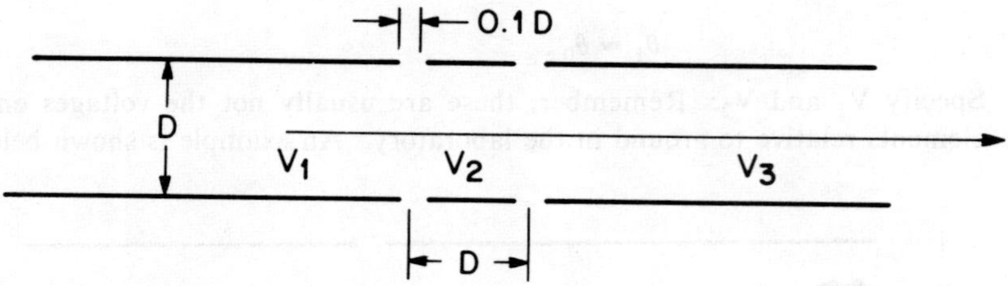

This lens is similar in appearance and function to a pair of two-tube lenses. However since the V_1/V_2 and V_2/V_3 gaps are only 1D apart, this combination of tubes must be treated as one lens. The reference midplane is in the middle of the central (V_2) element. Again we choose the entrance and exit planes to be D/2 on either side of the reference midplane for compatibility. It is possible to operate such a lens in a strongly accelerating mode ($V_3 > V_1$) with V_2 slightly negative. If V_2 is too negative, there will be insufficient "punch-through" from V_3 to pull the positrons through and the positrons will be reflected ("mirroring"). Since the criterion for mirroring is very complicated, the program listed here is only designed for $V_2 > 0$. Also, this subroutine should not be used when $V_1 \approx V_2 \approx V_3$ to within 10%. In this case the field-free transfer matrix is adequate.

Transfer Matrices for Other Einzel Lenses

When it is desirable to use a more compact version of the symmetric and asymmetric einzel lenses in which the central element has length D/2, the programs listed here can be used by simply changing the fitting constants. The list of constants to be changed and their new values can be obtained by comparing the constants listed by Harting and Read[2] with the ones listed here.

REFERENCES

1. I. J. Rosenberg, Alex H. Weiss, and K. F. Canter, Phys. Rev. Lett. **44** 1139 (1980).
2. *Electrostatic Lenses*, E. Harting and F. H. Read (Elsiever, Amsterdam, 1976).
3. K. F. Canter, W. S. Crane, P. H. Lippel and A. P. Mills, Jr., these proceedings.

4. C. E. Kuyatt, unpublished notes.

A large portion of these notes appear in the following book: *Building Scientific Apparatus*, J. H. Moore, C. C. Davis and A. Coplan (Addison-Wesley, London, 1983).

A discussion of the transfer matrix method and some pre-Harting and Read matrix elements are tabulated in the following: D. DiChio, S. V. Natali, C. E. Kuyatt and A. Galejs, Rev. Sci. Inst. **45**, 566 (1974).

5. Spherical abberation adds a spread of $\Delta r \approx 0.08d^3/D^2$ to the beam radius at an imaging point, where d is the beam diameter near the gap.

6. P. A. Sturrock, Static and Dynamic Electron Optics (Cambridge University Press, London, 1955).

Fortran Programs

```
C-----------------TWO TUBE LENS-----------------------------------------------
C
C         CALCULATES 2x2 ANALYTICAL TRANSFER MATRIX FOR A TWO-TUBE LENS
C         GIVEN TUBE VOLTAGES. THE MATRIX TM TRANSFERS a RAY FROM 1/2
C         DIAMETER BEFORE THE LENS GAP TO 1/2 DIAMETER AFTER THE LENS GAP
C
          SUBROUTINE TTL(V1,V2)
          DIMENSION  C(4,6), F(4)
          COMMON D,TM(2,2)                      !lens tube diameter,matrix elts)
          C(1,1)= 5.2904                        !(TM is returned)
          C(1,2)= 5.27 E-2
          C(1,3)= 8.4848 E-3
          C(1,4) =-2.4359 E-4
          C(1,5)= 2.8805 E-6
          C(1,6)= -1.2149 E-8                   !coefficients for first six terms
          C(2,1)= 4.6758                        !in power series for each element
          C(2,2)= 1.5878                        !of the transfer matrix
          C(2,3)= -8.2984 E-4
          C(2,4)= 1.3589 E-4
          C(2,5)= -1.5209 E-6
          C(2,6)= 6.0554 E-9
          C(3,1)= 4.9904
          C(3,2)= 7.6739 E-1
          C(3,3)= 5.9931 E-3
          C(3,4)= -1.5440 E-4
          C(3,5)= 1.7424 E-6
          C(3,6)= -7.1895 E-9
          C(4,1)= 5.5338
          C(4,2)= 5.5471 E-1
          C(4,3)= -1.7543 E-2
          C(4,4)= 2.8658 E-4
          C(4,5)= -2.6225 E-6
          C(4,6)= 9.5062 E-9
          DZ=D/2
          RF= -1                        !Accelerating lens
          V=V2/V1
          IF (V-2) 3,6,6                        !Strong lens or weak lens?
3         IF (V-0.5) 12,12,9                    !      (1/2 <weak <2)
12        RF= 1                         !Retarding
          V= 1/V
6         FF=V*D/((V-1)*(V-1))
          DO 10 I=1,4
          SUM = 0.0
          DO 20 J = 1,6
20        SUM = SUM + C(I,J)*(V**(J-1))
10        F(I) = FF*SUM
          TM(1,1) = (F(4)-DZ)/F(2)
          TM(1,2)= F(1)-(DZ-F(4))*(DZ-F(3))/F(2)
          TM(2,1)= -1/F(2)
          TM(2,2)= (F(3)-DZ)/F(2)
25        IF(RF)300,300,21
C         RETARDING LENS MATRIX ELEMENTS
21        DET = TM(1,1)*TM(2,2) - TM(1,2)*TM(2,1)
          TM11 = TM(1,1)
          TM12 = TM(1,2)
          TM21 = TM(2,1)
          TM22 = TM(2,2)
          TM(1,1) = TM22/DET
          TM(2,2) = TM11/DET
          TM(1,2) = TM12/DET
          TM(2,1) = TM21/DET
          GO TO 300
9         IF(V-1)18,18,15
18        RF = 1
          V = 1/V
C         ANALYTICAL TRANSFER MATRIX  FOR WEAK LENS
```

```
15      TM(1,1) = 0.6014+0.03285*(4.4833-V)**2
        TM(1,2) = 0.7158+0.0391*(3.6956-V)**2
        TM(1,2) = TM(1,2)*D
        TM(2,2) = 0.6704+0.0385*(3.926-V)**2
        TM(2,1) = 0.1242-0.00662*(V+3.332)**2
        TM(2,1) = TM(2,1)/D
        GO TO 25
300     CONTINUE
        RETURN
        END
```

```
C----------------------SYMMETRIC EINZEL LENS------------------------------
C       CALCULATES 2X2 ANALYTICAL TRANSFER MATRIX FOR SYMMETRIC EINZEL
C       LENS WITH CENTER TUBE LENGTH EQUAL TO TUBE DIAMETER.
C       THE MATRIX TM3S TRANSFERS A RAY FROM 1/2 DIAMETER BEFORE THE
C       SYMMETRY PLANE TO 1/2 DIAMETER PAST THE SYMMETRY PLANE
C
        SUBROUTINE SEL(V1,V2)
        COMMON D,TM3S(2,2)        !LENS TUBE DIAMETER, MATRIX ELTS
        DIMENSION C(4,7), F(4)    !(TM3S IS RETURNED TO CALLING PROGRAM)
        A=D
        V=V2/V1
C  TO AVOID DIVERGENCE OF THE VALUE OF FF WHEN V IS NEAR UNITY
        IF ( (v .gt. .9) .and. (v .LT. 1.1) ) GO TO 930
        P1 = -2.00
        P2 = 1.5
        IF ( (V .LT. 0.0) .OR. (V .GT. 14.0) ) GO TO 1000
        IF ( V .LE. 6.00 ) GO TO 325
         X1 = .6
         C(1,1) =  1.9332
         C(1,2) = -4.3330E-1
         C(1,3) =  2.6500E-1
         C(1,4) = -6.0627E-2
         C(1,5) =  7.3346E-3
         C(1,6) = -4.3132E-4
         C(1,7) =  9.8562E-6
         C(3,1) =  1.6328
         C(3,2) =  4.2632E-1
         C(3,3) = -3.4956E-1
         C(3,4) =  1.1481E-1
         C(3,5) = -1.6781E-2
         C(3,6) =  1.1077E-3
         C(3,7) = -2.7219E-5
         GO TO 900
325     IF ( (V .GE. .30) .AND. (V .LE. 3.8) ) GO TO 350
         X1 = .6
         C(1,1) =  1.9448
         C(1,2) = -6.1006E-1
         C(1,3) =  6.1246E-1
         C(1,4) = -2.9849E-1
         C(1,5) =  8.0495E-2
         C(1,6) = -1.0876E-2
         C(1,7) =  5.7447E-4
         C(3,1) =  1.5847
         C(3,2) =  1.0715
         C(3,3) = -1.8293
         C(3,4) =  1.2384
         C(3,5) = -3.8829E-1
         C(3,6) =  5.6981E-2
         C(3,7) = -3.1646E-3
         GO TO  900
```

```
350     X1 =.7
        C(1,1) =  1.5379
        C(1,2) =  2.0614E-2
        C(1,3) = -1.4098E-2
        C(1,4) =  4.2340E-2
        C(1,5) = -1.8810E-2
        C(1,6) =  3.7419E-3
        C(1,7) = -2.8505E-4
        C(3,1) =  1.4447
        C(3,2) =  4.0771E-1
        C(3,3) = -6.2306E-1
        C(3,4) =  5.1095E-1
        C(3,5) = -2.0603E-1
        C(3,6) =  4.0540E-2
        C(3,7) = -3.1196E-3
900     CONTINUE
        FF = ABS(V-1.0)**P1*(V+X1)**P2
        DO 920 I=1,3,2
         SUM = 0.0
         DO 910 J=1,7
910     SUM = SUM + C(I,J)*V**(J-1)
920     F(I) = FF*SUM*D
        F(2) = F(1)
        F(4) = F(3)
        TM3S(1,1) = ( F(4)-D/2.0) / F(2)
        TM3S(1,2) = ( F(1)*F(2) - (D/2.0-F(3))*(D/2.0-F(4)) ) / F(2)
        TM3S(2,1) = -1./F(2)
        TM3S(2,2) = ( F(3)-D/2.0 ) / F(2)
        GO TO 999
930     TM3S(1,1) = 1.0
        TM3S(1,2) = D
        TM3S(2,1) = 0.0
        TM3S(2,2) = 1.0
999     CONTINUE
        RETURN
1000    TYPE *,'ERROR-IMPROPER ARGUMENTS-FUNCTION OUT OF RANGE'
        RETURN
        END
```

```
C--------------------ASYMMETRIC EINZEL LENS-----------------------------
C       CALCULATES 2x2 ANALYTICAL TRANSFER MATRIX FOR AN ASYMMETRIC
C       (3-VOLTAGE) EINZEL LENS, WITH A CENTER TUBE LENGTH EQUAL TO
C       ONE LENS DIAMETER. THE MATRIX TM3A TRANSFERS A RAY FROM
C       1/2 DIAMETER BEFORE THE LENS CENTER TO 1/2 DIAMETER PAST
C       THE LENS CENTER.
C
C       THIS PROGRAM IS BASED ON 'PARAMETRIZATION OF FOCAL LENGTHS
C       OF ASYMMETRIC VOLTAGE LENSES,THREE CYLINDER LENS,A/D=1.0,
C       G/D=.1', HARTING AND READ,ELECTROSTATIC LENSES,APPENIDX A3.28, p 14
C       AND FITTING FORMULAS, SECTION IV.4, pp26-29
C
        SUBROUTINE AEL(V1,V2,V3)
C
C
        COMMON D,TM3A(2,2)        !LENS TUBE DIAMETER,MATRIX ELEMENTS
                                  !(TM3A IS RETURNED TO CALLING PROGRAM)
        DIMENSION B(5),C(5),DA(5),EF1(8,7),ECAPF1(8,7),ECAPF2(8,7)
        DIMENSION F(4)
C
```

```
C
      DELTA = .0010
      W = .8500
      A = 0.0
      B(1) =.16170
      B(2) =.50450
      B(3) =.02050
      B(4) =.00440
      B(5) =-.00063
      C(1) = .93112
      C(2) = .09758
      C(3) = -.12919
      C(4) =.04175
      c(5) =  -.00551
      DA(1) = 11.28091
      DA(2) =`-3.22069
      DA(3) = 2.01365
      DA(4) = -.45758
      DA(5) =  .08333

      EF1(1,1) = 1.03495
      EF1(1,2) = 1.11882e-01
      EF1(1,3) = 1.32174e-01
      EF1(1,4) = -1.62829e-02
      EF1(1,5) = 7.17307e-03
      EF1(1,6) = 1.37545e-02
      EF1(1,7) = 3.74351e-03

      EF1(2,1) = -3.23813e-01
      EF1(2,2) =-2.41281e-01
      EF1(2,3) =2.53292e-01
      EF1(2,4) =3.23788e-01
      EF1(2,5) =-9.60210e-02
      EF1(2,6) =-7.54862e-02
      EF1(2,7) =6.72792e-03

      EF1(3,1) =8.80456e-01
      EF1(3,2) =5.49850e-01
      EF1(3,3) =-1.35577
      EF1(3,4) =-8.86594e-01
      EF1(3,5) =5.67983e-01
      EF1(3,6) =2.01396e-01
      EF1(3,7) =-6.41804e-02

      EF1(4,1) =-1.07582
      EF1(4,2) =-6.99173e-01
      EF1(4,3) =1.98952
      EF1(4,4) =1.20228
      EF1(4,5) =-8.37688e-01
      EF1(4,6) =-2.62915e-01
      EF1(4,7) =9.98964e-02

      EF1(5,1) =6.90119e-01
      EF1(5,2) =4.77295e-01
      EF1(5,3) =-1.39458
      EF1(5,4) =-8.51186e-01
      EF1(5,5) =5.72321e-01
      EF1(5,6) =1.76314e-01
      EF1(5,7) =-7.14758e-02

      EF1(6,1) =-2.41165e-01
      EF1(6,2) =-1.78713e-01
      EF1(6,3) =5.10461e-01
      EF1(6,4) =3.25309e-01
      EF1(6,5) =-2.00270e-01
      EF1(6,6) =-6.19969e-02
      EF1(6,7) =2.70635e-02
```

```
EF1(7,1) =4.34255e-02
EF1(7,2) =3.44516e-02
EF1(7,3) =-9.41150e-02
EF1(7,4) =-6.30393e-02
EF1(7,5) =3.48010e-02
EF1(7,6) =1.04565e-02
EF1(7,7) =-5.45131e-03

EF1(8,1) = -3.15343e-03
EF1(8,2) = -2.65965e-03
EF1(8,3) =  6.92254e-03
EF1(8,4) =  4.82919e-03
EF1(8,5) = -2.40092e-03
EF1(8,6) = -6.29010e-04
EF1(8,7) = 4.77868e-04

ECAPF1(1,1) =1.03449
ECAPF1(1,2) =1.10706e-01
ECAPF1(1,3) =1.47563e-01
ECAPF1(1,4) =-1.44478e-02
ECAPF1(1,5) =-4.14182e-02
ECAPF1(1,6) =4.31242e-03
ECAPF1(1,7) =5.30027e-03

ECAPF1(2,1) =-2.95118e-02
ECAPF1(2,2) =-1.40352e-02
ECAPF1(2,3) =2.84927e-02
ECAPF1(2,4) =4.09893e-01
ECAPF1(2,5) =2.35488e-01
ECAPF1(2,6) =-1.11467e-01
ECAPF1(2,7) =-6.34556e-02

ECAPF1(3,1) =6.22674e-01
ECAPF1(3,2) =2.54525e-01
ECAPF1(3,3) =1.55313e-01
ECAPF1(3,4) =-8.92524e-01
ECAPF1(3,5) =-7.03318e-01
ECAPF1(3,6) =3.47238e-01
ECAPF1(3,7) =2.09922e-01

ECAPF1(4,1) =-5.52246e-01
ECAPF1(4,2) =-2.63768e-02
ECAPF1(4,3) =-4.05975e-01
ECAPF1(4,4) =1.08462
ECAPF1(4,5) =1.02859
ECAPF1(4,6) =-5.45072e-01
ECAPF1(4,7) =-3.36309e-01

ECAPF1(5,1) =2.84262e-01
ECAPF1(5,2) =-6.94823e-02
ECAPF1(5,3) =3.65071e-01
ECAPF1(5,4) =-7.45292e-01
ECAPF1(5,5) =-7.23181e-01
ECAPF1(5,6) =4.62471e-01
ECAPF1(5,7) =2.74640e-01

ECAPF1(6,1) =-7.97031e-02
ECAPF1(6,2) =4.15855e-02
ECAPF1(6,3) =-1.46338e-01
ECAPF1(6,4) =2.95768e-01
ECAPF1(6,5) =2.62573e-01
ECAPF1(6,6) =-2.13510e-01
ECAPF1(6,7) =-1.20403e-01

ECAPF1(7,1) =1.16114e-02
ECAPF1(7,2) =-8.62086e-03
ECAPF1(7,3) =2.68948e-02
ECAPF1(7,4) =-6.27541e-02
ECAPF1(7,5) =-4.73772e-02
ECAPF1(7,6) =5.02016e-02
ECAPF1(7,7) =2.69976e-02
```

```
ECAPF1(8,1) =-6.95395e-04
ECAPF1(8,2) =6.02917e-04
ECAPF1(8,3) =-1.79671e-03
ECAPF1(8,4) =5.49307e-03
ECAPF1(8,5) =3.39787e-03
ECAPF1(8,6) =-4.70848e-03
ECAPF1(8,7) = -2.45848e-03

ECAPF2(1,1) = 1.03449
ECAPF2(1,2) = 1.11090e-01
ECAPF2(1,3) = 1.47705e-01
ECAPF2(1,4) =-1.44012e-02
ECAPF2(1,5) =-4.14110e-02
ECAPF2(1,6) =4.30555e-03
ECAPF2(1,7) =5.29705e-03

ECAPF2(2,1) =-3.05871e-02
ECAPF2(2,2) =-3.98142e-01
ECAPF2(2,3) =-1.13979e-01
ECAPF2(2,4) =3.63081e-01
ECAPF2(2,5) =2.28275e-01
ECAPF2(2,6) =-1.04483e-01
ECAPF2(2,7) =-6.01810e-02

ECAPF2(3,1) =6.38233e-01
ECAPF2(3,2) =1.27290e-01
ECAPF2(3,3) =2.86785e-01
ECAPF2(3,4) =-6.39415e-01
ECAPF2(3,5) =-6.89409e-01
ECAPF2(3,6) =2.35140e-01
ECAPF2(3,7) =1.61768e-01

ECAPF2(4,1) =-6.06596e-01
ECAPF2(4,2) =8.01202e-02
ECAPF2(4,3) =-6.63164e-01
ECAPF2(4,4) =4.80334e-01
ECAPF2(4,5) =8.95776e-01
ECAPF2(4,6) =-2.61932e-01
ECAPF2(4,7) =-1.99538e-01

ECAPF2(5,1) =3.24521e-01
ECAPF2(5,2) =-1.86966e-01
ECAPF2(5,3) =5.51931e-01
ECAPF2(5,4) =-1.20187e-01
ECAPF2(5,5) =-5.14044e-01
ECAPF2(5,6) =1.60226e-01
ECAPF2(5,7) =1.17779e-01

ECAPF2(6,1) =-1.00700e-01
ECAPF2(6,2) =9.88908e-02
ECAPF2(6,3) =-2.09668e-01
ECAPF2(6,4) =-2.38575e-02
ECAPF2(6,5) =1.23775e-01
ECAPF2(6,6) =-5.85859e-02
ECAPF2(6,7) =-3.59603e-02

ECAPF2(7,1) =1.64378e-02
ECAPF2(7,2) =-2.28997e-02
ECAPF2(7,3) =3.58140e-02
ECAPF2(7,4) =1.68304e-02
ECAPF2(7,5) =-6.32244e-03
ECAPF2(7,6) =1.29699e-02
ECAPF2(7,7) =6.01850e-03

ECAPF2(8,1) =-1.11562e-03
ECAPF2(8,2) =1.99592e-03
ECAPF2(8,3) =-2.10751e-03
ECAPF2(8,4) =-2.09132e-03
ECAPF2(8,5) =-1.02304e-03
ECAPF2(8,6) =-1.40853e-03
ECAPF2(8,7) =-5.72327e-04
```

```
C
C
C       RETARDING OR ACCELERATING?
        RFLAG=-1
        IF(V3 .GT. V1) GOTO 6
        VTEMP=V3                     !IF RETARDING, INTERCHANGE V1,V3
        V3=V1
        RFLAG=1
        V1=VTEMP
6       VSUB3 = V3/V1 + DELTA
        USUB3 = ALOG(VSUB3)
        USUBM = 0.0
        TAU = 0.0
        H = 0.0
        DO 100 I=1,5
           USUBM = USUBM + B(I)*USUB3**(I-1)
           TAU = TAU + C(I)*USUB3**(I-1)
           H = H + DA(I)*USUB3**(I-1)
100     CONTINUE
        TAU = USUB3*TAU
        H = H*VSUB3/((VSUB3-1.)*(VSUB3-1.))
        VSUBM= (W-1.)*(VSUB3+1.)/(2.*W)
        Y = V2/V1 - VSUBM
        USUB2 = ALOG( ((Y*Y+A)**.5 + Y)/2. ) - USUBM
        CONST = ( H*TAU*TAU/4. )/( USUB2*USUB2+TAU*TAU/4. )
        F(1) = 0.0
        F(3) = 0.0
        F(4) = 0.0
        DO 150 I=1,7
           DO 125 J=1,8
            F(1) = F(1) + EF1(J,I)*(USUB3**(J-1))*(USUB2**(I-1))
            F(3) = F(3) + ECAPF1(J,I)*(USUB3**(J-1))*(USUB2**(I-1))
            F(4) = F(4) + ECAPF2(J,I)*(USUB3**(J-1))*(USUB2**(I-1))
125        CONTINUE
150     CONTINUE
        F(1) = CONST*F(1)*D
        F(2) = ( (V3/V1)**.5 ) * F(1)
        F(3) = CONST*F(3)*D
        F(4) = CONST*F(4)*D
C
C       compile with d-lines to printout focal lengths for comparison with
d       type *,f
C
        TM3A(1,1) = ( F(4)-D/2.0 ) / F(2)
        TM3A(1,2) = ( F(1)*F(2) - (D/2.0-F(3) )*(D/2.0-F(4)) )/F(2)
        TM3A(2,1) = -1.0/F(2)
        TM3A(2,2) = ( F(3)-D/2.0 ) / F(2)
        IF (RFLAG .NE. 1) RETURN
C       ELSE COMPUTE INVERTED ELEMENTS FOR RETARDING CASE
        DET=TM3A(1,1)*TM3A(2,2)-TM3A(1,2)*TM3A(2,1)
        TM11=TM3A(1,1)
        TM22=TM3A(2,2)
        TM12=TM3A(1,2)
        TM21=TM3A(2,1)
        TM3A(1,1)=TM22/DET
        TM3A(2,2)=TM11/DET
        TM3A(1,2)=TM12/DET
        TM3A(2,1)=TM21/DET
        VTEMP = V1
        V1 = V3
        V3 = VTEMP
        RETURN
        END
```

The Positron Trap — A New Tool for Plasma Physics

C. M. Surko
M. Leventhal
W. S. Crane
A. P. Mills, Jr.

AT&T Bell Laboratories
Murray Hill, New Jersey 07974

H. Kugel
J. Strachan

Princeton University
Princeton, New Jersey 08544

ABSTRACT

We discuss the possibility of using electrostatic trapping techniques to produce a collection of $\sim 10^{10}$ positrons in the form of a single-component, positron plasma with density n $\sim$ 10^7 cm^{-3}, temperature $\sim$0.2 eV, and particle containment time $\gtrsim 10^2$ sec. Possible uses of such a positron plasma are briefly discussed. They include the injection of bursts of positrons into a tokamak fusion plasma to study the transport of electron-mass charged particles in plasma and investigation of the uniques properties of an equal-mass plasma, such as evolution of the electron-positron beam-plasma instability.

I. Introduction

At a time when positrons continue to play an ever-increasing role in our understanding of the physical world, it is of interest to consider storing slow positrons and the possible uses of the resulting positron plasma. For a collection of classical particles, the criterion for collective behavior in $n\lambda_D^3 > 1$ and $L/\lambda_D >> 1$, where n is the positron density, L the smallest dimension of the system, and $\lambda_D = (kT/4\pi ne^2)^{1/2}$ is the Debye length in cgs units with T the positron-plasma temperature and n the density. A few numbers illustrate that such densities are plausible: If we choose L = 10 cm and T = 1 eV, then for $n \sim 10^6\ cm^{-3}$, $\lambda_D = 1$ cm, $L/\lambda_D = 10$ and $n\lambda_D^3 = 10^6$.

Single-component electron plasmas of these densities are now produced routinely using techniques developed recently by the group at the University of California at San Diego.[1,2] For example, in a room-temperature apparatus they achieve densities $n_e \sim 10^8\ cm^{-3}$, with temperatures $T_e \sim 0.5$ eV and plasma lifetimes $\gtrsim$300s.

The analogous positron plasma could have a number of interesting applications. For example, perhaps the simplest plasma experiment would be to study the interaction of the positron plasma with an electron beam. The electron beam would act as a means of excitation of positron plasma. It could also serve as a probe of the (non-linear) beam-plasma interaction in that measurement of the fluctuations and changes in the velocity distribution of the exiting beam can yield detailed information about the interaction of the beam with the plasma. Other consequences of an electron-positron plasma have been discussed by Wharton and Tsytovich.[3]

An interesting application of the positron trap described here would be to study the containment and transport of electron-mass charged particles in magnetically-confined fusion plasmas (e.g., a tokamak plasma). In magnetic fusion physics, one of the outstanding problems is the fact that the transport of electrons is much more rapid (e.g., by factors $\gtrsim 10^2$) than one would expect on the basis of classical Coulomb collisions. The electron heat transport is the most significant energy loss mechanism for such plasmas, and, to a large extent, it is likely to determine the criterion for ignition and useful fusion energy output. A positron trap could be used to store positrons from a radioactive source in the time (~5 minutes) between tokamak discharges. During the discharge, the positrons would be converted to a neutral positronium beam and injected into the tokamak plasma, where the positronium would be ionized. The positrons, thus injected and thermalized in times $\lesssim 100\ \mu s$, would then be transported out of the

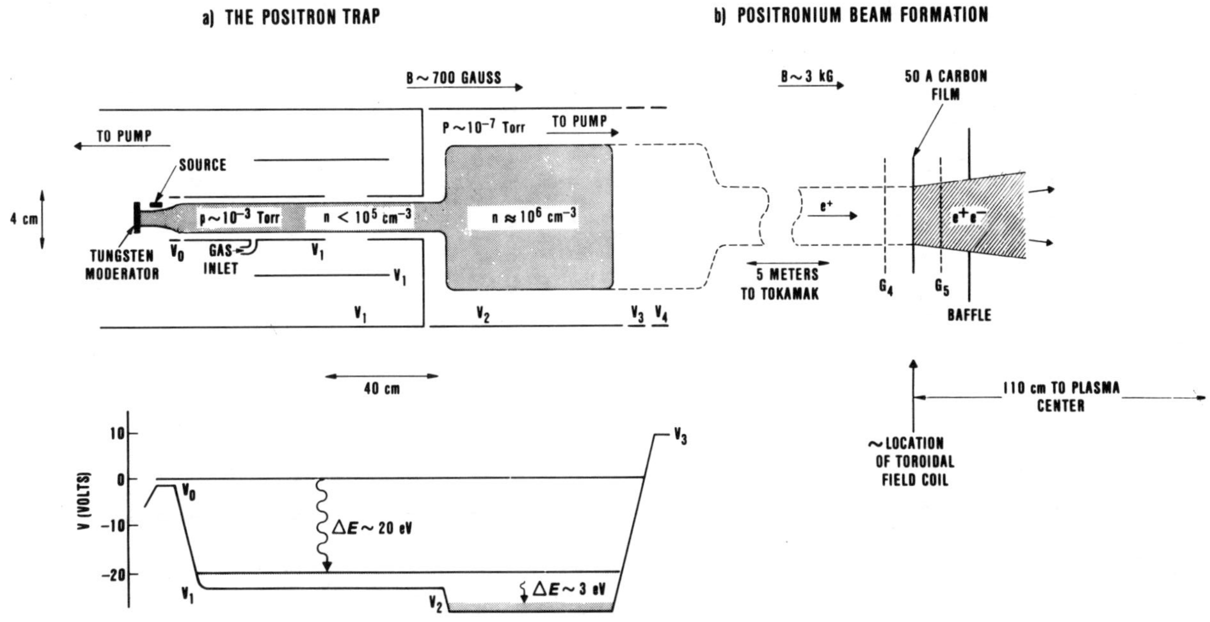

Fig. 1. Conceptual design of an apparatus for positron containment and the formation of a neutral positronium beam: (a) Positrons from a radioactive source (^{58}Co) are moderated in energy by a Tungsten crystal, and drift along the magnetic field into a potential well. Collisions in a high-pressure region ($p \sim 10^{-3}$ torr) provide sufficient energy loss to trap them between guard rings at voltages V_0 and V_3 in one round trip through the trap. Subsequent inelastic collisions trap them in the lower pressure region ($p \sim 10^{-7}$ torr) at potential V_2 in times <1 sec. In this low-pressure region, the containment time is expected to be $>10^2$ sec. (b) During tokamak discharge, the positrons in the trap are accelerated to ~ 1 keV by adjusting the voltages V_3 and V_4. Then the positrons travel to near the tokamak plasma (~ 100 cm from the plasma center), where they impinge on a 50 A carbon foil to form a positronium beam. The resulting burst of positronium atoms consists of a total of about 10^5 atoms with energies ranging from 10 to 10^3 eV in a time $\lesssim 10$ μsec.

plasma on the time-scale for electron transport (~30 – 100 ms). This transport time can be measured directly by the annihilation gamma rays emitted when the positrons strike a limiter. Such an experiment would provide the first direct measure of the transport of electron-mass particles in a tokamak plasma.

The remainder of this article is organized in the following way. In Sec. II we discuss the design of a containment device suitable for efficiently trapping positrons from a continuous positron source (which might be a radioactive source or an accelerator beam stop), and we also describe questions the properties of such a plasma. In Sec. III we describe questions the tokamak experiment might address. Finally, we conclude with some remarks about other possible applications of an efficient positron trap and the resulting positron plasma.

II. The Positron Trap

It is now well established that a uniform, linear magnetic field B ~ 1 kG with electrostatic potentials at the ends can contain a single-component electron plasma for times $\tau_e \gtrsim 10^2$ sec.[1,2] At high neutral gas pressures P, the containment time τ_e is proportional to 1/P and is limited by the combined effects of the radial space charge of the single-component plasma and the cross-field mobility due to electron-molecule scattering. At lowest pressures, τ_e is independent of P. While the limiting mechanism in this regime is not understood, τ_e is large and reproducible in a given geometry and magnetic field. The central point of this article is that such a trapping device will be useful to contain slow positrons, provided we can identify a suitable energy-loss mechanism to efficiently deposit positrons in the trap without disturbing the containment properties of the device.

The conceptual design of such a device is shown in Fig. 1. A ^{58}Co radioactive source emits positrons having a broad spectrum of energies up to 470 keV with a branching ratio for positron production of ~1/7. Thus, for example, a 7 Ci source will produce ~4×10^{10} positrons/sec. A key to this containment scheme is the production of slow positrons with energies ~1 eV. Fortunately, it is known that reflection of energetic positrons emitted by ^{58}Co source from a single-crystal Tungsten surface yields "slow" positrons[4] having energies of 2.5 ± 1 eV with an efficiency ~2×10^{-3}. For the 7 Ci source, this results in a total flux of 8×10^7 slow positrons per second.

As shown in Fig. 1, the slow positrons then enter a trap similar to that used[1] by Driscoll and Malmberg to contain single-component electron plasmas with a magnetic field (B ~ 10^3 Gauss) and guard rings at the ends to apply electrostatic potentials. An energy-loss mechanism is required to trap the continuous flow of positrons from the source. The trap in Fig. 1 shows two such stages: first a section with pressure $\sim 10^{-3}$ torr is used to initially trap the positrons; and then a deeper potential well at $\sim$10–7 torr is used to hold the positrons for long periods -- comparable, for example, to times $\sim$5 minutes between tokamak discharges. A more realistic apparatus would include intermediate stages to minimize the radial diffusion (by minimizing the time spent in the high-pressure regions) and to simplify the problems associated with differential pumping. Voltages on the cylindrical electrodes shown in Fig. 1 determine the electrostatic potentials in the different stages.

The gas species and pressure in the high-pressure region must be chosen so that positrons will be trapped in one transmit through the trap. Elastic scattering and the potential barrier establishing by V_0, could be used to contain the positrons for a number of bounces $N \sim \dfrac{V_1}{E - V_0} \sim 10$, where E is the energy of the injected positrons. Then electronic excitation (e.g., the inelastic cross section for positrons in Ar gas is $\sim 3 \times 10^{-16}\ \text{cm}^2$) can provide the required energy loss. In this case, there will be a loss due to positronium formation[5] of $\sim$2/3. Energy loss via vibrational excitation of molecules would be ideal, since then, the positron energy is below the positronium-formation threshold. However the cross sections for vibrational excitation of molecules are poorly known, and one can probably only count on[6] $\sigma_v \gtrsim 10^{-17}\ \text{cm}^2$. Thus vibrational excitation as an energy loss mechanism will probably be useful only after the first stage (e.g., in the 10^{-7} torr region shown in Fig. 1).

The working gas might be a mixture of Ar and N_2, although in an actual experiment a variety of gases and gas mixtures could be tried relatively easily. The voltage V_2 would be adjusted to inhibit positronium formation in the low-pressure region after the first inelastic collision. For the case of *electrons* in He gas, the lifetime in a low pressure region such as that shown in Fig. 1 is the order of 300s.[1] For the positron case, the efficiency of the trap is estimated to be $\gtrsim$20%, thus the total number of positrons trapped in 300 sec from a 7 Ci source is expected to exceed $\sim 5 \times 10^9$. Radial expansion will lead to a plasma diameter $\sim$20 cm, thus the volume plasma will be $\sim 3 \times 10^4\ \text{cm}^{-3}$. Such a plasma might be compressed electrostatically in the longitudinal direction by an order of magnitude and radially by a factor of more than two (by a suitably arranged magnetic field) resulting in a plasma density $\sim 10^7\ \text{cm}^{-3}$. The positrons are likely to cool via vibrational excitation of the gas molecules of energies $T \sim 0.3 - 0.5$ eV.[2] For a density $\sim 10^6\ \text{cm}^{-3}$, the Debye wavelength will be

$\lambda_D \sim 0.5$ cm, which is much less than the dimensions of the column of positrons. Thus the positrons will behave collectively as a plasma.

III. Measurement of Transport in a Tokamak Plasma

A. Design of an Experiment

During a tokamak discharge, we envision that the positrons from the trap would be accelerated to ~1 keV. They are then transported to the tokamak in a 3 kG guiding field (Fig. 1b) and made to impinge on a 50Å thick carbon foil, supported by a grid, where they form positronium (e^+e^-) at energies ~20 – 100 eV (see Fig. 2).[7] The neutral positronium will penetrate the tokamak plasma and be inonized. The deposition profile of the positrons with plasma is shown in Fig. 3.

The positrons will thermalize with the bulk plasma electrons within 1 μsec of injection. Due to motion of the positrons along the magnetic field lines and to the sheared magnetic field in the tokamak, the profile shown in Fig. 3 will also become symmetric about the magnetic axis on a time-scale of the order of 1 μsec.

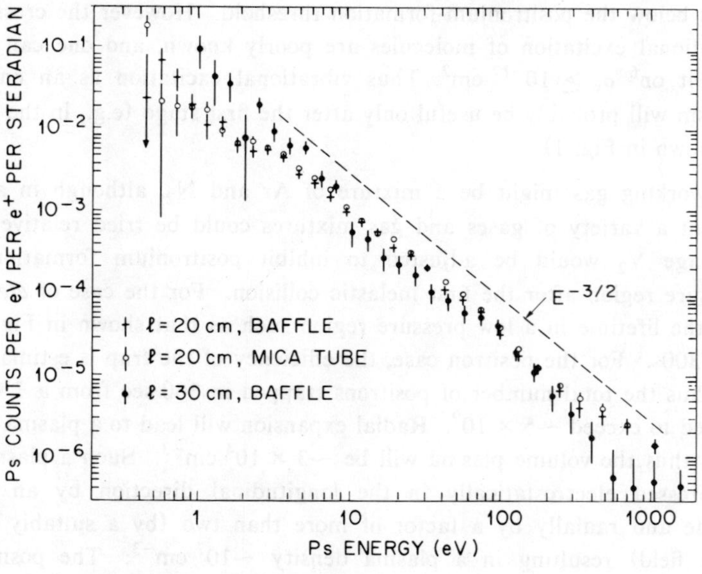

Fig. 2. Flux and energy distribution of positronium atoms produced by 1 keV positrons impinging on a 50Å carbon foil (from Ref. 7).

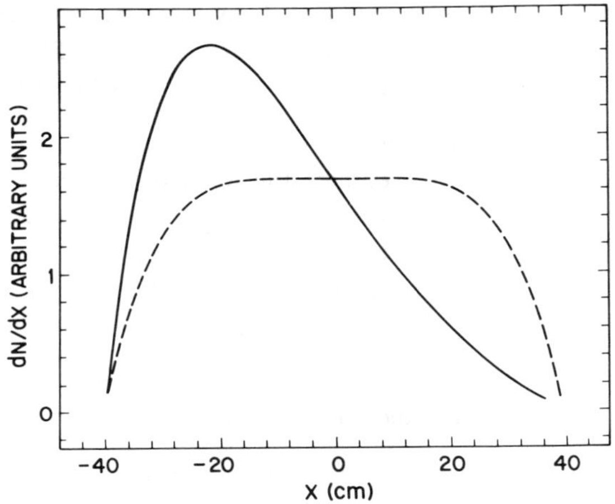

Fig. 3. Deposition profile of the positrons in the plasma as a function of distance from the plasma center. The solid curve assumes the positrons do not move after ionization. The dashed curve is the profile expected when the positrons average poloidally over a magnetic flux surface (in a time $\lesssim 1$ μsec). The positronium source (i.e., the carbon foil) is located at $x = -110$ cm. This calculation includes all positronium atoms emitted from the foil with energies >20 eV and takes into account the finite lifetime of the positronium. The cross-section is assumed to be $\sigma = 2 \times 10^{-16}$ cm^2: At 20 eV, σ is dominated by charge exchange, (i.e., $H^+ + (e^+e^-) \rightarrow H + e^+$), whole at energies $\sim$100 eV, ionization of the positronium by the plasma ions dominates.[8] We have also assumed $Z_{eff} = 2$, $\bar{n} = 2 \times 10^{13}$ cm^{-3} and a plasma with a parabolic density profile between $x = \pm 40$ cm.

The efficiency of the process (positrons) → (positronium) → (positrons in the plasma) is estimated to be $\sim 10^{-5}$, where the dominant loss mechanisms are the positronium conversion efficiency in the carbon foil ($\sim 10^{-4}$) and the 140 nsec lifetime of the positronium traveling the estimated 1 m distance from the foil to the plasma center. Approximately 10^4 positrons would be deposited in the plasma in a time <1 msec.

The transport of positrons out of the plasma can be measured by monitoring the positron flux to the limiter. This is accomplished by measuring the 511 keV γ radiation produced when a positron annihilates with an electron at the limiter. Since the positron lifetime for annihilation in a plasma of density $2 \sim 10^{13}$ cm^{-3} plasma is $\sim$10 sec and the diffusion tine to the limiter is $\sim 10^2$ msec, most of the positrons will impinge on the limiter. It is known from previous experiments that positrons with energies $\lesssim 100$ eV, impinging on a "dirty" metal plate, have a high probability ($\gtrsim 50\%$) of producing annihilation radiation, with the remainder of the positrons forming positronium atoms.[9] Thus the time between the deposition of the positron burst in the plasma and the arrival of the 511 keV annihilation radiation will be a measure of the cross-filed, positron transport rate.

The limiter and γ−ray detector system, prototypes of which are shown in Figs. 4 and 5, are designed to maximize the detector efficiency while protecting the detector from excessive heat loads. It is estimated that such a limiter/detector system can have an overall efficiency $\sim 10^{-1} - 10^{-2}$. Thus we expect $> 10^2$ γ−ray counts per tokamak discharge. This signal will arrive within a few particle containment times (i.e., $\sim 10^2$ msec) after the positrons are injected into the plasma.

B. Some Details on the Limiter and Detector System

The limiter surface which contacts the plasma must be a low Z, low sputtering material with excellent vacuum properties and have a high melting temperature and a high thermal conductivity. In addition, the limiter must be of sufficient thickness to provide the required thermal inertia to the high front-face temperatures. Shown in Figs. 4 and 5 are possible designs of such a limiter. The scintillators are positioned as close as possible to the limiter surfaces in

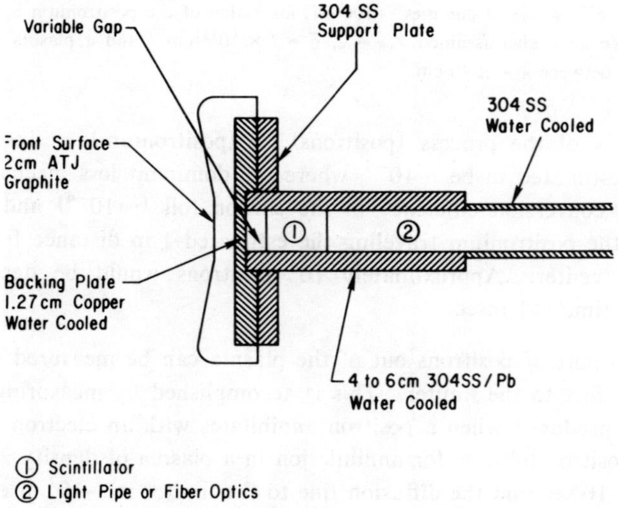

Fig. 4. A "rail" limiter with a scintillator for detecting the 511 keV annihilation radiation. The light pipe (or fiber optics) transmits the scintillator signal to a magnetically-shielded photomultiplier located in a low field region outside of the vacuum vessel.

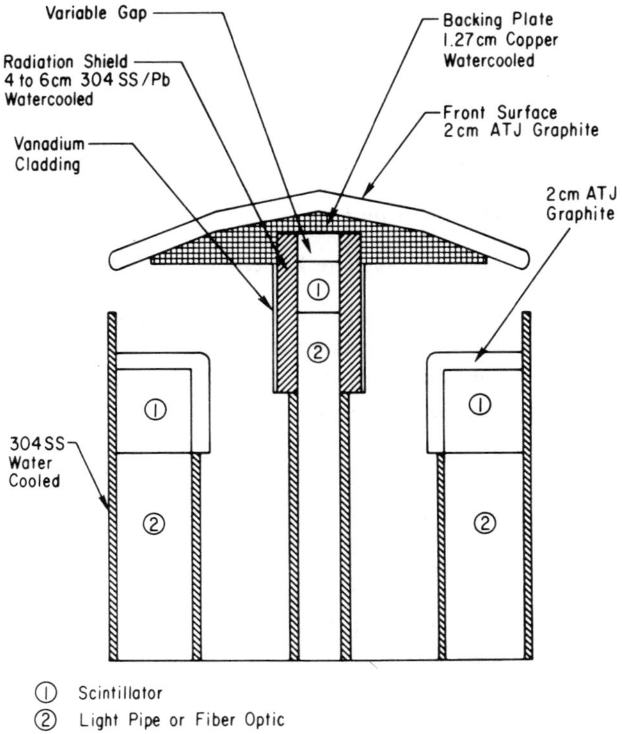

Fig. 5. A "scoop" limiter with three scintillator light pipe systems for detecting separately the 511 keV annihilation radiation at the center of the limiter and at each of the port edges. Additional detector systems could be positioned to view the vanadium neutralizer surfaces and the emission along the limiter surface.

water-cooled radiation shields. Optimal light pipes transmit the scintillator light from the high magnetic field region to a magnetically-shielded photomultiplier in a low-field region outside of the vacuum vessel. The scintillator material is NaI (Tℓ), which is chosen for its high photopeak efficiency at 511 keV, although other scintillators could also be considered.

For clean, low Z_{eff} discharges in plasmas which are not too low in density, there is expected to be no 511 keV background radiation. For example, for such discharges in the PLT tokamak, a detector volume 10^3 cm^3 located near the device was limited only by the cosmic-ray background.[10]

In some discharges, plasma-produced runaway electrons at MeV energies may present a special problem due to the large fluxes of hard x-rays and the annihilation radiation from pair production at the limiter. This effect of runaway electrons can be reduced by using a special runaway scrape-off limiter, located on the large major radius side of the plasma and slightly outshifted from the edge magnetic surface of the plasma. The energetic runaway electrons will be preferentially scraped off by this limiter, since their drift surfaces are outshifted from the magnetic surfaces, and runaway electrons are not expected to be observed at the top and bottom limiters.

C. Possible Physics Payoff of Such an Experiment

The distribution of arrival times of the γ−ray signal depends on both the diffusion and convection of the positrons and the initial positron deposition profile. It is expected that the deposition profile can be determined by the varying plasma density, the positronium beam energy, and the vertical angle of injection of the positrons. Thus, *the experiment described here provides a direct measure of the positron transport* rate. The radial profile of positron particle transport can be determined by varying the vertical angle or the positronium beam injection energy.

A fundamental issue in fusion physics is the transport of electrons in tokamak plasmas. It is well known that the electron thermal conductivity is large compared to that expected from Coulomb collisions.[11] However there are no direct measurements of electron particle transport. The positron transport measurements will undoubtedly shed light on this issue. Whether the electron and the positron transport coefficients are identical is, to our knowledge, an unanswered theoretical question. It will probably depend on which specific transport mechanism is dominant.

Variations of the positron transport experiment described above could be used to measure a variety of quantities of interest. For example, the positron "scrape-off" length at the plasma edge can be measured, either by varying the radial position of the detector/limiter relative to a fixed limiter or by measuring the annihilation rate along a limiter whose surface is at an angle to the toroidal direction. The net positron transport to the detector/limiter is given by the absolute intensity of the γ−ray signal.

Changes in transport during neutral-beam injection or magnetothydrodynamic (MHD) activity could be studied by suitable timing of the injection of the positron burst. For example, the effects of high β (i.e., high plasma pressure relative to B^2) and "fish-bone" (which is a kind of MHD activity particularly important for neutral beam heating) on the particle transport would be of interest.[12] The MHD-enhanced transport near rational magnetic surfaces could also be studied. Poloidal asymmetries in the transport to different limiters (e.g., top and bottom) could be studied by the use of two limiter/detector systems.

The positron velocity distribution in the plasma could be measured (at least in principle) by the angular spread in the γ—rays emitted during annihilation events from the bulk plasma. Finally, the spatial distribution of the positrons as a function of time could be measured directly by studying the bulk plasma annihilation events using a detector array and co-incidence techniques. However the anticipated count rates and the experimental tokamak geometry indicate that these last two experiments will be difficult.

IV. Concluding Remarks

We have described a device to produce a collection of $\sim 10^{10}$ positrons with a temperature ~ 0.5 eV and a containment time $> 10^2$s. We have also discussed one potential use of this device which is to study the transport of electron-mass charged particles in a tokamak fusion plasma.

If such a positron plasma is achieved, it is likely to be of use to study interesting features of the equal-mass, electron-positron plasma which could be obtained by adding electrons. Many of the interesting features of such a plasma require the two species to have equal temperature and densities.[3] Since the addition of electrons to the positron plasma is likely to degrade the containment time dramatically, experiments on such a compensated plasma will necessarily have to be pulsed in nature. In this case, the geometry of a double-plasma device would be convenient.[13] As we point out in the introduction, one relatively simple plasma experiment which might be carried out on a steady-state positron plasma would be to inject an electron beam to study the electron/positron beam-plasma instability. For example, the effects of "clump" and "hole" phenomena[14] are expected to be most pronounced for this case of equal masses. These effects lead to the growth of large-amplitude fluctuations at drift velocities well below the threshold for the usual (linear) ion-acoustic instability.

It is quite likely that there will be other uses of the positron trap described here. If this scheme is successful, a low-temperature, high-field version of the positron trap is a logical extension, based on the recent work of Malmberg et. al.[15] In the pure-electron plasma case, plasmas of densities $\sim 10^{10}$ cm^{-3} with temperature $< 10^\circ$K have been achieved. Other potential uses of the positron trap include study of the annihilation linewidth of the positrons in laboratory and fusion plasmas to model astrophysical processes and use of the trap as a source of intense positron bursts for atomic and solid-state physics experiments.

V. Acknowledgements

It is a pleasure to acknowledge the advice and encouragement of J. H. Malberg, C. F. Driscoll and W. D. White regarding the design of the positron trap based on their experience with the electron trapping techniques which they pioneered. We wish to thank B. L. Brown and O. R. Wood, II for useful discussions, and R. E. Olson for calculating the positronium ionization and charge-exchange cross sections relevant to the tokamak experiment.

REFERENCES

1. J. H. Malmberg and C. F. Driscoll, Phys. Rev. Lett. **44**, 654 (1980); C. F. Driscoll and J. H. Malmberg **50**, 167 (1983); and references therein.
2. J. S. Degrassie, Ph.D Thesis (University Microfilms, Ann Arbor, MI, 1977), unpublished.
3. V. Tsytovich and C. B. Wharton, Comments on Plasma Physics and Controlled Fusion **4**, (1978).
4. R. J. Wilson and A. P. Mills, Jr., Phys. Rev. **B27**, 3949 (1983) and P. J. Shultz, K. G. Lynn, W. E. Frieze and A. Vehanen, Phys. Rev. **B27**, 6626 (1983).
5. For the case of Argon, this is an estimate based on the energy dependence of the inelastic excitation and ionization cross sections. In H_2, survival fractions of 30% have recently been measured [B. L. Brown, Proceedings of the 7th International Conference on Positron Amihilation, edited by P. C. Jain and R. M. Singu (World Scientific, Singopore, 1985)].
6. J. W. Darewych (private communication).
7. A. P. Mills, Jr. and W. S. Crane, Phys. Rev. **A31** 593 (1985).
8. The relevant charge exchange and ionization cross-sections have recently been calculated by R. E. Olson (private communication).
9. K. F. Canter, A. P. Mills, Jr., and S. Berko, Phys. Rev. Lett. **33**, 7 (1974) (unpublished).
10. C. W. Barnes and J. Strachan, Phys. Fluids **26**, 2668 (1983). See, for example, the review article by H. P. Furth in *Fusion*, Vol. 1 (E. Teller, ed., Academic Press, New York, 1981).
11. K. McGuire and the PDX Group, Phys. Rev. Lett. **50**, 891 (1983); **51**, 1925 (1983) [E].

12. See, for example, R. J. Taylor, D. R. Baker, and H. Ikezi, Phys. Rev. Lett. **24**, 206 (1970).

13. R. H. Berman, D. J. Tetreault, and T. H. Dupree Phys. Fl. **28**, 155 (1985).

14. J. H. Malmberg, T. M. O'Neil, A. W. Wyatt, and C. F. Driscoll, Proc. of the 1984 Sendai Symposium on Nonlinear Plasma Phenomena.

APPENDIX

祝
BERKO教授
60歳誕生日

With my heartiest congratulations.

K. Fujiwara

Tokyo, Dec., 1984

BERKO, STEPHAN VITA

Birthdate: December 16, 1924; Oradea, Rumania

Education:

1946-47	Technische Hochschule, Munich Germany
1950	B.A. University of Virginia
1953	Ph.D. University of Virginia
1953-54	National Research Council Postdoctoral Fellow, Princeton University

Professional Record:

1954-57	Assistant Professor, University of Virginia
1957-58	Sloan Foundation Fellow, Bohr Institute, Copenhagen
1958-60	Associate Professor, University of Virginia
1960-61	Professor of Physics, University of Virginia
1961-	Professor of Physics, Brandeis University
1965-67	Chairman, Physics Department, Brandeis University
1968	Solid State Physics Institute, University of Paris, Orsay, France
1970-71	Chairman, Faculty Senate, Brandeis University
1971-74	Chairman, Physics Department, Brandeis University
1976-77	Guggenheim Foundation Fellow, Laue Langevin Institute, Grenoble, France
1978-	William R. Kenan Jr. Professorship

Honors and Awards:

Phi Beta Kappa
Sigma Xi
Fellow, American Physical Society
Virginia Academy of Science Award
Fellow, American Academy of Arts and Sciences

PhD Students of S. Berko

Prof. Norman A. Blum ('65)
Applied Physics Laboratory, The Johns Hopkins Universtiy, Laurel, MD 20810.

Dr. Paul H. Carr ('66)
Rome Air Dev. Ctr., Atn. RADC/EEAC, Hanscomb AFB, Bedford, MA 01731.

Dr. Steven Cushner ('71)
Dupont Photo Products Div., Cheesequake Rd., Parlen, NJ 08852.

Dr. William S. Farmer ('81)
Lincoln Laboratory, 244 Wood St., Lexington, MA.

Dr. Alan J. Greenberger ('73)
Room 2C310, AT&T Bell Laboratories, Murray Hill, NJ 07974.

Dr. Mohammad Haghgooie ('79)
Ford Motor Company, Scientific Research Lab., PO Box 2053, Dearborn, MI 48121.

Dr. Kenneth R. Hoffmann ('84)
University of Chicago, Franklin McQueen Institute, Chicago, IL 60637.

Dr. Chao-Kun Hu ('80)
IBM T. J. Watson Research Ctr., PO Box 218, Yorktown Heights, NY 10598.

Dr. G. Krithivas ('77)
Radiation Physics, University Hospital, 75 E. Newton St., Boston, MA 02118.

Dr. John J. Mader ('75)
Polaroid, Cambridge Optics Div., 38 Henry St.-2, Cambridge, MA 02138.

Dr. Allen P. Mills, Jr. ('67)
Room 1D338, AT&T Bell Laboratories, Murray Hill, NJ 07974.

Dr. David C. Schoepf ('85)
Physics Dept., Brandeis University, 415 South St., Waltham, MA 02154.

Prof. Suresh C. Sharma ('77)
Dept. of Physics, University of Texas at Arlington, PO Box 19059, Arlington, TX 76019.

Dr. Michael A. Schulman ('80)
Room S2075, Ford Motor Company, Scientific Research Lab., PO Box 2053, Dearborn, MI 48121.

Dr. Ilan Tal ('72)
Elscint, Ltd., Advanced Technology Center, PO Box 5258, Haifa, ISRAEL.

Dr. John H. Terrell ('65)
PO Box 134, Lincoln Ctr., MA 01773.

Dr. Howard L. Weisberg ('65)
R&D Associates, 4640 Admiralty Way, Marina del Rey, CA 90292.

PUBLICATIONS

1. "Multiple Scattering of Mu-Mesons in Magnetized Iron," Phys. Rev. *86*, 598 (1952).

2. "The Deflection of High Energy Electrons in Magnetized Iron" (with F. L. Hereford), Phys. Rev. *91*, 1127 (1953).

3. "Recent Experiments on the Ferromagnetic Deflection of Mu-Mesons and Electrons" (with F. L. Hereford), Va. J. of Science *4*, 101 (1953). Sheldon-Horsley Award for 1953.

4. "Free Radical Quenching of Positron Lifetimes" (with A. J. Zuchelli), Phys. Rev. *102*, 724 (1956).

5. "Effect of Melting on Positron Lifetime" (with E. H. Landes and A. J. Zuchelli), Phys. Rev. *103*, 828 (1956).

6. "Experimental Studies of Positron Interactions in Solids and Liquids" (with F. L. Hereford), Rev. Mod. Phys. *28*, 299 (1956).

7. "Angular Correlation of Annihilation Radiation for Oriented Graphite" (with J. S. Plaskett and R. E. Kelley), Phys. Rev. *106*, 824 (1957).

8. "Angular Distribution of Elastically Scattered 14.7 Mev Neutrons" (with W. D. Whitehead and B. C. Groseclose), Nucl. Phys. *6*, 210 (1958).

9. "Correlation of Annihilation Radiation in Oriented Single Metal Crystals" (with J. S. Plaskett), Phys. Rev. *112*, 1877 (1958).

10. "A Directional Scientillation Counter for Neutrons" (with R. F. Stetson), Nuc. Instrum. and Methods *6*, 94 (1959).

11. "Coulomb Excitation of Heavy Nuclei with 20 Mev Alpha Particles" Bull. Am. Phys. Soc. II, *4*, 414 (1959).

12. "Positronium Decay in Molecular Substances" (with W. Brandt and W. W. Walker), Phys. Rev. *120*, 1289 (1960).

13. "Band Structure and Positron Annihilation in Metals, Be and Mg" Phys. Rev. *128*, 2166 (1962).

14. "Differential Elastic Scattering of 15.2 Mev Neutrons by Ta, B, Th, and U" (with C. I. Hudson and W. S. Walker), Phys. Rev. *128*, 1271 (1962).

15. "Polarized Positron Annihilation in Ferromagnets" (with J. Zuckerman), Phys. Rev. Lett. *13*, 339A (1964).

16. "A Single Valued time to Pulse Height System Using Tunnel Diodes" (with H. L. Weisberg), IEEE Transact. Nucl. Sci. *II*, 406 (1964).

17. "Positron Annihilation and Positronium" (with M. Deutsch), *Alpha-, Beta- and Gamma-Ray Spectroscopy* (North Holland Publ., Amsterdam, 1965) 1583-1598.

18. "Two Photon Decay of Positrons in Transition Metals," *Optical Properties and Electronic Structure of Metals and Alloys* (North Holland Publ. Co., Amsterdam, 1966) 210-218.

19. "On Positron Lifetimes vs. Two Gamma Correlations in the Alkali Metals" (with J. H. Terrell and H. L. Weisberg), *Positron Annihilation)*, A. T. Stewart and L. O. Roelling, eds. (Academic Press, New York, 1967) 269.

20. "Positron Annihilation in Ferromagnetic Solids," *Positron Annihilation*, A. T. Stewart and L. O. Roelling, eds. (Academic Press, New York, 1967) 61-78.

21. "Positron Lifetimes in Metals" (with H. L. Weisberg), Phys. Rev. *154*, 249 (1967).

22. "Search for C Nonconservation in Electron-Positron Annihilation" (with A. P. Mills), Phys. Rev. Lett. *19*, 420 (1967).

23. Angular Distribution of Annihilation Radiation from Plastically Deformed Aluminum" (with J. Erskine), Phys. Rev. Lett. *19*, 307 (1967).

24. "Fermi Surface and Conventional Positron Annihilation Experiments: Copper as a Test Case" (with S. Cushner and J. Erskine), Phys. Lett. *27*A, 668 (1968).

25. "Polarized Positron Annihilation in Ferromagnetic Gadolinium" (with C. Hohenemser and J. Weingart), Phys. Lett. *28*A, 41 (1968).

26. "Anisotropic Positron Annihilation with the High Momentum Components of the Conduction Band in Copper" (with S. Cushner and J. Erskine), Phys. Rev. B*1*, 2852 (1970).

27. "Undergraduate Laser Diffraction Experiments Using Large Objects" (with Y. G. Lee, F. Wright and J. Rosenfeld), Am. J. Phys. *38*, 348 (1970).

28. "Investigation of the Fermi Surface of V_3Si by Means of Positron Annihilation" (with M. Weger), Phys. Rev. Lett. *24*, 55 (1970).

29. "Evidence for Positronium-Like Bloch States in Quartz Single Crystals" (with A. Greenberger, A. P. Mills, and A. Thompson), Phys. Lett. *32*A, 72 (1970).

30. "Spin Distribution Studies in Ferromagnetic Metals by Polarized Positron Annihilation Experiments" (with A. P. Mills), J. de Physique *32*, 287 (1971).

31. "Copper-Aluminum Alloy Fermi Surface Studies by Positron Annihilation" (with B. W. Murray and A. Thompson), Phys. Lett. *37*A, 461 (1971).

32. "Positron Annihilation Experiments and the Band Structure of V_3Si" (with M. Weger), *Computational Solid State Physics*, F. Herman, N. W. Dalton and T. Koehler, eds. (Plenum Press, New York, 1972) 59-77.

33. "Positron Lifetimes and $3\gamma/2\gamma$ Ratio in V_2O_3" (with A. P. Mills, A. Greenberger, S. C. Sharma), Bull. Am. Phys. Soc. *17*, 358 (1972).

34. "Angular Distribution of 2γ from Positron Annihilation in V_2O_3" (with A. Greenberger), Bull. Am. Phys. Soc. *17*, 358 (1972).

35. "Fermi Surface Studies of Disordered Alloys by Positron Experiments," *Charge Transfer/Electronic Structure of Alloys*, The Metallurgical Society of AIME L. S. Bennett and R. H. Willens, eds., New York (1974) 319-342.

36. "Efficient Positronium Formation by Slow Positrons Incident on Solid Targets" (with K. F. Canter and A. P. Mills), Phys. Rev. Lett. *33*, 7 (1974).

37. "Momentum Density Measurements by Positron Annihilation in Metals and Alloys; Recent Experiments with a Multicounter Two-Dimensional Angular Correlation Apparatus" (with J. Mader), Appl. Phys. *5*, 287 (1975).

38. "Observations of Positronium Lyman-α Radiation" (with K. F. Canter and A. P. Mills), Phys. Rev. Lett. *34*, 4 (1975).

39. "Comments on the Lock-Crisp-West Theorem with Experimental Examples" (with G. Beardsley, J. J. Mader and M. A. Shulman), Appl. Phys. *5*, 375 (1975).

40. "Fine Structure Measurements in the First Excited State of Positronium" (with A. P. Mills and K. F. Canter), Phys. Rev. Lett. *34*, 1541 (1975).

41. "Fermi Surface Studies of Non-Dilute Disordered Alloys by Positron Experiments" (with J. Mader), Phys. Cond. Matt. *19*, 405 (1975).

42. "Positron Annihilation in Germanium" (with M. A. Shulman and G. M. Beardsley), Appl. Phys. *5*, 367 (1975).

43. "A High Resolution Lifetime Study of Positron Trapping by Vacancies in Lead" (with S. C. Sharma and W. K. Warburton), Phys. Lett. *58*A, 405 (1976).

44. "Electronic Momentum Density by Two-Dimensional Angular Correlation of Annihilation Radiation in Aluminum" (with J. Mader, H. Krakauer and A. Bansil), Phys. Rev. Lett. *37*, 1232 (1976).

45. "Positron Annihilation," *The Compton Effect*, edited by Brian Williams (McGraw Hill, New York, 1977) 273-322.

46. "Review of Precision Positronium Experiments" (with A. P. Mills, Jr. and K. F. Canter), *Atomic Physics 5*, R. Marrus, M. Prior and H. Shugart, eds. (Plenum Publishing Corp., New York, 1977) 103-124.

47. "Momentum Density Measurements with a New Multicounter Two-Dimensional Angular Correlation of Annihilation Radiation Apparatus" (with M. Haghgooie and J. J. Mader), Phys. Lett. *63*A, 335 (1977).

48. "Electronic Momentum Densities by Two-Dimensional Angular Correlation of Annihilation Radiation" (with M. Haghgooie and J. J. Mader), Inst. Phys. Conf. Ser. No. *39*, 94 (1978).

49. "Direct Observation of the High-Momentum Components of the Conduction Band of Copper by Two-Dimensional Angular Correlation of Annihilation Radiation" (with M. Haghgooie and J. J. Mader) Phys. Lett. *69*A, 293 (1978).

50. "The Positron Annihilation Technique in Fermiology," J. Phys. *39*, C6-1568 (1978).

51. "Fermi Surface Studies in Disordered Alloys; Positron Annihilation Experiments" in *Electrons in Disordered Metals and at Metallic Surfaces*, edited by P. Phariseau, B. L. Gyoriffy, and L. Scheire, NATO Adv. Study Institutes Series B42 (Plenum Press, New York, 1979) 239-291.

52. "Positronium Experiments" (with K. F. Canter and A. P. Mills, Jr.) in *Progress in Atomic Spectroscopy, Part B*, edited by W. Hanie and H. Kleinpoppen (Plenum Press, New York, 1979) 1427-1451.

53. "Evidence for Microporosity and Dislocations in Pb(Cd) Alloys from Positron Annihilation Studies" (with C. K. Hu, G. R. Gruzalski and D. Turnbull), Solid State Comm. *31*, 65 (1979).

54. "Two-Dimensional Angular Correlation of Positron-Annihilation Radiation in Copper" (with S. Wakoh, M. Haghgooie and J. J. Mader), J. Phys. F.: Metal Phys. *9*, L231 (1979).

55. "Two-Dimensional Angular Correlation of Annihilation Radiation Experiments," *Positron Annihilation-Proc. of the Vth Int. Conf.*, R. R. Hasiguti and K. Fujiwara, eds. (The Japan Inst. of Metals, 1979) 65.

56. "A New System for the High Precision Measurement of the n=2 Fine Structure in Positronium" (with K. F. Canter, B. O. Clark and D. C. Schoepf), *Positron Annihilation-Proc. of the Vth Int. Conf.*, R. R. Hasiguti and K. Fujiwara, eds. (The Japan Inst. of Metals, 1979) 883.

57. "Positron Annihilation Lifetime and Doppler Profile Studies in Pb, Pb(Tl), and Pb(Cd)" (with C. K. Hu, G. R. Gruzalski and W. K. Warburton), *Positron Annihilation-Proc. of the Vth Int. Conf.*, R. R. Hasiguti and K. Fujiwara, eds. (The Japan Inst. of Metals, 1979) 231.

58. "Studies of the Fermi Surface of α-Phase CuZn Alloys" (with M. Haghgooie), *Positron Annihilation-Proc. of the Vth Int. Conf.*, R. R. Hasiguti and K. Fujiwara, eds. (The Japan Inst. of Metals, 1979) 291.

59. "Studies of the Fermi Surface of V_3Si by 2D ACAR Measurements" (with W. S. Farmer, F. Sinclair and G. M. Beardsley), Solid State Comm. *31*, 481 (1979).

60. "Positron Annihilation Experiments in Metals; Electronic Structure and Fermi Surface Studies," Scripta Metallurgica *14*, 23 (1980).

61. "Studies of the Fermi Surface of V_3Si by 2D ACAR Measurements" (with W. S. Farmer and F. Sinclair) in *Superconductivity in d- and f-band metals*, H. Suhl and M. B. Maple, eds. (Academic Press, New York 1980) 281-289.

62. "Positronium" (with H. N. Pendleton) Ann. Rev. Nucl. Part. Sci. *30*, 543 (1980).

63. "A Simple Positron Spin Analyzer using A Ge Detector" (with Takashi Akahane), *Positron Annihilation*, P. G. Coleman, S. C. Sharma, and L. M. Diana, eds. (North-Holland Publishing Co., New York, 1982) 874.

64. "Evidence of Excited State Ps Formation from Metal Surfaces in Ultra-High Vacuum" (with D. C. Shoepf, K. F. Canter, and A. H. Weiss), *Positron Annihilation*, P. G. Coleman,

S. C. Sharma, and L. M. Diana, eds. (North-Holland Publishing Co., New York, 1982) 165.

65. "Studies of the Fermi Surface of Tantalum by 2D ACAR Measurements" (with F. Sinclair and K. R. Hoffmann), *Positron Annihilation*, P. G. Coleman, S. C. Sharma, and L. M. Diana, eds. (North-Holland Publishing Co., New York, 1982) 316.

66. "The Autocorrelation Function $B^{2\lambda}$ ($\underline{r}$) in Angular Correlation Experiments" (with W. S. Farmer and F. Sinclair), *Positron Annihilation*, P. G. Coleman, S. C. Sharma, and L. M. Diana, eds. (North-Holland Publishing Co., New York, 1982) 319.

67. "Reconstructed 3D Positron-Electron Momentum Density in Cu: Fermi Surface and Many-Body Enhancement Studies" (with F. Sinclair and W. S. Farmer), *Positron Annihilation*, P. G. Coleman, S. C. Sharma, and L. M. Diana, eds. (North-Holland Publishing Co., New York, 1982) 322.

68. "Spin Dependent 2D ACAR Measurements in Gadolinium" (with K. R. Hoffmann), *Positron Annihilation*, P. G. Coleman, S. C. Sharma, and L. M. Diana, eds. (North-Holland Publishing Co., New York, 1982) 325.

69. "A Positron Annihilation Study of the Equilibrium Vacancy Ensemble in Aluminum" (with M. J. Fluss, B. Chakraborty, K. Hoffmann, P. Lippel, and R. W. Siegel, *Positron Annihilation*, P. G. Coleman, S. C. Sharma, and L. M. Diana, eds. (North-Holland Publishing Co., New York, 1982) 454.

70. "Momentum Density and Fermi-Surface Measurements in Metals by Positron Annihilation" in *Positron Solid-State Physics*, Proc. of the Int. School of Physics "E. Fermi" Course LXXXIII, W. Brandt and A. Dupasquier, eds. (North-Holland Publishing Co., New York, 1983) 64-145.

71. "Positron Annihilation Spectroscopy of the Equilibrium Vacancy Ensemble in Aluminum" (with M. J. Fluss, B. Chakraborty, K. R. Hoffmann, P. Lippel and R. W. Siegel), J. Phys. F: Met. Phys. *14*, 2831 (1984).

72. "A Monovacancy-Divacancy Model Interpretation of Positron Annihilation Measurements in Aluminum" (with M. J. Fluss, B. Chakraborty, P. Lippel and R. W. Siegel), J. Phys. F: Met. Phys. *14*, 2855 (1984).

73. "Fermi Surface and Conduction Electrons of $Na_{0.64}WO_3$ by Two Dimensional Angular Correlation of Positron Annihilation Radiation" (with T. Akahane, K. R. Hoffmann and T. Chiba), Solid State Comm. *54*, 823 (1985).

74. "A Study of the Equilibrium Vacancy Ensemble in Aluminum Using ID- and 2D-Angular Correlation of Annihilation Radiation" (with M. J. Fluss, B. Chakraborty, K. R. Hoffmann, P. Lippel and R. W. Siegel), *Positron Annihilation*, P. C. Jain, R. M. Singru, and K. P. Gopinathan, eds. (World Scientific Publ. Co., Singapore, 1985) 257.

75. "Positronium Formation in Voids of Vanadium" (with M. Hasegawa, Y. J. He, K. R. Hoffmann, R. R. Lee and T. Takeyama), *Positron Annihilation*, P. C. Jain,

R. M. Singru, and K. P. Gopinathan, eds. (World Scientific Publ. Co., Singapore, 1985) 260.

76. "The Fermi Surface of Cu-15.0% Ge ζ Phase Alloy by 2D ACAR Measurements" (with M. hasegawa, K. R. Hoffmann and R. R. Lee), *Positron Annihilation*, P. C. Jain, R. M. Singru, and K. P. Gopinathan, eds. (World Scientific Publ. Co., Singapore, 1985) 291.

77. "2D ACAR Momentum Density Study of the Nature of the Positron Surface State on Al(100)" (with K. F. Canter, K. G. Lynn, A. P. Mills, Jr., L. O. Roelling and R. N. West), *Positron Annihilation*, P. C. Jain, R. M. Singru, and K. P. Gopinathan, eds (World Scientific Publ. Co., Singapore, 1985) 951.

78. "An Experimental Method for Measuring Positronium Reflection from Surfaces" (with K. F. Canter, K. G. Lynn, A. P. Mills, Jr., L. O. Roelling, and M. Weber), *Positron Annihilation*, P. C. Jain, R. M. Singru, and K. P. Gopinathan, eds. (World Scientific Publ. Co., Singapore, 1985) 961.

79. "Positron or Positroniumlike Surface State on Al(100)?" (with K. G. Lynn, A. P. Mills, Jr., R. N. West, K. F. Canter and L. O. Roelling), Phys. Rev. Lett. *54*, 1702 (1985).

80. "Experimental Support for Physisorbed Positronium at the Surface of Quartz" (with P. Sferlazzo and K. F. Canter), Phys. Rev. B*32*, 6067 (1985)

Physics Demonstration Films

1. "Electron-Positron Annihilation," MIT Science Teaching Center and Educational Development Corporation Physics Demonstration Film (1970).

2. "Single Photon Polarization," MIT Science Teaching Center and Educational Development Corporation Physics Demonstration Film (1972).

Reprinted from *Positron Solid-State Physics*

Momentum Density and Fermi-Surface Measurements in Metals by Positron Annihilation.

S. BERKO

Brandeis University - Waltham, Mass. 02254

1. – Introduction.

1·1. *The early history.* – The study of positron interactions with electrons in condensed matter and the development of positrons as a probe of electronic structure has a long and well-documented history [1, 2]. This development spans four decades, from the earliest measurement of the angular correlation of annihilation radiation by BERINGER and MONTGOMERY in 1942 [3], to the latest high-precision reconstruction of the full three-dimensional momentum distribution from two-dimensional correlation data [4]. The fundamental techniques used in positron annihilation studies were developed rapidly, and by 1956 we find a review of the *Experimental studies of positron interactions in solids and liquids* by BERKO and HEREFORD [5] and a seminal work by FERRELL on the *Theory of positron annihilation in solids* [6]. By this time it was clear that, in some materials such as metals and semiconductors, positrons, after slowing down, annihilate from essentially « free » states, while in others, such as some insulators and organic liquids, a fraction of positrons form positronium—the positron-electron hydrogenlike bound state—prior to annihilation. In metals, many problems facing the use of positrons as a probe for electronic structure studies had been well understood by the late fifties. In 1958 we started work on oriented metal crystals, based on the pioneering studies on polycrystalline samples by DEBENEDETTI [7, 8] and STEWART [9]. Let us quote from the introduction from that paper [10], in order to present the flavor of our understanding of the field in those early years (the numbering of references has been changed so as to conform to our present order):

« Since the first measurements by DEBENEDETTI *et al.* [7] on the angular distribution of annihilation radiation in Au, a number of investigators have applied the same method to a variety of solids and liquids [5]. It has been pointed out by these workers that a study of annihilation of positrons in metals could provide direct information on the momentum distributions of valence electrons in these metals, since the probability of observing a certain angle between the two annihilation photons depends directly on the square of the Fourier transform of the positron-electron wave function. »

« Most existing measurements of the angular distributions have been performed on polycrystalline samples [7, 8], so that only the spherical average of the momentum distributions could be determined. The possibility of measuring momentum anisotropies in oriented single crystals was demonstrated by us in a previous paper on the angular correlation of annihilation radiation in oriented graphite [11]. In this paper we present more recent measurements on single crystals of Cu and Al, to examine how well this technique can be used in the determination of electron properties in metals, particularly the shape of the Fermi surface. »

« There seems to be no doubt that a positron has reached thermal equilibrium in a metal before it annihilates [12] (it has $\sim 10^{-10}$ s). We may, therefore, suppose that the electron-positron system annihilates from its ground state. A simple estimate of the positron lifetime using constant positron wave functions and annihilation with the valence electrons only yields values that are longer and vary more from metal to metal than the measured lifetimes. Calculations have been performed emphasizing the importance of strong correlations between the positron and the electrons [6, 13]. It is well known, however, that many properties of metals can be understood in terms of one-particle—uncorrelated—wave functions, and it seems to be worthwhile to investigate quite thoroughly whether this may not also be true for positron annihilation. Indeed before this is done, it will not really be clear what effects have to be attributed to correlation. For this purpose it is important to make calculations on several metals that have as different angular distributions of the annihilating photons as possible. »

« Roughly speaking, the angular distributions fall into two classes. There are those that appear very nearly parabolic with a fairly sharp cut-off at an angle θ given by hk'/mc, where k' is the wave vector corresponding to the Fermi surface of the valence electrons as calculated by the free-electron approximation; the alkalis, alkaline earths and aluminum are examples. The other class has angular distributions that extend way beyond the expected Fermi cut-off and do not show any prominent break; the noble metals are examples. »

« DeBenedetti and Primakoff [7], who first observed the wide angular distribution (in gold), attributed it to the sharp cut-off of the positron wave function at the core of the Au ion. This suggestion of the "excluded volume effect" was again proposed by Ferrell [6]. On the other hand, from an examination of the distributions obtained by Stewart from annihilation in the series Ni, Cu, Zn and Ga, Berko and Hereford [5] proposed that the large momenta were due to annihilation with the $3d$ electrons. »

As we shall see in these lectures, the physical picture emerging from these quoted paragraphs is not far from the present view of the situation. The most important effect not realized at that time was the great affinity of positrons to low-density regions of a metal, such as vacancies and dislocations. Papers presented at the first international positron conference held in 1965 [1] contain many of the ingredients required for the appreciation of this affinity (hindsight, of course): In a review paper on positron annihilation in molecular substances and ionic crystals, BRANDT [14] presented evidence for positron trapping in ionic crystals and developed formulae now called the « two-stage trapping model ». We quote from his conclusion: « Should these centers prove to be linked to lattice defects as considered here in a preliminary manner, an attractive complement to the method of color centers may develop for the characterization of defects in ionic crystals. » At the same meeting STEWART [15] reviewed the surprising results that the angular correlation changes substantially in several metals as a function of rising temperature *prior* to melting, rather than just in the liquid phase. To quote from his conclusion: « The behaviour of a positron in a hot crystalline metal and in a liquid metal is mysterious. » And yet in the same talk STEWART mentions the experiments by DEKHTYAR *et al.* [16], observing changes in the angular correlations in transition metal alloys induced by plastic deformation; the problem was that these authors attributed the observed effect (narrowing at low momenta) to a change in the *s vs. d* electrons upon plastic deformation. By 1967 we have reported the same narrowing effect upon plastic deformation in pure Al [17] and attributed it to positron trapping by dislocations. We quote the conclusion of our paper (again with reference numbers changed):

« In view of this new effect it is rather obvious that, irrespective of the purpose of the experiments, high-precision angular-distribution experiments will have to be performed with great care regarding the concentrations of impurities and dislocations. In particular, one must be very careful in interpreting deviations from parabolic behavior in terms of the properties of electrons and positrons in perfect crystals unless the samples are annealed carefully prior to experiment. It is also interesting to note that the peculiar dependence on temperature of the angular-distribution curves observed in some metals prior to melting [18, 19] might be explained in terms of the thermal generation of point defects and the above-postulated affinity of positrons to reside at vacancy sites. »

At the same time MACKENZIE *et al.* [20] reported that positron lifetimes in metals change with temperature and attributed this effect to trapping in thermally activated vacancies. These results led to the formulation of the trapping model for metals by CONNORS and WEST [21] and by BERGERSEN and STOTT [22], along the lines of the trapping model postulated earlier by BRANDT for ionic crystals. An explosive development followed, leading to the use of positron annihilation in metallurgical studies; this field has been amply reviewed [23], and will be discussed by several lecturers during the course of this Fermi School.

In our lectures we shall restrict ourselves to the discussion of positron annihilation studies used to understand in detail the interaction of positrons with electrons in metals with a « perfect lattice », *i.e.* in well-annealed samples showing (hopefully) little or no trapping at defects, and the use of positron annihilation experiments to study the electronic band structure of metals.

1·2. *The present physical picture: an overview.* – Before embarking on a detailed discussion of positron annihilation in metals, let us present a brief, qualitative, but up-to-date description of the overall behavior of positrons in a metal sample.

Positrons from a radioactive source are injected into the (single crystal) sample to be studied, having initially the typical energy distribution of a β^+ spectrum. The positrons lose energy rapidly (within $(10^{-12}\div10^{-11})$ s) by ionization, electron-hole pair creation and finally by positron-phonon scattering [24]. Within the sample the positrons reach thermal (or nearly thermal) equilibrium with the lattice, and end up spatially distributed along a characteristic « implantation profile » depending on the incident-energy spectrum and on the sample material [25]. Once in their low-energy states as delocalized Bloch waves (the positron forms its own « positron energy band » since there is no Pauli exclusion between the positron and the electrons), the positrons will diffuse with a mean free path characteristic of their energy. For details regarding positron diffusion see the lectures by Mills at this Fermi School. The important fact to remember is that the « implantation profile » in all materials is sufficiently deep so that the greatest fraction of thermalized positrons annihilates in the bulk of the samples to be studied.

In a metal the positron interacts strongly with the conduction electrons and forms a many-body system, with the electrons forming a characteristic screening cloud around the positron. Should a positron, during its diffusion stage, overlap a lower-than-average density region, such as a vacancy, a dislocation or a microvoid, it has a finite probability of being captured and thus being localized in such a region. The affinity for such regions is due to the repulsion between the positron and the positive atomic cores. Those (few) positrons that can diffuse to the sample surface have a finite chance of being expelled from the sample (thus a negative work function), an effect that that permits the production of variable-energy slow-positron beams [26]. The development of such high-intensity positron beams has led during the last several years to new applications of slow-positron physics to surface studies—a field that is being reviewed by Mills.

When the positron finally annihilates (with a lifetime of a few tenths of a nanosecond), the annihilation products are governed by QED selection rules: in spin singlet e^+-e^- overlap, annihilation can proceed via the 2γ channel, while spin triplet overlap produces three annihilation quanta. Since the 2γ annihilation cross-section is $\approx 10^3$ times larger than the 3γ cross-section [27], positrons annihilate in metals (no positronium formation) mainly via the 2γ channel. The angular correlation and the energy spectrum of the annihilation gammas reflect, by momentum and energy conservation, the momentum distribution of the annihilating e^+-e^- pair. Breaks in the angular correlations reflect the size and topology of the Fermi surface, while lifetime experiments yield information about the local electronic density at the positron site. In ferromagnetic materials, the $3\gamma/2\gamma$ yield from spin-aligned positrons combined with angular-correlation measurements can provide information about the spin-aligned electronic bands [28].

One can thus use positron annihilation as a tool in solid-state physics. However, as we have already mentioned, positrons, because of their strong Coulomb interaction with electrons and ions, do not form « ideal probes » of condensed matter, as do X-rays or neutrons. Yet these interactions have been sufficiently well studied and understood during the last two decades to allow the application of the positron annihilation technique to several important areas of solid-state physics. At the same time, positron annihilation experiments in well-understood samples such as in « nearly-free-electron » metals have provided theorists with a rich field for the application and testing of various techniques in many-body physics developed for the study of interacting charged fermions.

Our lectures are divided into five parts. After this introduction (first part) we present briefly in the second section the general theory of the 2γ angular-correlation experiments; we also outline the principles of Compton-profile measurements, another important technique used to study electronic momentum distributions. In the third part we discuss some general features of momentum distributions, particularly in crystalline solids, and present briefly the mathematical techniques permitting the reconstruction of the full three-dimensional momentum densities from its measured projections. In the fourth part we present a brief review of the experimental techniques, focusing our attention on the recently developed two-dimensional systems. Finally, in the fifth section, we present and discuss a few selected experimental results, in order to illustrate the main information obtainable from positron annihilation experiments. Our lectures will in no way correspond to a systematic review of all the published work on the subject—they represent rather a tutorial introduction to the field with some illustrative examples. For systematic studies we refer to the several comprehensive review articles in the literature [2, 29-34] and to other references cited therein.

2. – Theoretical considerations.

2·1. *Theory of* 2γ *angular-correlation experiments.* – The cross-section for 2γ annihilation of a slow free positron with a free, nonrelativistic electron is given by the Dirac formula $\sigma_{2\gamma} = \pi r_0^2 c/v$, where v is the relative e^+-e^- velocity ($v \ll c$) and $r_0 = e^2/mc^2$. Thus the annihilation rate $\lambda_{2\gamma}$ (inverse lifetime) of a slow positron in a noninteracting electron gas is $\lambda_{2\gamma} = \sigma_{2\gamma} nv = \pi r_0^2 cn$, with n being the density of electrons. In the derivation of this formula the positron-electron Coulomb interaction was not included. The changes required to take into account the full many-body (one positron-many electron) system encountered in real situations, such as in positron annihilation in an electron gas or in condensed matter, have been the subject of many theoretical investigations, and will form the topic of Carbotte's lectures at this Fermi School.

In his 1956 paper [6] FERRELL presented the second-quantization formulation of the theory; because of the large energies and momenta of the annihilation quanta ($E \sim mc^2$, $p \sim mc$) compared to the energy and momenta of the annihilating electron-positron pair, one obtains the Hamiltonian operator $H'(\boldsymbol{p})$ that produces a two-photon final state with total momentum $\boldsymbol{p}$ (we use $\hbar \equiv 1$):

$$H'(\boldsymbol{p}) = \sum_{p_1} \sum_{p_2} a_{\boldsymbol{p}_1} b_{\boldsymbol{p}_2} \delta_{\boldsymbol{p}_1+\boldsymbol{p}_2,\boldsymbol{p}}, \tag{2.1}$$

where $a_{\boldsymbol{p}_1}$ and $b_{\boldsymbol{p}_2}$ are the electron and positron annihilation operators, respectively. Using the point annihilation operators $\hat{\psi}_-(\boldsymbol{x}_1)$ and $\hat{\psi}_+(\boldsymbol{x}_2)$ given by

$$a_{\boldsymbol{p}_1} = \frac{1}{\sqrt{V}} \int \mathrm{d}\boldsymbol{x}_1 \exp[-i\boldsymbol{p}_1 \cdot \boldsymbol{x}_1] \hat{\psi}_-(\boldsymbol{x}_1) \tag{2.2}$$

and

$$b_{\boldsymbol{p}_2} = \frac{1}{\sqrt{V}} \int \mathrm{d}\boldsymbol{x}_2 \exp[-i\boldsymbol{p}_2 \cdot \boldsymbol{x}_2] \hat{\psi}_+(\boldsymbol{x}_2),$$

one obtains the Hamiltonian operator

$$H'(\boldsymbol{p}) = \int \mathrm{d}\boldsymbol{x} \exp[-i\boldsymbol{p} \cdot \boldsymbol{x}] \hat{\psi}_-(\boldsymbol{x}) \hat{\psi}_+(\boldsymbol{x}). \tag{2.3}$$

Thus the probability for a two-photon annihilation event with momentum $\boldsymbol{p}$, from an initial state $|i\rangle$ to a specific final state $|f\rangle$, is given by

$$\varrho_f^{2\gamma}(\boldsymbol{p}) = \text{const} \cdot \left| \langle f | \int \mathrm{d}\boldsymbol{x} \exp[-i\boldsymbol{p} \cdot \boldsymbol{x}] \hat{\psi}_-(\boldsymbol{x}) \hat{\psi}_+(\boldsymbol{x}) | i \rangle \right|^2, \tag{2.4}$$

where $|i\rangle$ stands for the initial state of the one-positron, n-electron system and $|f\rangle$ stands for the final $(n-1)$-electron and two-annihilation-quanta state after annihilation. Conservation of energy should, of course, be included by a δ-function. The « local » character of the annihilation is valid, since the annihilating particles do not change appreciably over the Compton wavelength

($\simeq 10^{-11}$ cm). If one does not measure the final states, such as in the standard angular-correlation experiments, one sums over final states, to obtain

$$\varrho^{2\gamma}(\boldsymbol{p}) = \sum_f \varrho_f^{2\gamma}(\boldsymbol{p}) . \tag{2.5}$$

In the approximation of neglecting the constraints of the energy-conserving δ-function, one can sum over the final states by completeness and thus eliminate the final state from the master formula for $\varrho^{2\gamma}(\boldsymbol{p})$, obtaining the master equation

$$\varrho^{2\gamma}(\boldsymbol{p}) = \text{const}\cdot \iint \mathrm{d}\boldsymbol{x}\,\mathrm{d}\boldsymbol{x}' \exp[i\boldsymbol{p}\cdot(\boldsymbol{x}-\boldsymbol{x}')]\langle i|\hat{\psi}_+^*(\boldsymbol{x})\,\hat{\psi}_+(\boldsymbol{x}')\,\hat{\psi}_-^*(\boldsymbol{x})\,\hat{\psi}_-(\boldsymbol{x}')|i\rangle . \tag{2.6}$$

With an appropriate choice of the multiplying constant (const $= \pi r_0^2 c/(2\pi)^3$), we obtain the momentum-dependent annihilation rate $\Gamma_{2\gamma}(\boldsymbol{p})$, *i.e.* the probability per unit time and per unit volume in momentum space for an annihilating positron to produce two photons with total momentum $\boldsymbol{p}$. The total annihilation rate (*i.e.* inverse lifetime) is then given by $\lambda_{2\gamma} = \int \mathrm{d}\boldsymbol{p}\,\Gamma_{2\gamma}(\boldsymbol{p})$.

CHANG LEE studied theoretically the annihilation process in some detail [35], showing why the Coulomb interaction between the positron and the electrons in the full many-body system can be treated separately from the interaction leading to annihilation. The exact nature of the approximations leading to eq. (2.6) has not been discussed thoroughly in the positron literature. The approximation leading to the elimination of final states from eq. (2.6) is somewhat similar to the impulse approximation in the theory of the Compton profiles. Thus one can apply the same reasoning used to justify the impulse approximation using the time-dependent correlation function representation—see, for example, the review of the Compton-profile theory by PLATZMAN and TZOAR [36].

One can go from the second-quantization language to configuration space, with a one-positron, many-electron initial wave function $\psi(\boldsymbol{r}_+, \boldsymbol{r}_1, \boldsymbol{\tau})$, properly antisymmetric in all electron co-ordinates, where $\boldsymbol{r}_+$ is the positron co-ordinate, $\boldsymbol{r}_1$ is an electron co-ordinate and $\boldsymbol{\tau}$ stands for the remaining electron co-ordinates $\{\boldsymbol{r}_2, \boldsymbol{r}_3, \dots, \boldsymbol{r}_n\}$, to obtain

$$\varrho^{2\gamma}(\boldsymbol{p}) = \sum_f \varrho_f^{2\gamma}(\boldsymbol{p}) = \text{const}\cdot \int \mathrm{d}\boldsymbol{\tau} \left| \int \mathrm{d}\boldsymbol{x} \exp[-i\boldsymbol{p}\cdot\boldsymbol{x}]\,\psi(\boldsymbol{x}, \boldsymbol{x}, \boldsymbol{\tau}) \right|^2 . \tag{2.7}$$

Finally, in the independent-particle approximation (IPM), for a positron in state $\psi_+(\boldsymbol{r})$ annihilating in a system of electrons described by a Slater determinant of single-particle states $\psi_{-j}(\boldsymbol{r})$, one obtains the IPM master formula

$$\varrho_{\text{IPM}}^{2\gamma}(\boldsymbol{p}) = \text{const}\cdot \sum_j \left| \int \mathrm{d}\boldsymbol{r} \exp[-i\boldsymbol{p}\cdot\boldsymbol{r}]\,\psi_+(\boldsymbol{r})\,\psi_{-j}(\boldsymbol{r}) \right|^2 . \tag{2.8}$$

Again, one can convert $\varrho^{2\gamma}_{\text{IPM}}$ to total IPM annihilation rates as above, obtaining

$$\lambda_{\text{IPM}} = \frac{\pi r_0^2 c}{(2\pi)^3}\int \mathrm{d}\boldsymbol{p}\sum_j \left|\int \mathrm{d}\boldsymbol{r}\exp[-i\boldsymbol{p}\cdot\boldsymbol{r}]\,\psi_+(\boldsymbol{r})\,\psi_{-j}(\boldsymbol{r})\right|^2. \tag{2.9}$$

2'2. *Geometry of the angular-correlation experiments.* – By momentum and energy conservation the two-photon momentum distribution $\varrho^{2\gamma}(\boldsymbol{p})$ results in an angular correlation and an energy distribution of the two annihilation quanta. In the center-of-mass frame the two photons are emitted at an angle of 180°, with total energy $E_{\text{T}} = 2mc^2 - E_{\text{B}}$, E_{B} being the binding energy of the electron and the positron in the system. In the laboratory frame the two photons are not anticolinear because of the finite momentum $\boldsymbol{p}$ to be conserved, and their energy is Doppler shifted. We have sketched in fig. 1 the ge-

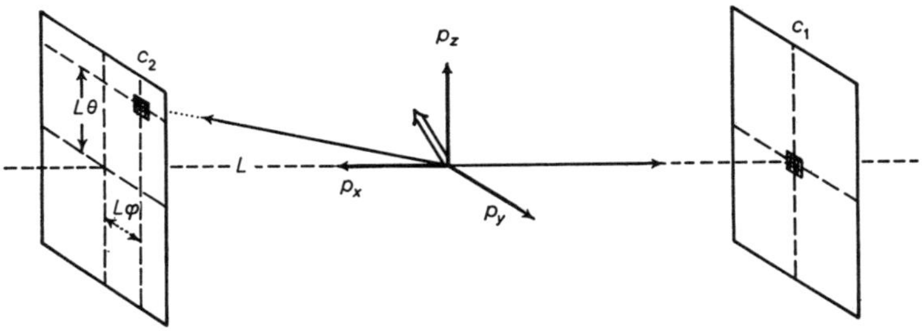

Fig. 1. – Geometry of the 2γ angular-correlation experiment. The double arrow represents the momentum $\boldsymbol{p}$ to be conserved by the two annihilation γ's.

ometry of the usual angular-correlation experiment. Given a typical atomic momentum $\boldsymbol{p}$ ($|\boldsymbol{p}| \sim 10^{-2}\,mc$), the angle between the two photons deviates from 180° only by a few milliradians. If we set a small detector at the origin of counter plane c_1 to detect one of the two photons, the second correlated photon will hit the counter plane c_2 at the co-ordinates θL and φL, since θ and φ are only a few milliradians; L is the distance of the counter plane from the annihilation center. Thus we obtain $p_z = \theta mc$ and $p_y = \varphi mc$, to a good approximation, since the magnitude of the photon momenta is mc within one percent. The third component, p_x, parallel to the main direction of the quanta can be measured by observing the energy of the two photons, E_1 or E_2. We obtain $E_{1,2} = mc^2 - E_{\text{B}}/2 \pm cp_x/2$ (again for $|\boldsymbol{p}| \ll mc$). Typical atomic momenta lead to Doppler shifts of a few keV; thus we can neglect in most measurements the smaller E_{B} term (we assume annihilations mainly with outer electrons). A momentum $\boldsymbol{p}$ corresponding to one atomic (momentum) unit (*i.e.* $|\boldsymbol{p}| = 1$ a.u., $m = 1$, $c = 137$) produces an angular deviation of 7.297 mrad, when $\boldsymbol{p}$ is $\perp$ to the 2γ axis, and an energy shift of ± 1.86 keV (from 511 keV) when ∥ to the 2γ axis. We can, therefore, measure momenta in milliradians, where one mrad $\equiv mc\cdot 10^{-3} = 0.137$ a.u.

We note that in an ideal experiment, using small detectors at c_1 and c_2 in coincidence, and measuing simultaneously the energy of both photons with high resolution, one can in principle obtain the partial three-dimensional momentum density $\varrho_j^{2\gamma}(\boldsymbol{p})$ of the 2γ pair restricted to a binding energy $E_{\mathrm{B}(j)}$. Add to this a simultaneous timing experiment, and one has the measurement of the partial rates $\Gamma_j^{2\gamma}(\boldsymbol{p})$. Usually, for counting rate reasons, this is not feasible, and one performs either angular correlation of annihilation radiation (ACAR) experiments, Doppler-shift measurements, or lifetime measurements, thus obtaining experimentally various integrals of $\Gamma_j^{2\gamma}(\boldsymbol{p})$.

In order to maximize the count rate, the usual *long-slit geometry* setup uses two long counters behind long collimating slits, thus integrating over one of the momentum components, say p_y (corresponding to long horizontal counters in fig. 1). Coincidences are measured as a function of the angle θ. The lack of energy measurement means an automatic integration over the p_x component. Thus, for a given momentum density $\varrho^{2\gamma}(\boldsymbol{p})$, the long-slit geometry setup measures, for perfect resolution, a coincidence rate proportional to the projection of the momentum distribution

$$N(p_z) = \iint \varrho^{2\gamma}(\boldsymbol{p})\,\mathrm{d}p_x\,\mathrm{d}p_y\,, \tag{2.10}$$

where $p_z = \theta mc$. The actual counting rate is obtained by folding the theoretical distribution of eq. (2.10) into the instrumental resolution of the apparatus. We call $N(p_z)$ the one-dimensional projection (1DP) of $\varrho^{2\gamma}(\boldsymbol{p})$, or the one-dimensional ACAR curve (1D ACAR).

More recently two-dimensional angular-correlation machines have been built [4], obtaining (without energy measurement and with perfect angular resolution) the two-dimensional ACAR correlation

$$N(p_y, p_z) = \int \varrho^{2\gamma}(\boldsymbol{p})\,\mathrm{d}p_x\,. \tag{2.11}$$

(Again, the actual measurement is a 2D convolution of $N(p_y, p_z)$ with the 2D resolution function of the apparatus.) Thus $N(p_y, p_z)$ is the two-dimensional projection (2DP) of $\varrho^{2\gamma}(\boldsymbol{p})$ (2D ACAR). In order to visualize the effect of these two geometries, one can apply eqs. (2.10) and (2.11) to a spherical momentum distribution corresponding to a (neutral) positron in a noninteracting electron gas, *i.e.* to $\varrho^{2\gamma}(\boldsymbol{p}) = \text{const}$ for $|\boldsymbol{p}| \leqslant p_{\mathrm{F}}$ and zero for $|\boldsymbol{p}| \geqslant p_{\mathrm{F}}$, where p_{F} is the Fermi momentum. One obtains for the « long-slit » geometry an inverted parabola cutting off at $p_z = \pm p_{\mathrm{F}}$. For the « point-slit » geometry, *i.e.* for the 2D ACAR surface, one predicts a hemisphere bounded in the (p_y, p_z)-plane by $p_y^2 + p_z^2 = p_{\mathrm{F}}^2$ (actually one obtains half an ellipsoid of revolution, since the normalization of the $N(p_y, p_z)$ ACAR surface is arbitrary).

2'3. *Theory of Compton-profile measurements.* – So far we have reviewed the theory of the ACAR experiments; we have seen that these experiments yield the projections of the momentum-space density of the electrons *as sampled by the positron, i.e.* the 2γ momentum density $\varrho^{2\gamma}(\boldsymbol{p})$. Our lectures would not be complete if we would not discuss, even briefly, an associated technique, based on the Compton effect, *i.e.* the technique of the Compton-profile measurements. Even in QED the Compton-scattering cross-section is related to the e^+-e^- annihilation cross-section, the Feynman diagrams of the two being related by a 90° rotation (crossing symmetry). The measurement of Compton-scattered energy profiles (CP) of X-rays or γ-rays provides information about the true electronic momentum distribution $\varrho(\boldsymbol{p})$ of a system, and is thus closely related to the ACAR technique. The CP measurements have a long history [37] carefully discussed by STUEWER and COOPER in an introductory article of a recent book on Compton scattering [38]. During the last decades the field has undergone a true renaissance due chiefly to the early efforts by R. J. WEISS in the U.S.A. and M. J. COOPER in England. Briefly, the Compton profile is the energy distribution of inelastically scattered X-rays or γ-rays at a fixed scattering angle; the Compton-scattered photons are Doppler broadened about the Compton wavelength due to the initial momentum distribution of the electrons producing the scattering. The pertinent geometry of the CP experiments is shown in fig. 2, with $\boldsymbol{k}_i$ and $\boldsymbol{k}_f$ being the incident and scattered

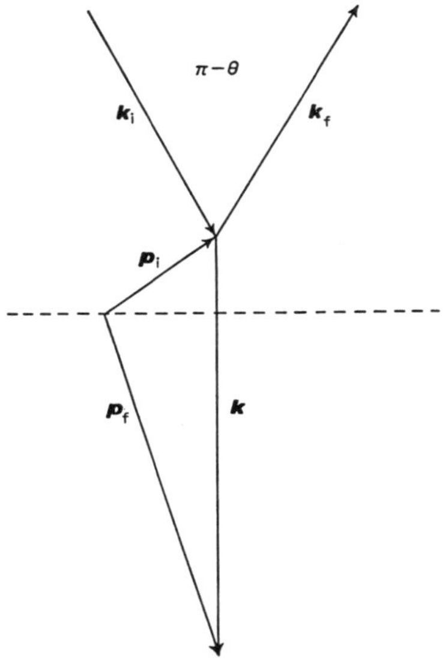

Fig. 2. – Geometry of the Compton-profile measurement. $\boldsymbol{k}_i$ and $\boldsymbol{k}_f$ are the incident and scattered photon momenta, $\boldsymbol{p}_i$ and $\boldsymbol{p}_f$ the corresponding electron momenta, θ is the scattering angle.

photon momenta, $\boldsymbol{p}_i$ and $\boldsymbol{p}_f$ the corresponding electron momenta; $\boldsymbol{k} = \boldsymbol{k}_i - \boldsymbol{k}_f$ is the scattering vector defining the z-axis of a co-ordinate system. The well-known Compton-wavelength shift, corresponding to $\boldsymbol{p}_i = 0$, is changed, in the simple case of a nonrelativistic free electron of finite $\boldsymbol{p}_i$, by the Doppler term depending on the p_z component of $\boldsymbol{p}_i$ in the $\boldsymbol{k}$ direction,

$$\lambda' - \lambda = \frac{2h}{mc}\sin^2\frac{\theta}{2} - \frac{2\lambda\sin(\theta/2)}{mc}\,p_z\,. \tag{2.12}$$

Since in a typical experiment one does not detect the recoiling electron, all momenta with component p_z will contribute for a given $\boldsymbol{k}$. For high-energy scattering with $|\boldsymbol{k}_i|$ much larger than the typical atomic momenta, the probability of measuring $\boldsymbol{k}_f$ should depend, to a good approximation, on the probability $J(p_z)\,\mathrm{d}p_z$ of having a momentum component between p_z and $p_z + \mathrm{d}p_z$, given a momentum-space (probability) density of $\varrho(\boldsymbol{p})$:

$$J(p_z) = \text{const}\cdot\iint\varrho(\boldsymbol{p})\,\mathrm{d}p_x\,\mathrm{d}p_y\,. \tag{2.13}$$

The integration is over a plane $\perp$ to $\boldsymbol{k}$, as indicated by the dashed line in fig. 2. This argument is, of course, strictly classical, based only on energy and momentum conservation.

Quantum mechanically, the cross-section for Compton scattering from a single electron with initial wave function ψ_i is given, in the nonrelativistic limit, by

$$\frac{\mathrm{d}^2\sigma}{\mathrm{d}\Omega\,\mathrm{d}\omega} = \left(\frac{\mathrm{d}\sigma}{\mathrm{d}\Omega}\right)_{\mathrm{T}}\frac{\omega_f}{\omega_i}\sum_f|\langle\psi_f|\exp[i\boldsymbol{k}\cdot\boldsymbol{r}]|\psi_i\rangle|^2\,\delta(E_f - E_i - \omega)\,, \tag{2.14}$$

where ω_i, E_i and ω_f, E_f are the initial and final photon and electron energies, $\omega = \omega_i - \omega_f$, ψ_f is the final electron wave function and $(\mathrm{d}\sigma/\mathrm{d}\Omega)_{\mathrm{T}}$ is the Thompson cross-section.

For the case in which the final electron energy is large with respect to its binding energy prior to scattering, one can use the « impulse approximation » [36], which then reduces eq. (2.14) to the desired form:

$$\frac{\mathrm{d}^2\sigma}{\mathrm{d}\Omega\,\mathrm{d}\omega} = f(\omega_i, \omega_f, \boldsymbol{k})\int\mathrm{d}\boldsymbol{p}\,\delta\left(\omega - \frac{k^2}{2m} - \frac{\boldsymbol{p}\cdot\boldsymbol{k}}{m}\right)\varrho(\boldsymbol{p})\,. \tag{2.15}$$

The 3D integral including the δ-function indicates the restriction of the Doppler shift on ω and reduces indeed to a 1D projection of $\varrho(\boldsymbol{p})$ equivalent to (2.13), where $\varrho(\boldsymbol{p})$ in second-quantized language is

$$\varrho(\boldsymbol{p}) = \text{const}\cdot\iint\mathrm{d}\boldsymbol{x}\,\mathrm{d}\boldsymbol{x}'\exp[i\boldsymbol{p}\cdot(\boldsymbol{x}-\boldsymbol{x}')]\langle i|\hat{\psi}^*_-(\boldsymbol{x})\,\hat{\psi}_-(\boldsymbol{x}')|i\rangle\,. \tag{2.16}$$

Here $|i\rangle$ corresponds to the many-electron state producing the Compton scattering.

In the impulse approximation the Compton-profile (CP) technique thus measures the 1D projection of the true momentum density $\varrho(\boldsymbol{p})$ onto a direction $p_z \parallel$ to the scattering vector $\boldsymbol{k}$. In the IPM $\varrho(\boldsymbol{p})$ becomes, for a Slater determinant of electron states $\psi_{-j}(\boldsymbol{r})$,

$$\varrho_{\text{IPM}}(\boldsymbol{p}) = N \sum_j \left| \int \mathrm{d}\boldsymbol{r} \exp[-i\boldsymbol{p}\cdot\boldsymbol{r}]\, \psi_{-j}(\boldsymbol{r}) \right|^2, \tag{2.17}$$

where N is an appropriate normalizing constant.

We thus see the great similarity *and* difference between Compton-profile and ACAR measurements, by comparing the master equations (2.16) and (2.17) for $\varrho(\boldsymbol{p})$ as measured by CP to the corresponding eqs. (2.6) and (2.8) for $\varrho^{2\gamma}(\boldsymbol{p})$ for the ACAR experiments. The main difference is, of course, the presence of the positron in (2.6) and (2.8); in ACAR one measures the projections of the « positron sampled momentum density » $\varrho^{2\gamma}(\boldsymbol{p})$, while in CP one obtains the projections of the « e^+ undisturbed » true $\varrho(\boldsymbol{p})$.

Note that in the IPM expressions one obtains $\varrho_{\text{IPM}}(\boldsymbol{p})$ from $\varrho^{2\gamma}_{\text{IPM}}(\boldsymbol{p})$ by simply setting $\psi_+(\boldsymbol{r}) = 1$. On the other hand, one cannot obtain $\varrho^{2\gamma}_{\text{IPM}}(\boldsymbol{p})$ from $\varrho_{\text{IPM}}(\boldsymbol{p})$ and the momentum density of the positrons alone, since eq. (2.8) for $\varrho^{2\gamma}_{\text{IPM}}(\boldsymbol{p})$ is the square of a convolution (using the Fourier-transform convolution theorem)

$$\varrho^{2\gamma}_{\text{IPM}}(\boldsymbol{p}) = \text{const}\cdot \sum_j \left| \int \varphi_+(\boldsymbol{p}')\,\varphi_{-j}(\boldsymbol{p}-\boldsymbol{p}')\,\mathrm{d}\boldsymbol{p}' \right|^2, \tag{2.18}$$

where the $\varphi(\boldsymbol{p})$'s are the single-particle momentum wave functions of the positron and the j-th electron.

We also note that the standard CP experiment measures the 1D projections of $\varrho(\boldsymbol{p})$, *i.e.* it is equivalent to the « long-slit » geometry in ACAR. (Compare eq. (2.13) with eq. (2.10).) If one could measure in coincidence the recoiling Compton electron, one could in principle measure the 2D projection of $\varrho(\boldsymbol{p})$, as in the 2D ACAR experiments. If not for inherent resolution problems in the present CP experiments, one would prefer to perform CP measurements over ACAR experiments, since $\varrho(\boldsymbol{p})$ is of more direct interest to solid-state physics than is $\varrho^{2\gamma}(\boldsymbol{p})$—we shall return to these questions in later sections.

3. – Momentum space in crystalline solids.

3·1. *General definitions.* – In the discussion of the theory of the Compton profile we have defined the momentum density $\varrho(\boldsymbol{p})$ for an interacting electron system in eq. (2.16). For a single particle the momentum density is the square of the momentum wave function $\varphi(\boldsymbol{p}) = \langle \boldsymbol{p}|\psi\rangle$ (where $\psi(\boldsymbol{r}) = \langle \boldsymbol{r}|\psi\rangle$) given by the usual Fourier-transform connection

$$\varphi(\boldsymbol{p}) = \mathrm{FT}_{-3}\{\psi\} \equiv (2\pi)^{-\frac{3}{2}}\int \exp[-i\boldsymbol{p}\cdot\boldsymbol{r}]\psi(\boldsymbol{r})\,\mathrm{d}\boldsymbol{r}\,. \tag{3.1}$$

We also use (3.1) to define our notation for the 3-dimensional Fourier transform FT_{-3}. The usual probability density in real space is $\varrho(\boldsymbol{r}) = \psi^*(\boldsymbol{r})\psi(\boldsymbol{r})$. It has become customary to call the corresponding $\varrho(\boldsymbol{p}) = \varphi^*(\boldsymbol{p})\varphi(\boldsymbol{p})$ the « momentum density » although this is somewhat deceiving; « probability density in momentum space » would be perhaps more adequate.

In diffraction theory one introduces the well-known « form factor » $F(\boldsymbol{p}) = \mathrm{FT}_3\{\varrho(\boldsymbol{r})\}$, since $F(\boldsymbol{p})$ is directly related to the measured diffraction intensities. One should not confuse $F(\boldsymbol{p})$ with $\varrho(\boldsymbol{p})$; given the wave function, one can compute $\varrho(\boldsymbol{r})$—and thus $F(\boldsymbol{p})$—as well as $\varrho(\boldsymbol{p})$. However, one *cannot* obtain $\varrho(\boldsymbol{p})$ from $\varrho(\boldsymbol{r})$ or $F(\boldsymbol{p})$; thus the measurement of $\varrho(\boldsymbol{p})$ provides additional information about the system. In case of a Slater determinant one can show that the total momentum density $\varrho(\boldsymbol{p}) = \sum_{j=1}^{N} \varrho_j(\boldsymbol{p})$, *i.e.* a sum over the one-electron momentum densities, a relation used already in eq. (2.17). For a general discussion of electron densities in an interacting n-electron system and for their relations to density matrices, see the lectures of SMITH jr. [39].

3·2. *The $B(\boldsymbol{r})$ function.* – In 1971, SMITH jr. and collaborators [40] introduced the Fourier transform of the momentum density, paralleling the definition of the form factor. Since then this new derived function has become very fashionable in connection with Compton-profile experiments [41-45], and it is usually denoted $B(\boldsymbol{r})$:

$$B(\boldsymbol{r}) = \mathrm{FT}_{-3}\{\varrho(\boldsymbol{p})\}\,. \tag{3.2}$$

We note that this function is symmetrical in definition to the X-ray form factor $F(\boldsymbol{p})$. Using the well-known Fourier-transform convolution theorem one can show that for IPM wave functions

$$B(\boldsymbol{r}) = (2\pi)^{-\frac{3}{2}} \sum_j \int \psi_j^*(\boldsymbol{r}+\boldsymbol{s})\,\psi_j(\boldsymbol{s})\,\mathrm{d}\boldsymbol{s}\,, \tag{3.3}$$

i.e. $B(\boldsymbol{r})$ is the autocorrelation function of the wave function in real space, and it is, therefore, connected to the off-diagonal matrix elements of the density matrix; the more general density matrix formulation has been discussed by BENESCH *et al.* [40]. The relationships between the various electron density functions we have defined so far in $\boldsymbol{r}$ and $\boldsymbol{p}$ space are shown in fig. 3, following the discussion by WEYRICH *et al.* [45]. We note that experimentally there *is* an asymmetry: whereas $F(\boldsymbol{p})$ is measured (by diffraction) and $\varrho(\boldsymbol{r})$ is computed,

$\varrho(\boldsymbol{p})$ can be measured (from its projections via the Compton-profile and positron experiments) and $B(\boldsymbol{r})$ is computed. It is perhaps amusing to note that the experimental results yield directly the momentum-space, and not the real-space information.

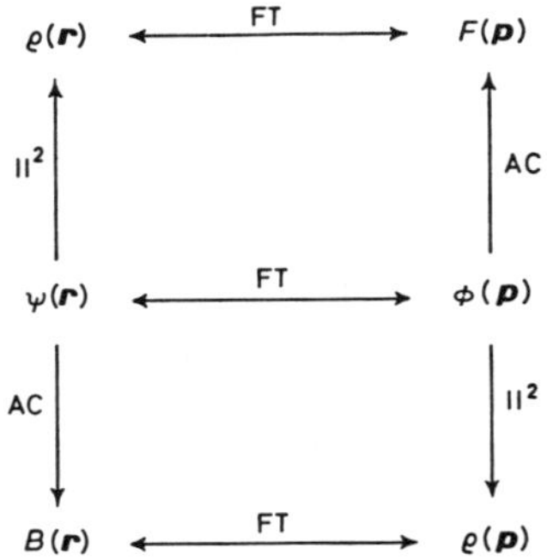

Fig. 3. – Relations between wave functions, densities and autocorrelation functions in real and momentum space, as discussed in the text. FT ≡ Fourier transform, $||^2$ ≡ « magnitude square », AC ≡ autocorrelation.

In analogy with the definition of $B(\boldsymbol{r})$ as the FT of $\varrho(\boldsymbol{p})$, we can define a corresponding $B^{2\gamma}(\boldsymbol{r})$ as

$$B^{2\gamma}(\boldsymbol{r}) = \mathrm{FT}_{-3}\{\varrho^{2\gamma}(\boldsymbol{p})\}\,. \tag{3.4}$$

In the IPM description $B^{2\gamma}(\boldsymbol{r})$ becomes the autocorrelation of the positron-electron wave function product, *i.e.* of $\psi_+(\boldsymbol{r})\psi_{-j}(\boldsymbol{r})$. We are unable, however, to further decompose in a simple way $B^{2\gamma}(\boldsymbol{r})$ in terms of a $B^+(\boldsymbol{r})$ and $B^-(\boldsymbol{r})$, *i.e.* into the separate autocorrelation functions of the positron and of the electron.

The $B(\boldsymbol{r})$ function has some extremely elegant and useful properties, some of purely mathematical importance, some of physical importance. For atomic and particularly for molecular systems, a certain type of orbital appears much more clearly identifiable in $B(\boldsymbol{r})$ than in $\varrho(\boldsymbol{p})$—see, for example, the discussion by WEYRICH *et al.* [45]. For the periodic-potential problem in solid-state physics several theorems derived by SCHÜLKE [43, 44] turn out to be most useful. The most important application of $B(\boldsymbol{r})$ is, however, in the mathematical reconstruction of the full three-dimensional momentum density $\varrho(\boldsymbol{p})$ from its measured one- or two-dimensional projections. We shall return to these applications after discussing the general behavior of momentum densities in periodic systems.

3·3. *Momentum density in periodic systems.* – We proceed to apply the basic IPM formula for $\varrho^{2\gamma}(\boldsymbol{p})$, *i.e.* eq. (2.8), to the case of a single crystal, *i.e.* a periodic potential $V(\boldsymbol{r}) = V(\boldsymbol{r}+\boldsymbol{R})$, where $\boldsymbol{R}$ is a lattice vector. In this case $\psi_+(\boldsymbol{r})$ and $\psi_{-j}(\boldsymbol{r})$ become Bloch waves:

$$(3.5) \qquad \psi_{\boldsymbol{k}^+}(\boldsymbol{r}) = u_{\boldsymbol{k}^+}(\boldsymbol{r}) \exp[i\boldsymbol{k}^+\cdot\boldsymbol{r}] \quad \text{and} \quad \psi_{\boldsymbol{k}^-,j}(\boldsymbol{r}) = u_{\boldsymbol{k}^-,j}(\boldsymbol{r}) \exp[i\boldsymbol{k}^-\cdot\boldsymbol{r}],$$

where j now is a band index. For a single-electron Bloch wave we have

$$(3.6) \qquad \varrho^{2\gamma}_{\boldsymbol{k}^+,\boldsymbol{k}^-,j}(\boldsymbol{p}) = \text{const}\cdot\left|\int d\boldsymbol{r} \exp[-i\boldsymbol{p}\cdot\boldsymbol{r}]\psi_{\boldsymbol{k}^+}(\boldsymbol{r})\psi_{\boldsymbol{k}^-,j}(\boldsymbol{r})\right|^2.$$

The Fourier transform appearing in eq. (3.6) is sometimes denoted by $A_j(\boldsymbol{p}, \boldsymbol{k}^+, \boldsymbol{k}^-)$—see, for example, the review by MIJNARENDS [31]. At a finite temperature we have [7]

$$(3.7) \qquad \varrho^{2\gamma}(\boldsymbol{p}) = \text{const}\cdot \sum_{\boldsymbol{k}^+,\boldsymbol{k}^-,j} F_+(\boldsymbol{k}^+, T)F_-(\boldsymbol{k}^-, T)\varrho^{2\gamma}_{\boldsymbol{k}^+,\boldsymbol{k}^-,j}(\boldsymbol{p})$$

with F_+ being a Boltzmann distribution for the positron near the bottom of its band and F_- the usual Fermi distribution for electrons.

During the last several years the behavior of $\varrho^{2\gamma}(\boldsymbol{p})$ as a function of temperature has come under investigation. One should note that in eq. (3.7) the $\varrho^{2\gamma}_{\boldsymbol{k}^+,\boldsymbol{k}^-,j}(\boldsymbol{p})$ can also be temperature dependent, since the wave functions have to be computed at the finite temperature of the sample—a most complex task achievable only under severe approximations [46].

For zero temperature we set $\boldsymbol{k}^+ = 0$, $\boldsymbol{k}^- = \boldsymbol{k}$, to obtain

$$(3.8) \qquad \varrho^{2\gamma}_{\boldsymbol{k},j}(\boldsymbol{p}) = \text{const}\cdot \sum_{\boldsymbol{G}} |A^{+-}_j(\boldsymbol{k}, \boldsymbol{G})|^2 \delta(\boldsymbol{p}-\boldsymbol{k}-\boldsymbol{G}),$$

where the $A^{+-}_j(\boldsymbol{k}, \boldsymbol{G})$'s are the coefficients in the expansion of the periodic function $u^+(\boldsymbol{r})\,u_{\boldsymbol{k},j}(\boldsymbol{r}) = \sum_{\boldsymbol{G}} A^{+-}_j(\boldsymbol{k}, \boldsymbol{G}) \exp[i\boldsymbol{G}\cdot\boldsymbol{r}]$; the $\boldsymbol{G}$'s are the reciprocal-lattice vectors of the crystal. Then

$$(3.9) \qquad \varrho^{2\gamma}(\boldsymbol{p}) = \sum_{\boldsymbol{k},j} n_j(\boldsymbol{k})\,\varrho^{2\gamma}_{\boldsymbol{k},j}(\boldsymbol{p}),$$

where $n_j(\boldsymbol{k})$ is the step function equal to one for occupied states and zero for unoccupied states, *i.e.* with the step at the Fermi $\boldsymbol{k}_F$. We note that the important dependence of $\varrho^{2\gamma}(\boldsymbol{p})$ on the Fermi surface appears in (3.9) via the $n_j(\boldsymbol{k})$'s. It is clear from eq. (2.18) that the $A^{+-}_j(\boldsymbol{k}, \boldsymbol{G})$'s of the master equation (3.8) can be written as a convolution of the separate Fourier coefficients A^+ and A^- of the u^+ and u^- wave functions as

$$(3.10) \qquad A^{+-}(\boldsymbol{k}, \boldsymbol{G}) = \sum_{\boldsymbol{G}'} A^+(\boldsymbol{G}')A^-(\boldsymbol{k}, \boldsymbol{G}-\boldsymbol{G}').$$

The discussion of the application of eq. (3.8) to the case of a lattice with a basis was given by BRANDT *et al.* [47].

We note that for the case of the true momentum density as measured by the Compton-profile experiments, *i.e.* without the positron wave function, we obtain the IPM master equation at $T = 0$ for $\varrho(\boldsymbol{p})$ by simply setting $A_j^{+-}(\boldsymbol{k}, \boldsymbol{G})$ equal to $A^-(\boldsymbol{k}, \boldsymbol{G})$ in eqs. (3.8) and (3.9). The following discussion of the qualitative dependence of $\varrho^{2\gamma}(\boldsymbol{p})$ on various band behaviors holds, therefore, equally well for $\varrho(\boldsymbol{p})$.

As we pointed out a long time ago [10], eq. (3.8) shows the fundamental difference between real momentum $\boldsymbol{p}$ and crystal momentum $\boldsymbol{k}$: an electron with quantum number $\boldsymbol{k}$ will contribute not only at $\boldsymbol{p} = \boldsymbol{k}$ but at all $\boldsymbol{p} = \boldsymbol{k} + \boldsymbol{G}$, with amplitude $|A_j^{+-}(\boldsymbol{k}, \boldsymbol{G})|^2$, as sketched in fig. 4. The $\boldsymbol{G} \neq 0$ contributions are the « high-momentum components » (HMC) of the momentum density.

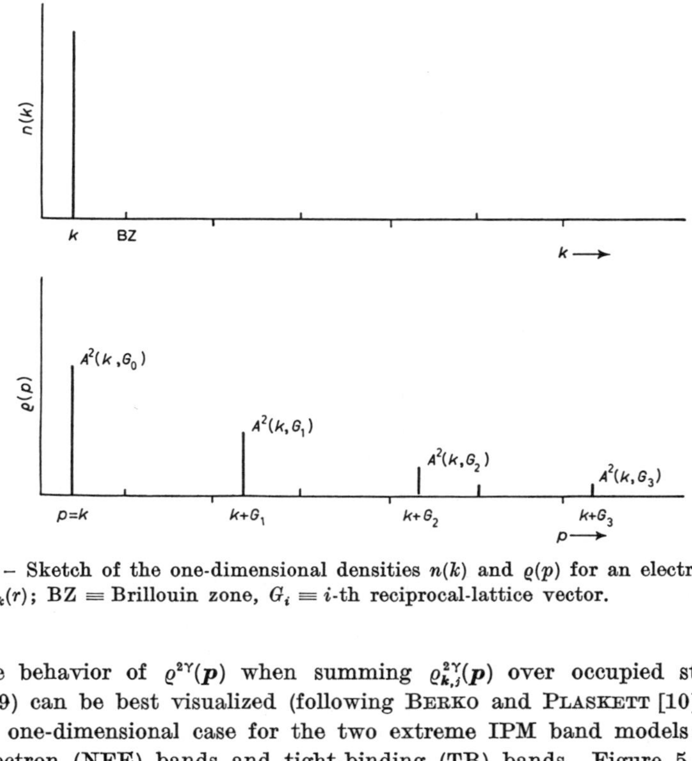

Fig. 4. – Sketch of the one-dimensional densities $n(k)$ and $\varrho(p)$ for an electron in the state $\psi_k(r)$; BZ $\equiv$ Brillouin zone, $G_i \equiv i$-th reciprocal-lattice vector.

The behavior of $\varrho^{2\gamma}(\boldsymbol{p})$ when summing $\varrho_{\boldsymbol{k},j}^{2\gamma}(\boldsymbol{p})$ over occupied states by eq. (3.9) can be best visualized (following BERKO and PLASKETT [10]) in the simple one-dimensional case for the two extreme IPM band models: nearly-free-electron (NFE) bands and tight-binding (TB) bands. Figure 5 exhibits the $\boldsymbol{k}$ and the $\boldsymbol{p}$ one-dimensional distributions for the following three cases: *a*) Fermi surface in first zone, *b*) no Fermi surface, *i.e.* $\boldsymbol{k}$ distribution touching Brillouin zone (BZ), and *c*) Fermi surface in the second band (plotted in the

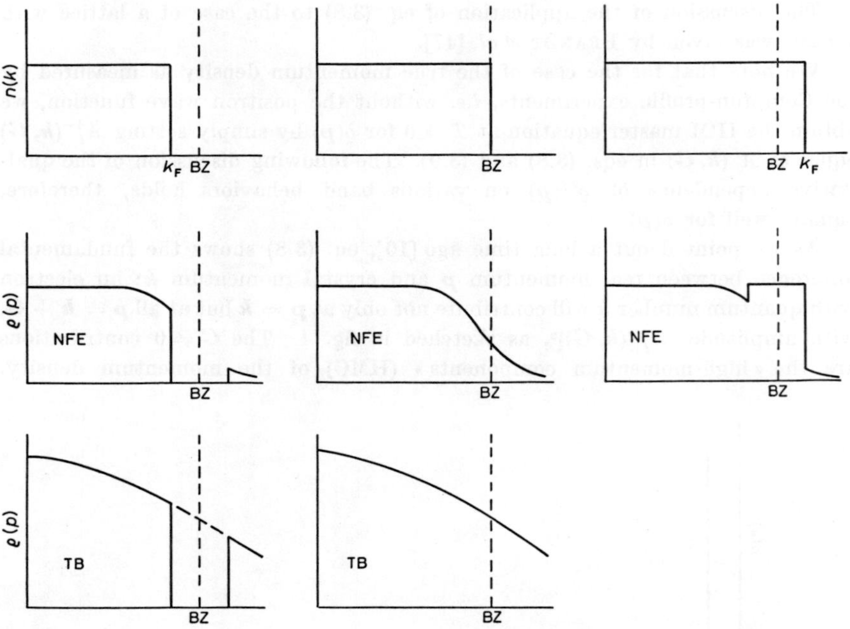

Fig. 5. – **k** *vs.* **p** space distribution in a one-dimensional nearly-free-electron (NFE) and tight-binding (TB) model for the three cases discussed in the text (shown only for $p \geqslant 0$).

k-space extended-zone scheme). The following features are noteworthy, and do not depend on the approximation used: 1) The momentum distribution will exhibit breaks not only at the Fermi surface in the first zone ($\boldsymbol{G} = 0$), but, due to the HMC's, these breaks reoccur in all zones, $\boldsymbol{G} \neq 0$ (the so-called « Umklapp annihilations », for lack of a better word, for the positron case). 2) In the case in which the **k** distribution just touches the BZ, the **p** distribution becomes continuous; in fact, in the first-nearest-neighbor TB approximation, this continuous distribution is equal to the atomic momentum distribution used to form the TB wave function (for the case of constant ψ^+, *i.e.* the « true » $\varrho(\boldsymbol{p})$), as proved in ref. [10]. 3) It is not necessary for the largest FS break to occur in the first zone, as shown in case *c*) of fig. 5.

Mijnarends [48, 49] has analyzed the behavior of the three-dimensional density $\varrho^{2\gamma}(\boldsymbol{p})$ in oriented single metal crystals, using group-theoretic arguments regarding the symmetry property of the $\varrho_{\boldsymbol{k},j}(\boldsymbol{p})$'s; he concludes that not all electronic bands have to contribute to the momentum density in each zone. Thus the reconstruction of the Fermi surface from high-precision momentum density measurements has to be performed with great care, searching for

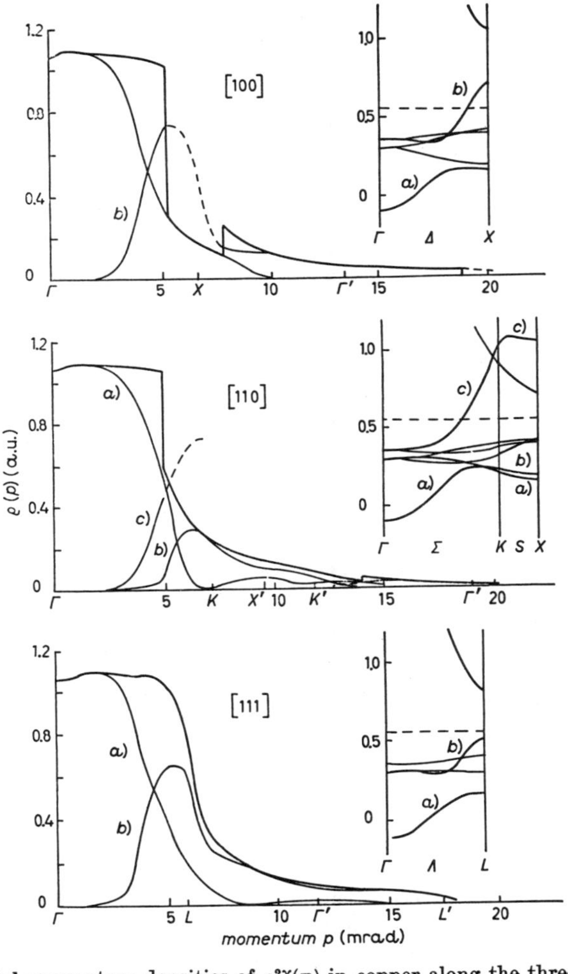

Fig. 6. – Radial momentum densities of $\varrho^{2\gamma}(\boldsymbol{p})$ in copper along the three main crystallographic directions, as computed by MIJNARENDS [48]; the inset shows the relevant band structure, with bands that cannot contribute for symmetry reasons marked by thin lines.

« FS breaks » in higher zones as well. As an example of the behavior of a realistic $\varrho^{2\gamma}_{\text{IPM}}(\boldsymbol{p})$, we reproduce in fig. 6 the copper $\varrho^{2\gamma}(\boldsymbol{p})$ momentum density computations of Mijnarends [48] along three major axes. The IPM copper density was obtained using a modified KKR computation, and includes the effect of a $\boldsymbol{k}^+ = 0$ positron wave function. Although the effect of the various bands is quite complex, the net momentum density is quite similar to the NFE model of fig. 5. We note that, along the ⟨111⟩ directions where there is no FS in copper (the « necks »), $\varrho^{2\gamma}(\boldsymbol{p})$ becomes continuous at point L on the BZ as predicted for the second case in fig. 5. For a more detailed discussion of symmetry relations, and for the applications of band theory to evaluate $\varrho^{2\gamma}(\boldsymbol{p})$ in various metals, see the lectures by MIJNARENDS in this volume, as well as ref. [31].

We note that the general behavior of $\varrho^{2\gamma}(\boldsymbol{p})$ as indicated in fig. 5 and the distinction of $\boldsymbol{p}$ *vs.* $\boldsymbol{k}$ space hold equally for the true momentum density $\varrho(\boldsymbol{p})$. $\varrho(\boldsymbol{p})$ thus has a specific topology in $\boldsymbol{p}$ space: for an unfilled band it is delimited in each zone by the Fermi surface (FS), like the $\boldsymbol{k}$-space distribution in the repeated-zone scheme, but it is nonperiodic, being modulated at each $\boldsymbol{p} = \boldsymbol{k} + \boldsymbol{G}$ by the weights $|A_j(\boldsymbol{k}, \boldsymbol{G})|^2$. The topology of $\varrho^{2\gamma}(\boldsymbol{p})$ is similar.

3·4. *Reconstruction of the three-dimensional momentum densities from the measured projections.* – We have seen in the previous subsections that the ACAR and CP experiments provide projections of the momentum densities $\varrho^{2\gamma}(\boldsymbol{p})$ or $\varrho(\boldsymbol{p})$. How can one obtain the three-dimensional momentum densities from their projections? The first mathematical technique used to reconstruct $\varrho^{2\gamma}(\boldsymbol{p})$ from its « long-slit » projections was introduced by MIJNARENDS [50, 51], and is still much used in 1D ACAR as well as CP work. When studying the literature, it becomes clear that the field of image reconstruction from projections is extremely active, and has been independently developed in many parts of science. With the advent of X-ray and γ-ray tomography, and more recently the PET technique (positron emission tomography) [52], image reconstruction has become a veritable industry [53-55]. In pure mathematics RADON was the first to discuss the projection transform in some detail already in 1917 [56]. The first use in physics was apparently in radioastronomy [57] in the fifties.

Essentially, most reconstruction techniques are based on a fundamental Fourier-transform theorem: the FT of a projection of a function is equal to a section through the origin of the Fourier transform of the function. The theorem was apparently used for reconstruction in different fields at different times by many scientists. In the Compton-profile field it forms the basis of a reconstruction technique introduced by MUELLER in 1977 [58]. The Fourier projection theorem is almost self-evident; we write it in terms of $\varrho(\boldsymbol{p})$ and $B(\boldsymbol{r})$ (defined in subsect. **3**·1 and **3**·2) with a projection $J(p_z)$ onto the z-axis:

$$B(x, y, z)\Big|_{\substack{x=0\\y=0}} = (2\pi)^{-\frac{3}{2}}\iiint \varrho(\boldsymbol{p})\exp[-i\boldsymbol{p}\cdot\boldsymbol{r}]\,\mathrm{d}\boldsymbol{p}\Big|_{\substack{x=0\\y=0}} = \tag{3.11}$$

$$= (2\pi)^{-\frac{3}{2}}\int\left[\iint \varrho(\boldsymbol{p})\,\mathrm{d}p_x\,\mathrm{d}p_y\right]\exp[-ip_z z]\,\mathrm{d}p_z = (2\pi)^{-\frac{3}{2}}\int J(p_z)\exp[-ip_z z]\,\mathrm{d}p_z\,.$$

Thus we have the theorem

$$\mathrm{FT}_{-1}\{J(p_z)\} = 2\pi B(0, 0, z)\,. \tag{3.12}$$

We can prove eq. (3.12) in a more elegant form, in terms of a projection onto a general direction $\hat{e}_z$, to obtain

$$\text{FT}_{-1}\{J(p_z;\hat{e}_z)\} = \text{FT}_{-1}\left(\int \varrho(\boldsymbol{p})\,\delta(\boldsymbol{p}\cdot\hat{e}_z - p_z)\,\mathrm{d}\boldsymbol{p}\right) = 2\pi\int B(\boldsymbol{r})\,\delta(\boldsymbol{r}-\hat{e}_z z)\,\mathrm{d}\boldsymbol{r}\,. \tag{3.13}$$

Clearly, if one has enough projections $J(p_z;\hat{e}_z)$ measured along various directions $\hat{e}_z$, one can Fourier transform each and then interpolate $B(\boldsymbol{r})$ onto a cubic grid; the inverse Fourier transform then yields the full reconstructed 3D $\varrho(\boldsymbol{p})$.

The above holds for ACAR data also; simply replace $\varrho(\boldsymbol{p})$ by $\varrho^{2\gamma}(\boldsymbol{p})$ and $J(p_z)$ by $N(p_z)$. We obtain $B^{2\gamma}(\boldsymbol{r})$ as the FT of $\varrho^{2\gamma}$ (see eq. (3.4)). In the case of 2D ACAR measurements $N(p_y, p_z)$ one obtains the corresponding important 2D equivalent theorem:

$$\text{FT}_{-2}(N(p_y, p_z)) = (2\pi)^{\frac{1}{2}} B^{2\gamma}(0, y, z)\,. \tag{3.14}$$

This means that the 2D Fourier transforms of the 2D ACAR results are proportional to the Fourier transform of $\varrho^{2\gamma}$ evaluated in a plane going through the origin and with a normal parallel to p_x, the integration direction for obtaning $N(p_y, p_z)$.

The main problem in using these theorems (eqs. (3.13) and (3.14)) directly lies in the proper interpolation in reciprocal space. The various image reconstruction techniques developed specifically for ACAR or CP data expand $\varrho(\boldsymbol{p})$ *or* $B(\boldsymbol{r})$ into various orthonormal functions and solve the inversion formula for the expansion coefficients. Thus MIJNARENDS expands $\varrho(\boldsymbol{p})$ into cubic harmonics [50, 51] and obtains a solution for the expansion coefficients using the Fourier-Hankel transform. MAJUMDAR developed reconstruction formulae using the Mellin transform, also applicable to reconstruction from 2D ACAR data [59]. Mueller's Fourier-transform method [58] involves first subtracting from $\varrho(\boldsymbol{p})$ an isotropic term and then approximating the anisotropic part by a Fourier series. Pecora and Ehrlich's reconstruction of $\varrho(\boldsymbol{p})$ in separate planes from $N(p_y, p_z)|_{p_y=0}$ ACAR data uses an expansion in truncated Fourier series in polar co-ordinates [60]. More recently HANSEN [61] developed an inversion technique where the Compton profiles J are first Fourier transformed into $\boldsymbol{r}$ space and then the $B(\boldsymbol{r})$ function is expanded into lattice harmonics, prior to reconstruction of $\varrho(\boldsymbol{p})$. In all these techniques the most important problem is to follow carefully the noise propagation. Since the symmetry of the reciprocal lattice is taken into account, great care has to be taken with noise: pure noise can also lead to nice-looking patterns due to the use of symmetry.

Another important question has to do with the number of projections required for reconstruction. In principle the use of a finite number of projections introduces an « effective resolution » in the reconstructed $\varrho(\boldsymbol{p})$, which worsens with increasing $|\boldsymbol{p}|$ circumferentially, and is given roughly by $|\boldsymbol{p}|\alpha$, where α is the angle between projections. (See a brief discussion of this point by MIJNARENDS [31].) Very often poorer experimental resolution and more projections would have led to better results. This fact has been often overlooked so far; many workers use few projections with good momentum resolution and statistics, rather than optimizing the momentum resolution and total counts per projection with the number of projections.

At Brandeis we have used a reconstruction method borrowed from medical physics (tomography), the so-called filtered back-projection (FBP) technique [62]. Mathematically the technique is equivalent to the Fourier technique. We reconstruct $\varrho^{2\gamma}(\boldsymbol{p})$ from 2D ACAR $N(p_y, p_z)$ data one plane at a time (*i.e.* $p_y = \text{const}$), but the method can be used for simultaneous reconstruction of the full 3D $\varrho^{2\gamma}(\boldsymbol{p})$ also. The simplest description of the back-projection method involves the reconstruction of a two-dimensional density function $\varrho(p_x, p_z)$ from its one-dimensional projections $N(p_{z'}; \varphi_j)$, onto the $\hat{e}_{z'}(j)$ directions rotated by an angle φ_j with respect to the axis $\hat{e}_z$ attached to the crystal $\big(i.e.\ \hat{e}_{z'}(j)\cdot\hat{e}_z = \cos\varphi_j\big)$. We define the discrete back-projection as

$$(3.15)\qquad \varrho_{\text{BP}}(p_x, p_z) = \sum_{j=1}^{m} \Delta\varphi \int N(p_{z'}; \varphi_j)\,\delta(p_{z'} - p_z\cos\varphi_j - p_x\sin\varphi_j)\,\mathrm{d}p_{z'}\,,$$

where $\Delta\varphi$ is the angle between the projection axes ($\Delta\varphi = \pi/m$), using m equally spaced projections. It can be shown that ϱ_{BP} is *not* the correct density $\varrho(\boldsymbol{p})$ even for the integral limit $\varrho_{\text{BP}}(\boldsymbol{p}) = \int_0^\pi N(p_{z'}; \varphi)\,\mathrm{d}\varphi$; by taking the FT of ϱ_{BP} one finds that it corresponds to the FT of the true ϱ divided by the magnitude of the «frequency», in our case by $|\boldsymbol{r}|$. Thus one has to first multiply the FT of ϱ_{BP} by $|\boldsymbol{r}|$ and then obtain, by inverse FT, the true ϱ. In practice one uses a filter $f(r)$ that behaves as $|\boldsymbol{r}|$ below a certain limit and goes to zero above it, the limit being determined by the resolution function and the «Nyquist frequency». We do not have to go, however, to $\boldsymbol{r}$ space: a convolution of the projections $N(p_{z'}; \varphi_j)$ with the FT of the «frequency» filter, $F(p_z) = \text{FT}\{f(r)\}$, is an equivalent operation. After convolution, $N_{\text{conv}}(p_{z'}; \varphi_j) = N(p_{z'}; \varphi_j) * F(p_{z'})$, one back-projects to obtain the discrete «filtered back-projection» ϱ_{FBP}:

$$(3.16)\qquad \varrho_{\text{FBP}}(p_x, p_z) = \sum_{j=1}^{m} \Delta\varphi \int N_{\text{conv}}(p_{z'}; \varphi_j)\,\delta(p_{z'} - p_z\cos\varphi_j - p_x\sin\varphi_j)\,\mathrm{d}p_{z'}\,.$$

Following CHESLER and RIEDERER [62], we use at Brandeis the FT of a ramp filter multiplied by a Hanning window [63]. For $F(\boldsymbol{r})$ we shall show the results of such a reconstruction in connection with 2D ACAR experiments on Cu and V_3Si.

3'5. *The behavior of $B(\boldsymbol{r})$ for periodic systems.* – In the previous subsection we discussed the use of the $B(\boldsymbol{r})$ function in the reconstruction of $\varrho(\boldsymbol{p})$. In addition, $B(\boldsymbol{r})$ has been found to have several important mathematical properties in the case of periodic potentials (*i.e.* Bloch wave functions), properties that have been found useful in analyzing Compton-profile measurements (and, within a certain approximation, also ACAR results).

The first such property, as discussed by PATTISON and WILLIAMS [41], is specific to a simple noninteracting electron gas, *i.e.* when $\varrho(\boldsymbol{p}) = n(|\boldsymbol{k}|)$, with $n = 1$ for $|\boldsymbol{k}| \leqslant k_f$ and $n = 0$ for $|\boldsymbol{k}| > k_f$ (see eq. (3.9)). We know that in that case the Compton profile $J(p_z)$ is an inverted parabola—and so is the 1D ACAR curve, for a noninteracting positron in a free-electron gas. In this simple case one obtains for $B(r) = \mathrm{FT}_{-1}\{J(p_z)\}$

$$B(r) = \frac{3}{p_f^2 r^2}\left[\frac{\sin p_f r}{p_f r} - \cos p_f r\right], \tag{3.17}$$

where $J(p_z)$ was normalized to unit area, and p_f is the free-electron Fermi momentum ($\equiv k_f$).

Thus it follows that $B(r)$ has zeros whenever $p_f r = 4.493$, 7.725, 10.904, etc. The important point to notice is that the resolution of the apparatus does not shift these zeros; the resolution function that convolutes $J(p_z)$ becomes an envelope function in Fourier space (*i.e.* multiplies $B(r)$), thus damping the oscillations in eq. (3.17), without shifting the zeros. Therefore, it becomes possible to determine the Fermi momentum even for the very poor resolution of the Compton-profile data. The simple formula (3.17) does not hold, however, for anisotropic Fermi surfaces, nor for an interacting Fermi gas, and the theorem thus has very limited applicability, restricted to simple metals.

A more general set of properties which is applicable, within the IPM, to all periodic systems has been derived by SCHÜLKE [43, 44]. One can show that for the case of Bloch waves (see notation of subsect. **3**˙1 and **3**˙2) with quantum number $\boldsymbol{k}$ and band index j

$$B(\boldsymbol{r}) \propto \sum_{j,\boldsymbol{k}} n_j(\boldsymbol{k}) \exp[-i\boldsymbol{k}\cdot\boldsymbol{r}] \int \mathrm{d}\boldsymbol{s}\, u^*_{\boldsymbol{k},j}(\boldsymbol{r}+\boldsymbol{s})\, u_{\boldsymbol{k},j}(\boldsymbol{s})\,. \tag{3.18}$$

The extension of (3.18) to $B^{2\gamma}(\boldsymbol{r})$ for positron experiments is self-evident, with $u_{\boldsymbol{k},j}(\boldsymbol{r})$ being replaced by $u^+(\boldsymbol{r})\, u_{\boldsymbol{k},j}(\boldsymbol{r})$. From eq. (3.18) one can easily prove the following theorems:

1) For filled bands, *i.e.* for an insulator, $B(\boldsymbol{r})$ has zeros for all $\boldsymbol{r} = \boldsymbol{R}$, where the $\boldsymbol{R}$'s are the lattice vectors of the periodic system, except for $\boldsymbol{R} = 0$.

2) If we first sum eq. (3.18) over $\boldsymbol{k}$ space, we obtain, for filled bands,

$$B(\boldsymbol{r}) \propto \sum_j \int W_j^*(\boldsymbol{r}+\boldsymbol{s})\, W_j(\boldsymbol{s})\, \mathrm{d}\boldsymbol{s}\,, \tag{3.19}$$

where $W_j(\boldsymbol{r})$ is the Wannier function for band j.

3) For a metal we obtain

$$B(\boldsymbol{R}) = (2\pi)^{-\frac{3}{2}} \sum_{j,\boldsymbol{k}} n_j(\boldsymbol{k}) \exp[-i\boldsymbol{k}\cdot\boldsymbol{R}]\,. \tag{3.20}$$

Thus one can obtain the Fermi surface of the system by simply inverting eq. (3.20); $n_j(\boldsymbol{k})$ is the occupation number for the j-th band, and, as usual, $\boldsymbol{k}$ is defined in the first BZ. One can replace in eq. (3.2) the summation over $\boldsymbol{k}$ by the usual integration over a Brillouin zone (BZ). The above theorems are true within the IPM for Compton profiles. For ACAR data they are only approximately correct, even within the IPM, because of the presence of $\psi^+(\boldsymbol{r})$. More theoretical work will be required to understand the nature of these approximations.

3·6. *The* LCW *construction theorem.* – In 1972, LOCK, CRISP and WEST [64] derived a theorem that allows—within a certain approximation—a simple transformation of the momentum density obtained either by Compton-profile or ACAR measurements into the projections of the «occupation number» $n_j(\boldsymbol{k})$ in $\boldsymbol{k}$ space. If $N_{ijk}(p_z)$ represents the 1D projection of the momentum density along the direction normal to the (ijk)-planes of the reciprocal lattice, the LCW theorem states that the function

$$F_{ijk}(p_z) = L_{ijk} \sum_{n=-\infty}^{\infty} N_{ijk}(p_z + nL_{ijk}) \qquad (n \text{ integer}) \tag{3.21}$$

reproduces the functional variation of the cross-sectional areas of the Fermi surface in a repeated-zone scheme; the contribution to $F_{ijk}(p_z)$ from filled zones is a constant. In eq. (3.21) L_{ijk} is the spacing of the (ijk)-planes. The easiest way to prove the LCW theorem is to go to the full 3D $\varrho^{2\gamma}(\boldsymbol{p})$ (or $\varrho(\boldsymbol{p})$) momentum density. We form the «folded momentum density» $\tilde{\varrho}^{2\gamma}(\boldsymbol{p}) = \sum_{\boldsymbol{G}} \varrho^{2\gamma}(\boldsymbol{p}-\boldsymbol{G})$. $\tilde{\varrho}^{2\gamma}$ is also the convolution of $\varrho^{2\gamma}(\boldsymbol{p})$ with the «sampling function» $G(\boldsymbol{p}) = \sum_i \delta(\boldsymbol{p}-\boldsymbol{G}_i)$; $\tilde{\varrho}^{2\gamma}(\boldsymbol{p}) = \varrho^{2\gamma}(\boldsymbol{p}) * G(\boldsymbol{p})$. Using the IPM definition of $\varrho^{2\gamma}(\boldsymbol{p})$ for a periodic system (eqs. (3.8) and (3.9)), we obtain

$$\tilde{\varrho}^{2\gamma}(\boldsymbol{p}) = \text{const}\cdot \sum_{\boldsymbol{k},j,\boldsymbol{G}} n_j(\boldsymbol{k})\,\delta(\boldsymbol{p}-\boldsymbol{k}-\boldsymbol{G})\Big\{\sum_{\boldsymbol{G}'} |A_j^{+-}(\boldsymbol{k},\boldsymbol{G}')|^2\Big\}. \tag{3.22}$$

(Remember: $\boldsymbol{k}$ is restricted to the first zone.)

Thus in the approximation that the $\boldsymbol{G}'$ sum in the brackets is $\boldsymbol{k}$ independent, $\tilde{\varrho}^{2\gamma}(\boldsymbol{p})$ is proportional to the occupation number $n_j(\boldsymbol{k})$ summed over all bands, *in the extend-zone scheme*; this is the LCW theorem in 3D momentum space, *i.e.*

$$\tilde{\varrho}^{2\gamma}(\boldsymbol{p}) = \text{const}\cdot \sum_{\boldsymbol{k},j,\boldsymbol{G}} n_j(\boldsymbol{k})\,\delta(\boldsymbol{p}-\boldsymbol{k}-\boldsymbol{G})\,. \tag{3.23}$$

For Compton-profile measurements the above equation for $\tilde{\varrho}(\boldsymbol{p})$ holds exactly within the IPM, since A^{+-} in eq. (3.22) gets replaced by A^{-} and $\sum_{\boldsymbol{G}} |A_j^{-}(\boldsymbol{k}, \boldsymbol{G})|^2 = 1$ by wave function normalization.

One obtains the 1D or the 2D form of the LCW theorem by properly projecting eq. (3.22) onto an axis, thus obtaining eq. (3.21), or onto a plane for the equivalent 2D ACAR equation. Care has to be taken to limit the range of integration of the projections, in order not to obtain infinities.

Thus the LCW theorem allows one to obtain the $\sum_j n(\boldsymbol{k}, j)$'s or the projections thereof, and thus obtain the Fermi surface (or projections of the FS) of the system, without having to analyze the momentum density first. As we have seen, the theorem holds for the Compton profiles in the IPM; for ACAR analyses, in addition to IPM, the assumption that $\sum_{\boldsymbol{G}} |A_j^{+-}(\boldsymbol{k}, \boldsymbol{G})|^2$ is $\boldsymbol{k}$ independent is needed (a simple plane-wave positron wave function would do). In a subsequent paper LOCK and WEST [65] gave corrections to this approximation for ACAR data. The applicability of the theorem to 1D ACAR as well as CP experiments was discussed by BEARDSLEY *et al.* [66], with Cu and Ge data as examples.

The LCW theorem was introduced before the importance of the $B(\boldsymbol{r})$ function was recognized. The theorem turns out to be directly connected with

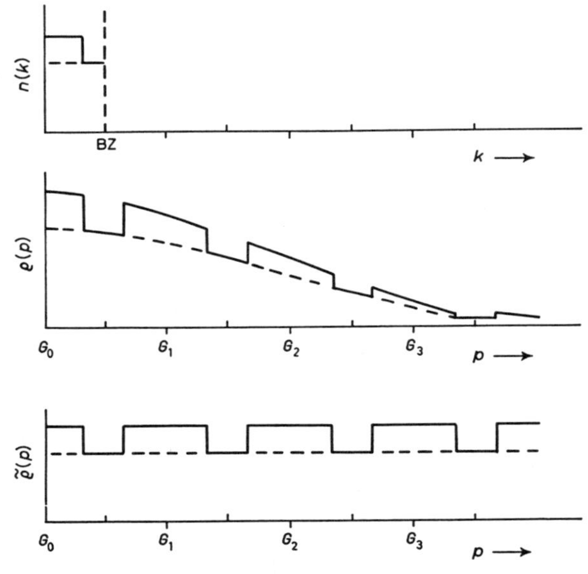

Fig. 7. – Sketch of the LCW construction for a one-dimensional metal with filled (core) electrons and an unfilled conduction band. BZ ≡ Brillouin zone, G_i ≡ i-th reciprocal-lattice vector (densities shown only for $p \geqslant 0$); all vertical scales are in arbitrary units.

properties of $B(\boldsymbol{r})$; it is, in fact, the $\boldsymbol{p}$-space equivalent to the $\boldsymbol{r}$-space formula of eq. (3.20). One inverts eq. (3.20) to obtain

$$\sum_j n_j(\boldsymbol{k}) = \text{const}\cdot\sum_{\boldsymbol{R}} B(\boldsymbol{R}) \exp[i\boldsymbol{k}\cdot\boldsymbol{R}] = \\ = \text{const}\cdot\int B(\boldsymbol{r})\sum_{\boldsymbol{R}}\delta(\boldsymbol{r}-\boldsymbol{R})\exp[i\boldsymbol{k}\cdot\boldsymbol{r}]\,d\boldsymbol{r}\,. \tag{3.24}$$

We can extend the range of $n_j(\boldsymbol{k})$ to all of $\boldsymbol{p}$ space by multiplying eq. (3.24) by $\delta(\boldsymbol{p}-\boldsymbol{k}-\boldsymbol{G})$ and summing over $\boldsymbol{k}$ and $\boldsymbol{G}$. One then recognizes in the last expression the FT of the product between $B(\boldsymbol{r})$ and $R(\boldsymbol{r}) = \sum_{\boldsymbol{R}} \delta(\boldsymbol{r}-\boldsymbol{R})$; using the convolution theorem and the fact that the sampling functions $R(\boldsymbol{r})$ and $G(\boldsymbol{p})$ are FT's of each other, one obtains the convolution const$\cdot\varrho(\boldsymbol{p}) * G(\boldsymbol{p})$, *i.e.* $\tilde{\varrho}(\boldsymbol{p})$, thus the 3D LCW theorem of eq. (3.23).

In fig. 7 we indicate the effect of the LCW construction by having sketched the behavior of $n(\boldsymbol{k})$, $\varrho(\boldsymbol{p})$ and $\tilde{\varrho}(\boldsymbol{p})$ for a simple one-dimensional system consisting of a filled band and an unfilled one (we sketch a TB conduction band to indicate $\varrho(\boldsymbol{p})$ in several zones).

4. – Experimental techniques.

4·1. *One-dimensional* (« *long-slit* ») ACAR *measurements*. – In subsect. 2·2 we have discussed the geometry of the « long-slit » apparatus measuring the 1D projections $N(p_z)$. A typical apparatus is shown schematically in fig. 8.

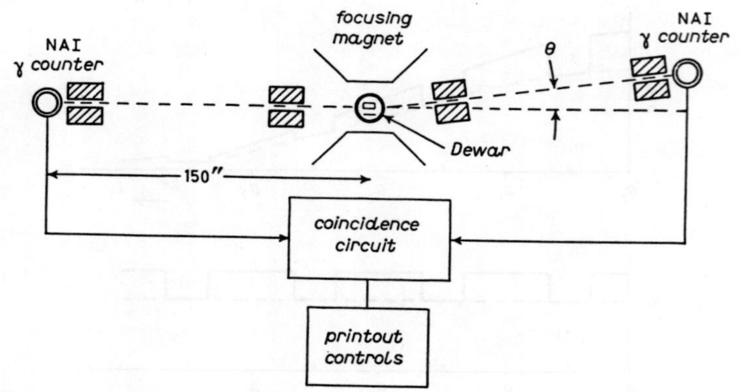

Fig. 8. – Sketch of apparatus used for one-dimensional (« long-slit ») ACAR measurements; the NaI γ counters consist of long cylindrical scintillators positioned perpendicular to the plane of the sketch. The shaded areas are the « long » lead slits, defining the angle θ and the effective resolution $\Delta\theta$.

The γ-ray detectors are usually long, cylindrically shaped NaI(Tl) scintillators, ~ 5 cm in diameter, behind lead slits that define the resolution width Δp_z of the apparatus. In order to measure true 1D projections, the length of the crystals has to be sufficient to submit an angle φ much larger than the typical width of $\sim (15 \div 30)$ mrad of an angular distribution. Usually φ is over 100 mrad. Depending on the experiment, Δp_z usually varies from $(0.2 \div \div 0.5)$ mrad for Fermi-surface work to $(2 \div 3)$ mrad for defect studies. One detector is mounted on a moving arm and is stepped automatically for measuring the coincidence rate as a function of the angle θ. At Brandeis University we built a triple pair detector system; the resolving time of the electronic coincidence circuit is $(20 \div 40) \cdot 10^{-9}$ s. The samples (usually oriented single crystals) are mounted in a variable-temperature Dewar, between the pole pieces of a (~ 20 kG) magnet used to focus the positrons from a ^{58}Co or ^{22}Na source onto the face of the sample. In our machine, micrometer screws enable us to translate the magnet and Dewar assembly perpendicular to the (2γ) axis of the apparatus, and to rotate it independently of the arms holding the slits which are pivoted under the magnet; this feature is very useful for the alignment of the sample.

Care has to be taken to correct the data for the finite slit length of the detectors. For these correction techniques see ref. [67]. In our machine one detector subtends a smaller angle φ than the other, by about the width of an ACAR distribution; this has the effect of making the effective resolution along the length of the detector flat topped, thus decreasing somewhat the end corrections. The usual long-lived positron sources used for ACAR work are ^{22}Na (~ 2.6 y) and ^{58}Co (~ 71 d). Up to 100 mCi of ^{22}Na or 500 mCi of ^{58}Co have been used. A complete list and description of β^+ sources can be found in a paper by LORCH [68]. More details about these 1D ACAR experimental techniques, as well as experimental questions in general, can be found in the many general review articles mentioned earlier (see also the review by SINGRU [69] and the lecture notes by MACKENZIE at this school).

4.2. *Doppler-broadening measurements.* – In subsect. **2.2** we saw that that the energy of the annihilation gamma-rays is Doppler shifted by the p_x component of the momentum parallel to the gamma-ray (by $\pm cp_x/2$) and that typical atomic (or solid-state) momenta lead to Doppler shifts of a few keV. The energy measurement of these gammas yields, therefore, the projection of the momentum distribution of the positron-electron system onto the direction of the gamma measured. In our notation, the energy measurement of one of the annihilation γ's is, therefore, equivalent to a 1D ACAR measurement, yielding $N(p_x) = \int\int \varrho^{2\gamma}(\boldsymbol{p})\, \mathrm{d}p_y\, \mathrm{d}p_z$. In principle, the energy measurement can be performed by a bent-crystal spectrometer to a high precision; the efficiency of such spectrometers is, however, very low and one thus requires a very large β^+ source. In practice, Doppler spectrometry is performed more commonly with the use of Ge(Li) or intrinsic Ge γ-ray spectrometers. The advantage of such spectrometers over performing 1D ACAR measurements is in their efficiency; one can accumulate the equivalent of a full ACAR spectrum simultaneously by using a multichannel energy analyzer. On the other hand, the essential drawback in the Doppler-broadening technique is the limitation in the obtainable energy resolution. At present the best Ge spectrometers have an effective resolution of ~ 1 keV, corresponding in an ACAR setup to an effective momentum resolution of $\approx 4mc \cdot 10^{-3}$, *i.e.* an angular resolution of ≈ 4 mrad. Thus the effective momentum resolution is about a factor of ten worse than the usual ACAR resolution. For this reason Doppler broadening is not used usually for Fermi-surface work; it is used, however, quite regularly for defect studies, where the high efficiency is of paramount importance. At Brandeis we find, however, that with the multiple long-slit setup, using wide angular slits (~ 2 mrad), we can obtain an ACAR counting rate approaching that of a Ge Doppler spectrometer (albeit with a much stronger source).

We have seen that, in principle, one could combine angular-correlation geometry with Doppler spectrometry, in order to obtain directly the 3D $\varrho^{2\gamma}(\boldsymbol{p})$. Such an experiment was actually performed by Singru [70] as a demonstration of the technique on a Cu sample; because of the extremely low counting rate, however, such measurements are not very practical.

Another interesting application of Doppler-broadening spectroscopy has been introduced by Lynn *et al.* [71, 72]. The principle of the technique, called by them the « two-dimensional Doppler-broadening technique », is to use two Ge(Li) detectors in colinear geometry and observe in time coincidence the energy of both annihilation gammas, E_1 and E_2. By forming electronically $E_1 + E_2$ and $E_1 - E_2$, they obtain the 1D momentum distribution via the Doppler shift *vs.* the binding energy of the electrons. Given the poor energy resolution of the Ge(Li) detectors, one can only distinguish the annihilation of tightly bound electrons (several keV) from those of conduction electrons, but one cannot separate the annihilation of various bands of the outer electrons. « Pile-up » and incomplete charge collection also cause severe problems. An important advantage of using such a coincidence technique over the usual 1D Doppler-broadening technique is the very low background it achieves; one can thus obtain the momentum distribution at very high momenta. Typical 1D ACAR results are usually presented to $\sim \pm 20$ mrad, while the momentum distributions obtainable by the 2D Doppler technique can be extended to an equivalent of ± 60 mrad. At these momenta, the distribution in Al, for example, is down by five orders of magnitude from the zero-momentum peak!

Finally we should mention another combination of multiparameter measurement: that of momentum distribution *vs.* time. By combining a Ge(Li) Doppler spectrometer with a lifetime setup using a ^{22}Na source, one can obtain a 1D momentum distribution *vs.* positron lifetime [73, 74]. The effective time resolution of such a system ($\sim (300 \div 500)$ ps) has not been made as short as the best dedicated lifetime setup can achieve (< 200 ps), and the application of this technique has been limited mainly to nonmetallic systems. If the time resolution could be improved, many interesting experiments could be performed, particularly in systems exhibiting multiple lifetimes due to positron trapping.

4'3. *Compton-profile measurements.* – We shall now present a brief description of the experimental techniques used in Compton-profile (CP) measurements, since they are quite similar to the Doppler-broadening measurements in positron physics. We have seen in subsect. **2'3** that the energy distribution of the inelastically (Compton) scattered X-rays or γ-rays is directly related to the 1D projection of the momentum density $\varrho(\boldsymbol{p})$ of the electrons involved in scattering. Many of the early Compton-profile (CP) measurements were performed with X-ray beams. A typical arrangement for such measurements involves a X-ray tube with a high-Z target such as Mo, a Soller slit collimator, and the energy measurement of the scattered X-rays at a fixed large-angle scattering ($\sim 170°$), using a crystal spectrometer. The system produces the Compton broadened spectra of the $K\alpha_1$ and $K\alpha_2$ lines superimposed; after separation of the two lines, the energy spectrum is converted to a momentum distribution $J(p_z)$ (see eq. (2.13)). The advantage of the method is the reasonably good momentum resolution of the crystal spectrometer ($\sim (0.15 \div 0.3)$ a.u.). The disadvantages are the strong absorption of the X-rays in the sample, limiting the measurements to low-Z materials, and the slow data accumulation due to the low efficiency of the crystal spectrometer. The introduction of Ge(Li) detectors and nuclear γ sources instead of X-rays has permitted the measurement of the CP from high-Z samples with a substantial increase in data

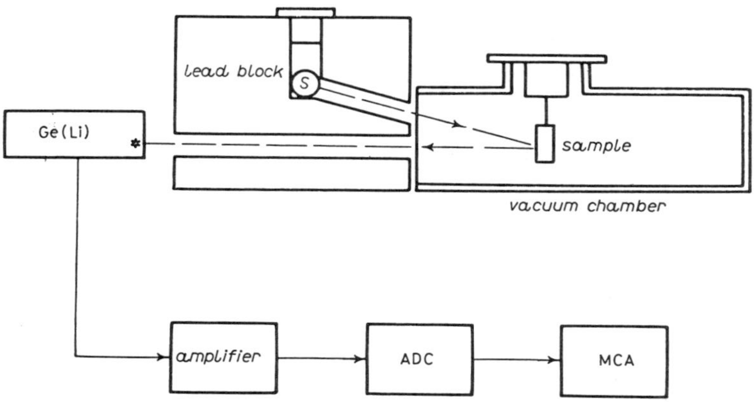

Fig. 9. – Schematic sketch of a typical Compton-profile setup; S is the radioactive γ source, such as ^{198}Au.

accumulation rate—a development paralleling the use of Doppler broadening in positron experiments. The effective momentum resolution of such Ge(Li) setups is strongly limited, however, by the energy resolution of the Ge(Li) spectrometer. A sketch of a simple Compton-profile setup using γ sources is shown in fig. 9. For a thorough review of Compton-scattering experiments up to 1976, see the review book edited by WILLIAMS [75]; the CP experimental techniques have been reviewed by WEISS *et al.* [76]. More recently COOPER reviewed the γ-ray sources for Compton scattering [77]. A substantial improvement in counting rate has been achieved recently by SCHNEIDER *et al.* [78] using the 412 keV radiation from extremely-high-intensity ^{198}Au sources (over 100 Ci!). The effective momentum resolution, however, remains poor ($\sim$ 0.4 a.u.; 0.137 a.u. corresponds to 1 mrad angular resolution in ACAR work).

The possibility of using monochromatized synchrotron radiation has also been recently explored [79]; the available intensity is superior to the usual X-ray setup, but at present one is still limited to low energies and thus low-Z samples. The new generation of synchrotron sources incorporating « wigglers » to enhance the high-energy part of the spectrum might permit the application of CP to high-Z samples with the good resolution provided by the crystal spectrometers.

We have mentioned previously that the CP measurements are easier to interpret theoretically than the ACAR results, because of the presence of the positron in the latter; on the other hand, the effective momentum resolution of a 1D ACAR setup is substantially better than that of the CP experiment (by a factor of three to ten). Thus at present one can obtain more information about FS shapes by 1D ACAR experiments than by CP measurements. Another advantage of positron annihilation for FS experiments is the possibility of 2D momentum distribution measurements.

4'4. 2D ACAR *measurements*. – In the long-slit geometry the integration of $\varrho^{2\gamma}(\boldsymbol{p})$ over a plane results in a substantial loss of important detail. It had been recognized some time ago that the ACAR technique lends itself to the measurement of 2D distributions corresponding to an integration over a single momentum component, by going from long-slit to a nearly « point-slit » geometry. In order to compensate for the substantial reduction in counting rate, multiple detector arrays and, more recently, position-sensitive γ-ray detectors have been developed and successfully used to obtain 2D ACAR data.

The early 2D ACAR experiments and their specific application to Fermi-surface (FS) studies in metals and alloys have been discussed by us in an address at the *III International Conference on Positron Annihilation* (ICPA) in 1973. At that time we also reported the first results obtained with a 22-detector angular-correlation apparatus [80] and their use to obtain Fermi surfaces of disordered alloys. Those results were restricted to a few slices in the full 2D correlation space, but clearly demonstrated the power of 2D over

the 1D (« long-slit ») ACAR experiments. Based on these results we completed a larger multicounter system (64 NaI detectors), and described this apparatus at the 1976 IV ICPA in Denmark [81]. Since that conference the new system has permitted us to obtain the first full 2D ACAR surfaces in several materials [4, 82, 83]. Several other 2D ACAR systems have been proposed, designed and built since then.

The multidetector systems mentioned above were, of course, not the first to provide some 2D information in positron experiments. The first point-detector experiment by COLOMBINO *et al.* [84] showed interesting 2D effects in graphite already in 1963; the intrinsic low counting rate of 2D experiments was compensated by the use of *in situ* sources by FUJIWARA and SUEOKA [85] in the first « crossed-slit » experiments that demonstrated the measurability of the Cu ⟨111⟩ Fermi-surface necks. Similar experiments were reported by WILLIAMS and coworkers [86]. Another important step towards the development of the full 2D machines was also the success of the « rotating specimen » method using colinear counter geometry as investigated independently by SUEOKA [87] and WILLIAMS *et al.* [88]. By 1981 several groups have built or are in the process of building 2D ACAR machines. These machines differ from each other by the nature of the position-sensitive detectors used. At Brandeis University we have used the multidetector system with 64 individual NaI(Tl) scintillators. Other laboratories have built machines based on proportional wire chambers [89], Anger cameras [90], or position-sensitive Ge detectors [91]. The ultimate resolution of these detectors will be limited by source strength and by the thermal motion of the positron—$(0.2 \times 0.2)\ \text{mrad}^2$ resolution seems feasible in the near future.

4·5. *General design considerations.* – Figure 10 contains the geometry of the typical 2D ACAR setup. Areas L and R (rectangles in this case) represent

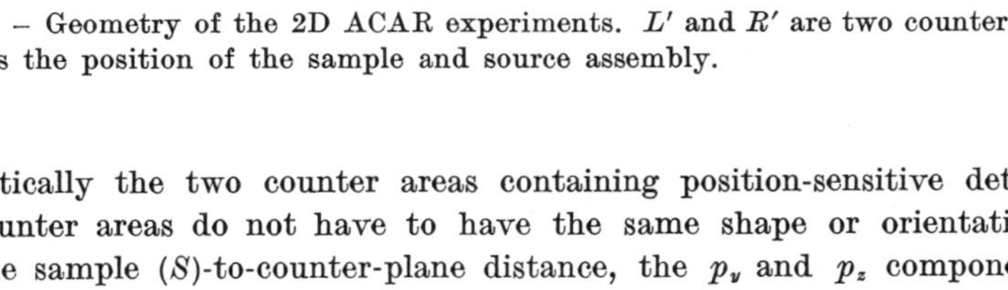

Fig. 10. – Geometry of the 2D ACAR experiments. L' and R' are two counter planes, and S is the position of the sample and source assembly.

schematically the two counter areas containing position-sensitive detectors. The counter areas do not have to have the same shape or orientation. If L is the sample (S)-to-counter-plane distance, the p_y and p_z components of the momentum $\boldsymbol{p}$ carried by the two annihilation γ's detected in coincidence by the detector elements L_{ij} and R_{kl} centered at co-ordinates $(y_i z_j)$ and $(y_k z_l)$ are given by

$$p_y = (y_i + y_k) mc/L\,, \qquad p_z = (z_j + z_l) mc/L\,. \tag{4.1}$$

Equation (4.1) is, of course, only true in the small-angle approximation $y/L \ll 1$ and $z/L \ll 1$, well fulfilled in ACAR experiments. The momentum resolution $f_{ij,kl}$ of the individual detector element pair is given by the two-dimensional convolution of the « effective areas » of the two detectors (this can include a variable efficiency across the face of the detector). Besides these « pair resolution functions » $f_{ij,kl}$, another parameter, which we call the momentum sampling function $S(p_y p_z)$, is of major importance. $S(p_y p_z)$ measures the number of effective « detector pairs » that contribute to the determination of the momentum density at (p_y, p_z). For the case of equal efficiency across the detector planes, $S(p_y p_z)$ is simply proportional to the overall 2D convolution of the two counter areas. Two examples might be instructive: For the case of a set of detectors positioned along a line Y_L and Z_R (two $\perp$ « lines »), $S(p_y p_z)$ corresponds to a uniform rectangular sampling surface covering momentum space $(Y_L * Z_R) mc/L$; for each measured momentum pair one counter pair contributes. For two square arrays, S has a tentlike shape with all diagonal counter pairs contributing at the peak of S at $p_y = p_z = 0$. Douglas and Stewart suggest using two rectangular counter planes (multiwire proportional counters) rotated by 90° and resulting in a flat-topped $S(p_y p_z)$ [92].

Ideally one could store all counter pair coincidences separately in a matrix $(ij; kl)$. If the efficiency of the detectors is known and is constant in time, one can save greatly in computer memory by storing all counter pairs with equal $(i + k, j + l)$ at one memory location. A separate measurement of counter pair resolutions $f_{ij,kl}$ and of $S(p_y p_z)$ is important. We also note that adding all counter pairs contributing to the same $(p_y p_z)$ as defined by eq. (4.1) introduces a slight integration in $\boldsymbol{p}$ space depending on the angles subtended by the counter areas (usually a few degrees maximum). The effective momentum resolution also depends on the finite size of the sample irradiated by the e^+ beam and the effect of the finite temperature on the positron. The minimum size of the source dimension (a few mm) governs the choice of L to be within $(5 \div 15)$ m. We shall now describe the 64-detector apparatus used by our research group at Brandeis.

4.6. *The* 64-*detector* 2D ACAR *system*. – The most important consideration guiding us in the design of the system was cost and manpower. The decision to use individual detectors was made before Anger cameras had been developed with sufficient resolution, and before the development to medium-efficiency γ converters for proportional counters. The system consists of two arrays of

32 NaI detectors each with the sample-detector distance L of 10 m, with interchangeable lead collimators in front of the detectors. The detectors are conventional NaI scintillators (5 cm long, 3.8 cm diameter) coupled to RCA 6199 phototubes. (We did not use smaller-diameter detectors and phototubes because of cost considerations.) The 32 detectors on either side are mounted in two layers, one behind the other, in offset arrays. The NaI crystals are surrounded by a lead structure to further minimize counter-to-counter scattering. The collimation system consists usually of a set of crossed-slit rectangles corresponding to 0.5 mrad by 1.5 mrad. Smaller circular resolution collimators have also been used. If stationary, the counter system measures data on a square grid 3.6 mrad apart, corresponding to the closest distance between detectors. The 32×32 cross-coincidences correspond to 109 unique (p_z, p_y) pairs. In order to fill in the gap between these points, the whole counter assemblies are movable.

Each detector array is mounted on a modified milling machine base, which allows for precision positioning by stepping motors. One array, on the scanning side, is mounted on a precision rotary table positioned on the moving milling machine platform. Both motions (linear and rotational) are actuated by stepping motors controlled by the on-line computer in order to point the collimators towards the sample. The reproducibility of an angle is estimated to be within 0.0025 mrad. The sample and the e^+ source are both mounted on a cold finger cooled to liquid-nitrogen temperature in a Dewar. The e^+ beam from the source is focused by a 24 kG magnet. Magnet and Dewar assembly can be positioned to a high accuracy for alignment.

The block diagram of the data collection system is shown in fig. 11. The output of every detector is discriminated against noise and shaped to a 40 ns pulse for coincidence timing and to a 1 μs pulse for counting single events

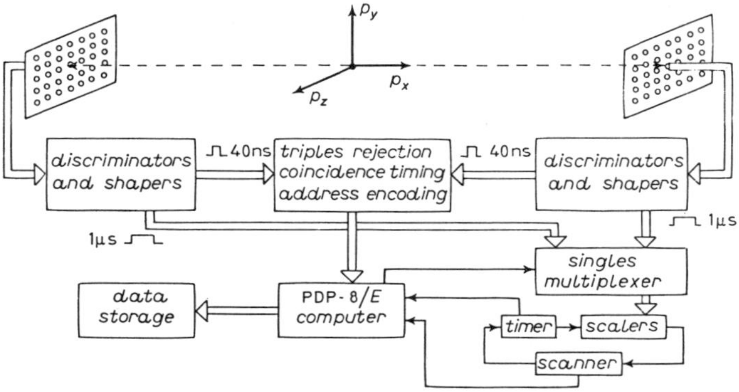

Fig. 11. – Block diagram of data-handling electronics used in the Brandeis 2D ACAR multicounter system.

(« singles »). The « singles » of all detectors are multiplexed and counted in eight scalers to continuously monitor stability and to be used for accidental rate correction. The fast (40 ns) pulses are routed to a digital circuit which performs the following functions: *a*) scattering rejection: any multiple signal in either detector array is rejected as having not a unique « address »; *b*) coincidence timing (a single coincidence circuit provides this service); *c*) address encoding: the identification number of both detectors participating in the time-coincident event is encoded into a 10-bit binary word acceptable by the computer. Valid coincidences are read by the computer (PDP-8/E) through program interrupts, whereas the « singles » are read in batches of eight after a preset time has elapsed.

All system control functions are performed by the PDP-8/E computer through specially designed interfaces and software. The control functions include *a*) motion control through the stepping motors, *b*) magnetic-field switching for ferromagnetic polarization experiments, *c*) scan control of scalers and other units connected in a « daisy-chain » configuration. The actual position of the moving array and the magnetic-field status are monitored continuously by the computer. Data are taken by moving one of the counter sets automatically in the p_z direction (the direction of the smaller dimension of the collimating slit), thus filling in the p_z co-ordinate (usually in steps of 0.2 or 0.3 mrad). Subsequent change in p_y allows one to fill in the full (p_z, p_y)-plane with the desired density of measured points. Although in principle we have to move the detectors only between adjacent 3.6 mrad points, the apparatus can move over a wider p_y and p_z range, thus permitting the comparison of a wide range of individual pair efficiencies. The stability of the system is continuously monitored by counting the singles rates in each detector via the multiplexer system. Typically data are obtained in steps of 0.2 mrad in the p_z direction over a range of ± 40 mrad and in 0.3 mrad steps for the p_y direction over a $(0 \div 20)$ mrad range. A typical coincidence counting rate for copper at $p_y = p_z = 0$, using a 500 mCi ^{58}Co source, is ≈ 0.2 c/s per counter pair; the total coincidence rate is approximately 50 c/s. The full 2D ACAR surface can be measured with 0.4 % accuracy at the peak in $10 \div 14$ days of counting time. Although at present we keep all 1024 detector pair counts in the computer, we find the efficiencies very stable in time. Because of our moving system as many as 200 counter pairs contribute to a typically measured momentum pair, thus averaging out small differences in counting efficiencies. For the purpose of displaying the data we sometimes use a Gaussian smoothing routine (a 2D Gaussian with FWHM of $(0.5 \div 0.8)$ mrad gives good results).

(The above two subsections **4˙5** and **4˙6** are nearly *verbatim* transcriptions from a lecture given in 1979 at ICPA-5 [4].)

4˙7. *Other* 2D ACAR *systems.* – Before describing some experimental ACAR results we briefly mention several other position-sensitive detectors used for

2D ACAR work. Positron physicists have recently discovered that, while they were developing ACAR techniques for condensed-matter applications, a parallel development in annihilation radiation detection took place in medical physics, where annihilation γ's have been used for years for localization and more recently for detailed PET [52]. Although in these applications the deviation from anticolinearity is not measured, the requirements for high-efficiency position-sensitive γ-ray detectors are very similar to those of 2D ACAR work.

The technique of using a large NaI crystal optically coupled to several phototubes for γ localization was introduced by ANGER as early as 1958 [93], and was used by him for positron medical application in 1966 [94]. WEST and coworkers [90] have used a pair of modern Anger cameras for a new high-efficiency 2D ACAR system. Their Anger cameras consist of two 50.8 cm diameter, 1.27 cm thick NaI(Tl) crystals, optically coupled to an array of 37 photomultipliers, 7 cm in diameter, producing two position co-ordinates for each pulse detected (via an analog electronic system). The effective 2D resolution reported by WEST is (0.65×0.65) mrad^2 FWHM.

One of the earliest proposed 2D ACAR systems was to use a block of Ge for γ localization [95]. Apparently the project encountered several technical difficulties in the past; a successful 9 cm $\times$ 6 cm high-purity position-sensitive Ge detector has been developed and its performance has been described by TRIFTSHÄUSER *et al.* [91].

Perhaps the most promising detectors for future 2D work are multiwire proportional chambers. They owe their success to the development of the high-resolution lead converter-proportional chamber combination by JEAVONS and coworkers [89, 96] in a collaborative effort by CERN and the University of Geneva. A 10 cm $\times$ 10 cm prototype was used successfully to obtain 2D ACAR data from several oriented crystals [97, 98]. The second, larger version of this highly successful multiwire chamber has been completed at the University of Geneva, and is the subject of a lecture at this school by MANUEL. The efficiency of the new cameras is reaching 10% and a 30 cm $\times$ 30 cm chamber is about to be completed by MIJNARENDS in Holland. There is little doubt that with this size and efficiency, and with a possible momentum resolution of (0.2×0.2) mrad^2, these systems will surpass, in the near future, the results obtained by the multicounter systems or by the Anger cameras.

5. – Experimental ACAR results.

5'1. *General considerations.* – As we mentioned in the introduction, we shall not attempt to present an exhaustive and systematic review of all ACAR work in metals; instead we shall try to illuminate some of the important results by describing a few representative experiments. Whenever possible we shall discuss the newer 2D ACAR results; we shall use mainly Brandeis 2D data, simply because we are more familiar with the details.

We have seen in the theoretical sections that in principle ACAR data can provide information about the band structure of the sample studied; one obtains projections of the wave-function-dependent $\varrho^{2\gamma}(\boldsymbol{p})$ and, by studying their discontinuities, the measure of the Fermi surface (FS). A difficult and important problem facing the interpretation of the ACAR results concerns the effect of the e+-e− many-body interactions on $\varrho^{2\gamma}(\boldsymbol{p})$. Perhaps the most important result of the many theoretical studies dealing with this problem is the conclusion that the e+-e− many-body interactions do not shift the discontinuities in $\varrho^{2\gamma}(\boldsymbol{p})$ from the position of the FS of the sample [99]. Thus one can obtain the FS size and topology without having to rely heavily on a complete theoretical understanding of the e+-e− correlations.

The ACAR experiments divide themselves naturally into two classes: experiments performed on simple, well-understood metals, in order to study the detailed effect of the e+-e− many-body correlations, and/or to confirm the applicability of the ACAR technique, and Fermi-surface experiments on systems where the more standard FS solid-state techniques cannot be applied—such as FS measurements of substitutional disordered alloys, of systems undergoing structural phase transitions, etc. In order to compare the ACAR results with theory, it is usual to first compute $\varrho^{2\gamma}(\boldsymbol{p})$ in the IPM approximation (*i.e.* using eqs. (3.7)-(3.9)), by computing the electronic wave functions as well as the positron wave function using various band-theoretic techniques. For details of such computations, see the lectures by MIJNARENDS at this Fermi School. A detailed comparison of the 1D or 2D projections of the theoretical $\varrho^{2\gamma}(\boldsymbol{p})$ with 1D or 2D ACAR data requires either deconvolution of the momentum resolution from such data or convolution of the theoretical projections by the resolution of the experimental apparatus. Data deconvolution is known to be highly problematic [100], and we prefer data comparison with resolution-folded theoretical projections, whenever possible. It is customary to assign the lack of agreement with IPM computations to e+-e− many-body effects. Great care has to be taken, therefore, to ensure a reliable, detailed IPM computation based on « first principle » band computations, such as APW, KKR or OPW. Pseudowavefunction computations can lead to discrepancies, since they usually do not include properly the « high momentum » components of the conduction electrons, and thus underestimate the « Umklapp annihilation » contributions.

As mentioned earlier, comparison of ACAR experiments with theory that assumes delocalized Bloch wave functions for the positron requires the use of metallurgically carefully characterized, well-annealed samples, which should have a low concentration of vacancies, dislocations and other defects, to ensure no appreciable positron localization due to trapping at such defects.

The use of ACAR theory in analyzing the experimental data also involves the understanding of the various thermal effects encountered, the most important being the question of positron thermalization prior to annihilation.

5'2. *Temperature effects.* – Equation (3.7) contains the formula for $\varrho^{2\gamma}(\boldsymbol{p})$ at a finite temperature T. The most obvious thermal effect is due to the statistical distributions F_+ and F_- of the positron and the electrons, respectively. In addition, the momentum density itself will change due to electron-phonon and positron-phonon interactions. We have for the positron the Boltzmann distribution

$$F_+(p_+, T) = (2\pi m^* k_B T)^{-\frac{3}{2}} \exp[-p_+^2/2m^* k_B T] , \tag{5.1}$$

where m^* is the positron's effective mass and k_B is the Boltzmann constant. The electron F_- is the usual Fermi distribution that becomes the step function $n(\boldsymbol{k})$ for zero temperature. For a simple free-electron system we convolute these two distributions; the effect of the F's will appear strongest at p_F, the Fermi momentum « cut-off ». We note, however, that in momentum space the effect of F_+ dominates over that of F_-, since the positron is thermally « smeared » about $p_+ = 0$, while the electron is « smeared » about p_F. One easily estimates that this « smearing » is $(\Delta p)_- \sim (k_B T/2E_F) p_F$ due to the electron and $(\Delta p)_+ \sim$ $\sim (3k_B T/2E_F)^{\frac{1}{2}} p_F$ due to the positron. At room temperature we obtain $(\Delta p)_- \sim$ $\sim p_F/300$, while $(\Delta p)_+ \sim p_F/10$; for a typical Fermi momentum of $p_F \simeq 5mc \cdot$ $\cdot 10^{-3} = 5$ mrad, this corresponds to $(\Delta p)_- \sim 0.016$ mrad and $(\Delta p)_+ \sim 0.5$ mrad. With an angular resolution of $\sim (0.1 \div 0.3)$ mrad the thermal smearing due to positron motion can thus be observed, while that of the electrons is negligible.

STEWART and coworkers have explored these relationships to study the thermal effects in alkali and other simple metals; in particular, they have studied the question of positron thermalization prior to annihilation.

5'3. *Positron thermalization.* – In a metal, high-energy positrons slow down to energies of the order of the Fermi energy by ionization and by plasmon excitation in times that are a small fraction of their observed lifetime, as already predicted in the early thermalization work by LEE-WHITING [101]. The positrons proceed to lose energy by creating particle-hole pairs in the conduction band and by interaction with phonons.

CARBOTTE and ARORA [102] studied the slowing-down process by Green's function technique, and WOLL and CARBOTTE [103] used the Boltzmann equation to calculate this relaxation process. More recently MIKESKA [104], HEDE and CARBOTTE [105] and PERKINS and CARBOTTE [106] analyzed the importance of the positron-phonon interaction.

A review by BERGERSEN and PAJANNE [107] discusses most of the relevant theoretical work on thermalization. Experimentally, KIM *et al.* [108] suggested, in an analysis of their experiment on the Fermi momentum cut-off in alkali metals, that there is a minimum temperature, on the order of $(100 \div 200)$ K, below which the positron will not have time to thermalize prior to annihilation. The theoretical computations predicted, however, a lower minimum temperature of the order of ≈ 50 K, particularly when the positron-phonon interaction was taken into account [106]. This discrepancy seems to have been resolved recently in favor of theory: KUBICA and STEWART [109] have reported high-precision experiments in K, Mg and Al, that indicate thermalization prior to annihilation down to $(10 \div 30)$ K. This result is important, since positron thermalization means that the effective momentum resolution is not as restricted by the positron motion as suggested earlier [108]; at 4.2 K, momentum resolutions of the order of 0.1 mrad, *i.e.* of 0.014 a.u., could be achieved.

5·4. *Alkali metals.*

5·4.1. Many-body effects: lifetime measurements. Most theoretical discussions regarding the e^+-e^- correlations have focused on the interacting electron gas with one positron present, *i.e.* on the positron-electron jellium model; only a few computations have tried to solve the much harder problem of the presence of the periodic lattice. These computations have been compared with experimental results on metals where the lattice plays a small role, *i.e.* the « nearly-free-electron » metals such as some of the alkali metals and aluminum. These metals are known to possess a nearly spherical Fermi surface; Na is a particularly good candidate for nearly-free-electron-band behavior. (For a review of the Fermi surface of alkali metals, see [110].) The many-body theories had to explain two distinct experimental results, that at first sight seem almost contradictory:

a) The experimental lifetimes in all metals indicate that the effective density at the site of the positron is greatly « enhanced », *i.e.* the annihilation rates observed are almost an order of magnitude larger (shorter lifetimes) than predicted by IPM computations.

b) On the other hand, for nearly-free-electron metals the 1D ACAR curves show a reasonably small deviation from « parabolic » behavior, as expected from a simple noninteracting electron gas.

The search for the successful explanation of these two trends made positron annihilation in simple metals a most challenging field for many-body computations. The first detailed computation to predict both lifetimes and momentum-dependent annihilation probabilities in detail was the work by KAHANA and KAHANA and CARBOTTE [111], based on a Green's function formalism of eq. (2.6). The electron-positron Green's function was solved by summing over ladder diagrams, including the effects of the electron and positron self-energies; $\varrho^{2\gamma}(\boldsymbol{p})$ as well as the total annihilation rate $\lambda_{2\gamma}$ were computed as functions of the effective r_s, *i.e.* the electron density n_0 of the jellium ($r_s = (4/3\,\pi n_0)^{-\frac{1}{3}}$). Their lifetime computation agreed well with the experimental lifetimes measured by WEISBERG and BERKO [112] in alkali metals and in other simple metals, except for the low-density alkali metals. Since these annihilation rate measurements have become so important in testing various computations, we have plotted them in fig. 12; we also plotted several theoretical computations (by no means all) for comparison.

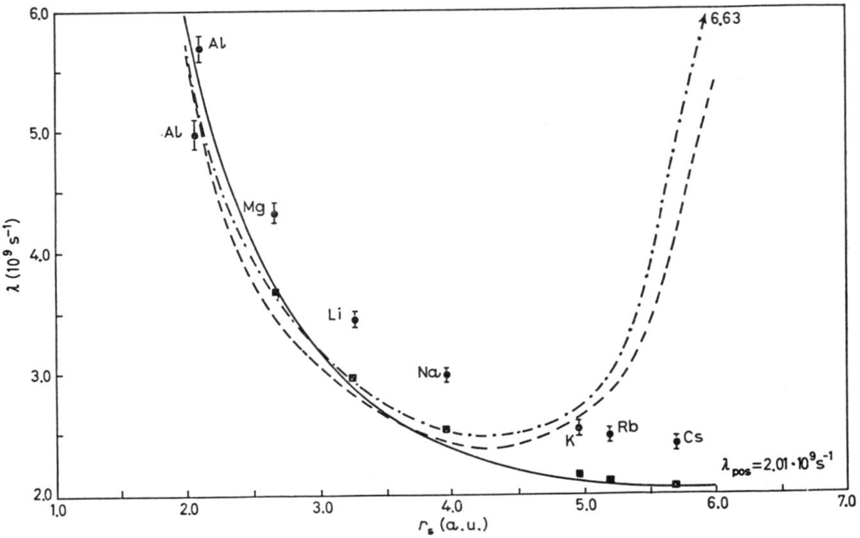

Fig. 12. – Experimental annihilation rates (inverse lifetimes) in alkali metals and some other nearly-free-electron metals [112], plotted against r_s. The two Al rates indicate the typical range of rates obtained in Al samples with different amount of defects (higher rate corresponds to well-annealed sample). The square data points are rough estimates of Battacharyya and Singru [113] of the contribution of the conduction electrons to the total rate, after the estimated core annihilation rate was removed. The theoretical predictions plotted are: —·—·— KAHANA [111], — — — SJÖLANDER and STOTT [114], ——— BATTACHARYYA and SINGWI [113]. λ_{pos} is the positronium annihilation rate with rapid spin exchange.

It was noticed by CROWELL, ANDERSON and RITCHIE [115] that Kahana's ladder approximation breaks down for $r_s \geqslant 5$; the more recent semi-empirical computations by BATTACHARYYA and SINGWI [113] eliminate this divergent behavior. Another theoretical computation by SJÖLANDER and STOTT is also plotted [114]. Using a new formalism based on an interacting-boson description, ARPONEN and PAJANNE [116] also predict a properly convergent behavior for large r_s (*i.e.* low density). For a more complete set of references regarding the many-body computations, see the lectures at this school by CARBOTTE and the references contained in the recent papers on the subject by MALDAGUE [117] and by BORONSKI, SZOTEK and STACHOWIAK [118], who describe an exact solution of the Kahana equation. While discussing briefly positron lifetime results, we should like to also mention the remarkable success of Brandt, Isaacson and Paulin [119] in explaining the detailed (centroid shift based) lifetime measurements of MacKenzie, Jackman and Thrane [120] for a host of metals using the semi-empirical density enhancement factor developed by BRANDT and REINHEIMER [121]. The observed pressure dependence of the total annihilation rates in alkali metals by MACKENZIE *et al.* [122] has been explained theoretically by KUBICA and STOTT [123].

5·4.2. Many-body effects; 1D ACAR measurements. The many-body computations discussed above predict a momentum-dependent annihilation rate $\Gamma^{2\gamma}(\boldsymbol{p})$ usually expressed as a product of a many-body enhancement factor $\varepsilon(\boldsymbol{p})$ and the IPM annihilation rate $\Gamma^{2\gamma}_{\text{IPM}}(\boldsymbol{p})$, *i.e.* $\Gamma^{2\gamma}(\boldsymbol{p}) = \varepsilon(\boldsymbol{p})\Gamma^{2\gamma}_{\text{IPM}}(\boldsymbol{p})$. KAHANA and CARBOTTE find that, as one approaches the Fermi surface, the annihilation rate increases; above the Fermi momentum the annihilation rate is essentially zero, due to a surprisingly effective cancellation of terms, as compared to the well-known high-momentum many-body tail of an interacting electron gas without the positron present. It has become usual to parametrize the isotropic enhancement as

$$\varepsilon(p) = a + b\gamma^2 + c\gamma^4, \qquad \text{where } \gamma = p/p_F. \tag{5.2}$$

For tables of the coefficients a, b and c *vs.* r_s, see ref. [29, 116]. We have plotted in fig. 13 $\varrho(p)$ for the noninteracting Fermi gas (« e_{free} ») and for the

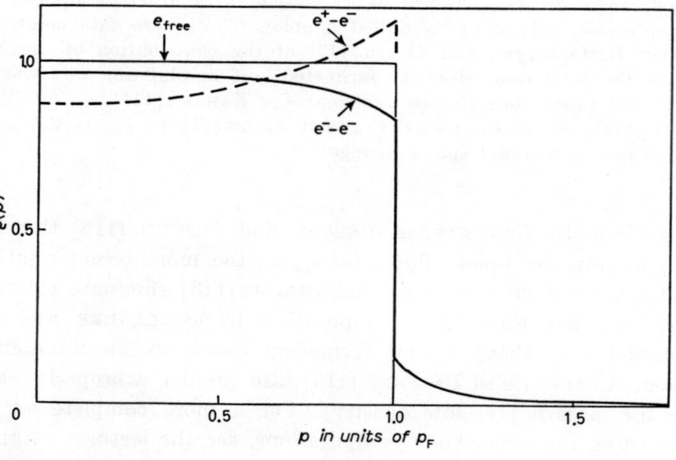

Fig. 13. – The effect of many-body interactions on $\varrho(p)$. Details discussed in the text; curves are renormalized to equal volume in p space.

interacting electron gas (curve marked e⁻-e⁻) as well as $\varrho^{2\gamma}(p)$ (curve marked e⁺-e⁻), normalized to equal volume in $\boldsymbol{p}$ space, for easy comparison. The curves were computed for an electron density equivalent to that of the conduction band of Al.

The study of the e⁺-e⁻ many-body effects on the 1D ACAR curves in alkali metals was pioneered by STEWART and his coworkers. For a spherical Fermi surface the momentum-dependent $\varepsilon(p)$ of eq. (5.2) produces a 1D ACAR curve which shows a small but distinct « bulge » from the inverted parabolic shape

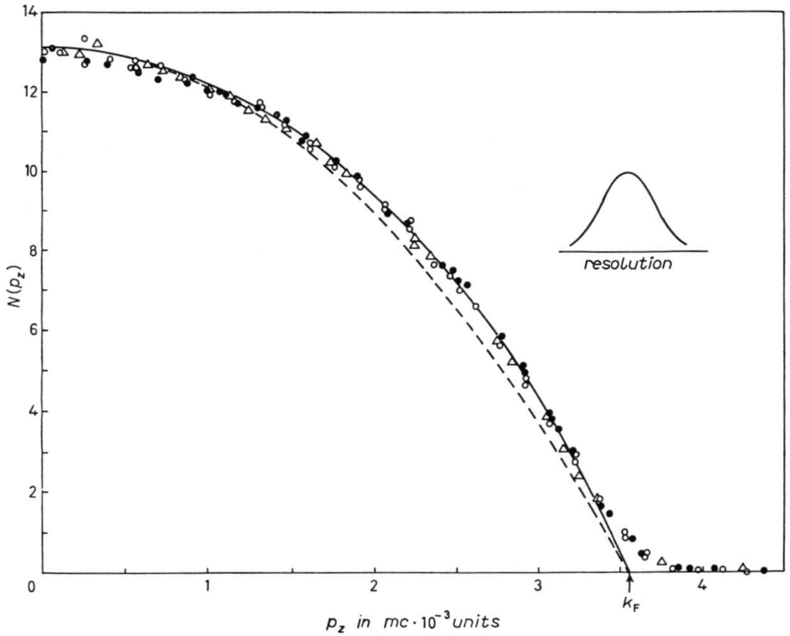

Fig. 14. – Long-slit angular correlation in sodium by DONAGHY and STEWART [124], with $N(p_z)$ along three orientations: ••• [110], ∘∘∘ [100], △△△ [111]. The experimental points are compared to the theoretical prediction of Kahana (———) and to a free-electron parabola (— — —); a low-intensity Gaussian has been subtracted from the data, assumed to be due to core annihilation.

expected for a noninteracting electron gas. A careful study of this shape permits the experimental measurement of the enhancement parameters b/a and c/a of eq. (5.2). In fig. 14 we show the first result of this many-body effect as observed by DONAGHY and STEWART [124] on single crystals of Na. A low-intensity Gaussian was fitted to the high momentum of the experimental curve and subtracted from the data; this Gaussian was meant to represent the momentum distribution produced by the annihilation with the core electrons. We see that the distribution is indeed isotropic and that the experimental points agree well with the theoretical curve predicted by KAHANA [111].

In a more detailed set of experiments KIM and STEWART [108] have measured the 1D ACAR curves in Li, Na, K and Rb over a wide temperature range ((15 ÷ 650) K). After core subtraction the experimental curves were fitted to a parametrized theoretical function involving the convolution of the « effective resolution » due to the actual resolution of the apparatus and the thermal motion of the positron (eq. (5.1)) with the many-body 1D ACAR curve resulting from the 2D integration of eq. (5.2). Thus the fit provided a measure of b/a and c/a of eq. (5.2) and of the effective positron mass m^* of eq. (5.1). In such fits the resultant values of b/a and c/a are highly correlated, and a linear combination such as $(b + c)/a$ is better determined experimentally ($\pm\sim 10\,\%$ in

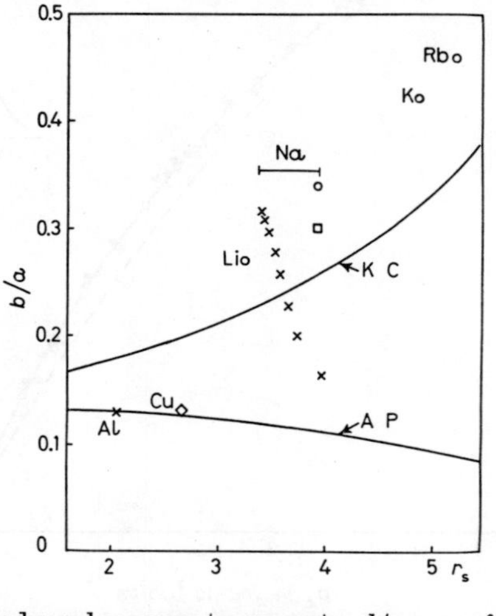

Fig. 15. – The many-body enhancement parameter b/a as a function of r_s; the KC and AP curves are the theoretical predictions of Kahana and Carbotte [111] and Arponen and Pajanne [116], respectively. The experimental values are from analyses of Donaghy and Stewart (□, ref. [124]), Kim and Stewart (○, ref. [108]), Gustafson and Willenberg (×, ref. [125]) and Sinclair, Farmer and Berko (◇, ref. [126]).

Kim and Stewart's case). The Kim and Stewart values obtained for b/a are plotted in fig. 15, against r_s; several other more recent experimental results are also plotted, and so is a new set of theoretical values by ARPONEN and PAJANNE [116]. We note that these new results seem in substantial disagreement with those of Kahana and Carbotte [111]: as r_s increases (electron density decreases), ARPONEN and PAJANNE predict a decrease in the « peaking » in $\varrho^{2\gamma}(\boldsymbol{p})$ as $\boldsymbol{p}$ approaches $\boldsymbol{p}_F$, while the Kahana theory predicts the opposite.

(The c/a values follow effectively the same trend.) Another difference has to do with the « many-body HM tail »: unlike KAHANA *et al.*, ARPONEN and PAJANNE obtain a finite albeit small « tail » above p_F.

GUSTAFSON and WILLENBERG [125] have performed a 1D ACAR *vs.* pressure experiment on Na. Their enhancement parameters indicate a surprisingly steep and so far unexplained dependence on pressure, also plotted (as b/a) in fig. 15. We note that the Na value at zero pressure is in substantial disagreement with the value of Kim and Stewart, and fits better the theoretical prediction of Arponen and Pajanne. We also include data for Cu [126], to be discussed later. Earlier results by ARIAS-LIMONTA and VARLASHKIN [127] in Rb and Cs also seem to favor the Arponen and Pajanne theory (they observe essentially a parabolic behavior), in disagreement with the Kim and Stewart data. We conclude, therefore, that a new study of the ACAR (1D as well as 2D) shapes in alkali metals would be highly desirable; in particular, careful pressure measurements would be most important, in order to check the unusual dependence of b/a reported in Na [125].

5·4.3. Effective-mass results. As mentioned above, the study of 1D ACAR curves *vs.* T also led STEWART and coworkers to a measurement of the effective positron mass m^* [128]. The positron effective mass m^* in eq. (5.1) can deviate from the positron mass m due to several interactions. The lattice interaction produces a nearly parabolic positron band with a positron band mass of about $1.1m$ [129]. A much more complex effect is the contribution to the effective mass of the e^+-e^- interactions, another fundamental many-body problem. The existing computations using the random-phase approximation (RPA) [130] yield effective masses of 1.08 for Al to 1.25 for K (in units of m); a more recent computation by BALDO and PUCCI [131] goes beyond RPA to yield somewhat larger values for the alkali metals. The positron-phonon interaction effect on m^* is small; on the other hand, the phonons can substantially influence the momentum distribution which deviates from the Boltzmann shape, with the result that a Boltzmann fitting can give larger « apparent » masses [132]. This effect could be responsible for the fact that the experimentally measured m^* are larger than the theoretical values. KIM and STEWART [108] obtain 1.8 ± 0.3 for Li and Na, 2.1 ± 0.3 for K and 2.3 ± 0.3 for Rb.

KUBICA and STEWART [109], using a much improved long-slit apparatus with excellent resolution (~ 0.16 mrad), performed high-precision temperature measurements in K, Na, Mg and Al. They obtain « apparent » effective masses of 1.6 ± 0.1 in K and Na, 1.2 ± 0.1 in Mg and 1.55 ± 0.1 in Al, in fair agreement with the theory that includes the positron-phonon change in the e^+ momentum distribution.

In the above experiments the shape of the distribution assumed for core annihilation (and Umklapp annihilation) is a Gaussian. The influence of the exact shape of these contributions on the extracted b/a and c/a parameters has not been studied sufficiently. The use of more detailed band computations for the core and Umklapp shapes would be of great interest. Another problem arising with Na and Li at low temperatures is their martensitic (structural) transformation. Further work on these problems is presently pursued in several laboratories.

5·4.4. Fermi-surface measurements. We note that in these 1D ACAR experiments another free parameter determined by fitting procedure is the Fermi momentum p_F. Using their high-resolution long-slit apparatus Stewart's group reported recently [133] new Fermi diameters in the [100], [110] and [111] directions of Li at temperatures just above the martensitic transition (78 K). Their data show about a 2 % departure from the free-electron sphere, with small bulges in the [110] directions. These high-precision experiments follow a long line of Li 1D ACAR experiments as well as some crossed-slit measurements by WILLIAMS and collaborators [134], and clearly demonstrate the applicability of the ACAR technique to FS measurements: because of the martensitic transition standard FS techniques are unable to yield information on bulk samples of Li. The various Li band computations have to include nonlocal exchange and correlation terms to explain the near sphericity of Li [135]. The new high-resolution 2D ACAR systems will shed, hopefully, more light on this old problem.

5·5. *Divalent metals.* – After the first long-slit experiments in oriented crystals of Be by STEWART [136] and BERKO [137], many experiments were performed on Mg, Zn and Cd. Notably, one should mention the 3D reconstruction of $\varrho^{2\gamma}(\boldsymbol{p})$ in Zn using the Mijnarends method, from three 1D projections obtained by MOGENSEN *et al.* [138], by KONTRYM-SZNAJD and STACHOWIACK [139]. They find many details in $\varrho^{2\gamma}$, but do not discuss the « effective resolution » brought about by the use of only three projections—see our subsect. **3·4** on this problem. More recently PAWLOWSKA [140] computed theoretically the momentum densities in Zn and presented group-theoretic selection rules for the h.c.p. structure. The most recent divalent-metal study is the Mg 1D ACAR experimental work by SHIOTANI *et al.* [141], who compare their result with the APW computations (including the e^+ wave function) by WAKOH. They find that the observed crystalline « anisotropies » (*i.e.* difference between $N(p_z)$'s along two orientations) could be explained well by the IPM computations. On the other hand, the actual overall shapes of the individual $N(p_z)$'s required the introduction of empirical enhancement factors. A good fit was obtained by enhancing the conduction band by an isotropic Kahana-like momentum-dependent term, and multiplying the IPM core contribution by a momentum-independent enhancement factor. This pattern seems to be typical in almost all other ACAR studies on metals. The momentum-independent many-body enhancement behavior for tightly bound electrons is roughly in agreement with the computations of Carbotte and Salvadori [142].

Perhaps the highest-precision Fermi-momentum measurement by positron annihilation to date is that by KUBICA and STEWART [109] in Mg, a reference quoted earlier regarding effective-mass measurements. They quote the value of p_F in Mg along the crystallographic c-axis as $(5.29 \pm 0.005) \cdot 10^{-3}\, mc$, corresponding to $(0.725 \pm 0.001)\, a_0^{-1}$, where a_0 is the Bohr radius: this is in excellent agreement with the value $p_F = 0.722\; a_0^{-1}$ obtained from magneto-acoustic

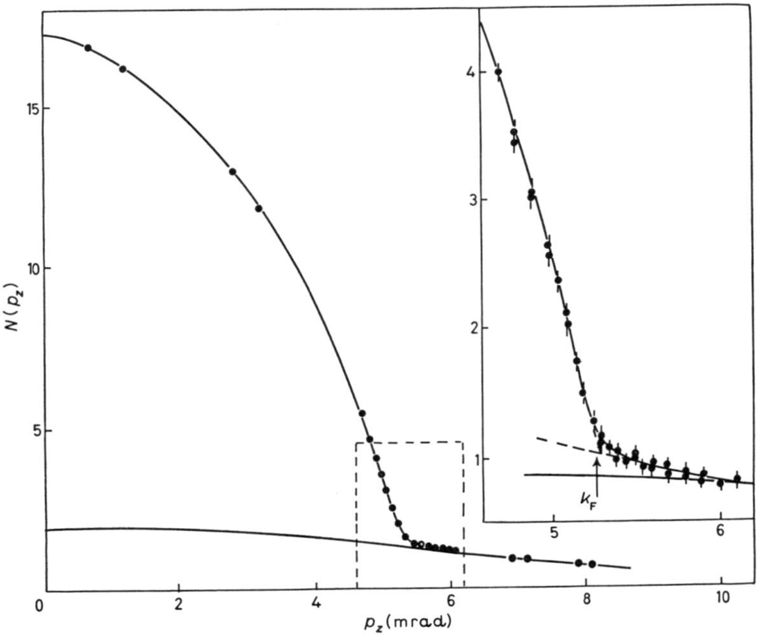

Fig. 16. – High-resolution 1D ACAR curve in Mg at 4.2 K, p_z along the c-axis, by KUBICA and STEWART [109]; $\Delta p \sim 0.16$ mrad. Inset shows $N(p_z)$ in the region of the Fermi momentum $p_z = k_F$.

attenuation measurements. The precision of such 1D ACAR experiments thus approaches the precision of more standard solid-state FS techniques. We plot in fig. 16 their 1D ACAR from a Mg crystal at 4.2 K to exhibit the high precision required around the Fermi momentum, in order to obtain the above-discussed results (the smearing corresponds to a free-positron temperature of (10 ± 10) K).

5·6. *Aluminum.*

5·6.1. 1D ACAR measurements. Ever since our first 1D ACAR experiment on oriented single crystals [10], aluminum, being a nearly-free-electron metal with well-known band structure, has provided, together with copper,

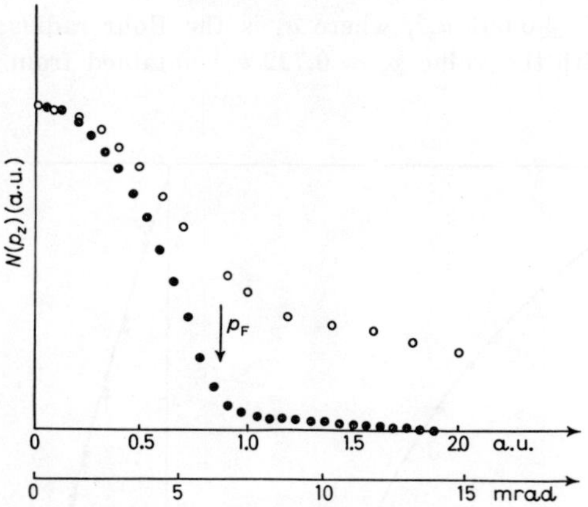

Fig. 17. – Comparison of Compton-profile (○) [143] and 1D ACAR (●) [144] results in Al.

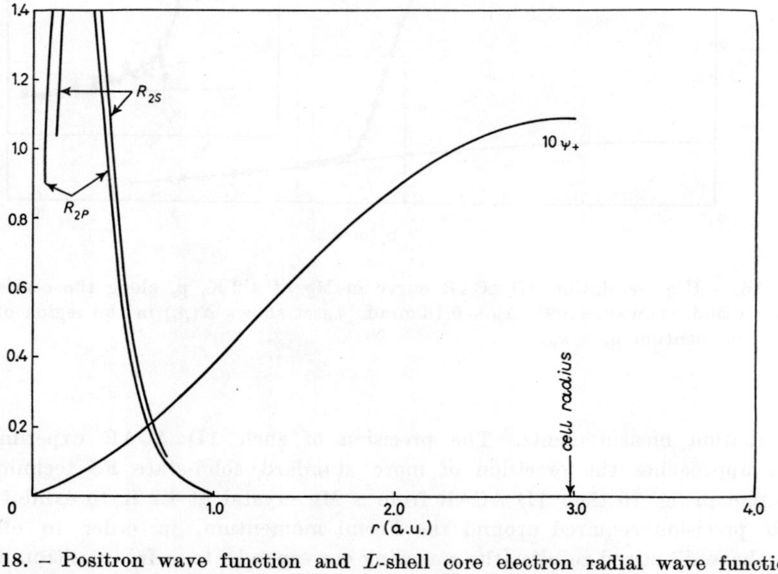

Fig. 18. – Positron wave function and L-shell core electron radial wave functions in Al [10].

ideal tests for ACAR theory. Aluminum has also been studied extensively by the Compton-profile (CP) technique, discussed in subsect. **2**.3 and **4**.3. In fig. 17 we compare the aluminum CP result [143] with the 1D ACAR data [144]. We note the poorer resolution of the typical CP result, as discussed in subsect. **4**.3. A more striking observable difference is the much larger high-momentum (HM) tail of the CP curve. It is clear that the presence of the positron wave function diminishes substantially the core contribution to $\varrho^{2\gamma}(\boldsymbol{p})$ as compared to the « true » momentum-space density $\varrho(\boldsymbol{p})$, whose projection is given by the CP data. In fig. 18 we show the result of our early positron wave function computation [10] (using a Wigner-Seitz method), indicating the small ψ^+ overlap with the $2p$ and $2s$ core electrons. The early spherically symmetric Wigner-Seitz $\psi_+(\boldsymbol{r})$ has been replaced by the proper anisotropic computation using either a plane-wave expansion [145] or by a modified positron pseudo-wavefunction computation [146]. However, these anisoptropic wave functions do not change substantially the behavior of $\psi_+(r)$ around the atomic cores.

5.6.2. Core electron enhancement. If one is interested in the e^+-e^- many-body effects on the HM components of $\varrho^{2\gamma}(\boldsymbol{p})$, one can write the actual $\varrho^{2\gamma}(\boldsymbol{p})$ in terms of the IPM densities as

$$\varrho^{2\gamma}(\boldsymbol{p}) = \sum_j \varepsilon_j(\boldsymbol{p})\, \varrho^{2\gamma}_{\mathrm{IPM},j}(\boldsymbol{p}) \,, \tag{5.3}$$

where $\varepsilon_j(\boldsymbol{p})$ represents the many-body enhancement function for the band j. A detailed fit of the projections of eq. (5.3) to the actual 1D or 2D ACAR data can then be used to obtain the enhancement factors $\varepsilon_j(\boldsymbol{p})$ for the various bands. This procedure is, of course, not unique; by going to large $|\boldsymbol{p}|$ one hopes to avoid the contribution of the conduction electrons, and thus obtain the core enhancement $\varepsilon_{\mathrm{core}}(\boldsymbol{p})$.

The small HM component in the 1D ACAR from Al can be more clearly studied with the Doppler-profile (DP) technique using two Ge detectors in coincidence, as discussed in subsect. **4**.2. In fig. 19 we show the DP data obtained in Al by LYNN *et al.* [71, 72] as a Doppler-shift energy distribution plotted on a semi-logarithmic scale (ΔE of 1 keV in « 2D » DP corresponds to $\Delta\theta = 1.96$ mrad in 1D ACAR). The theoretical core annihilation fractions with $1s$, $2s$ and $2p$ electrons are also indicated. A comparison of the total experimental annihilation rate and the IPM partial core annihilation rate with the fit of the core fractions in fig. 19 led LYNN *et al.* to conclude that the many-body enhancement for core electrons (see eq. (5.3)) at large momenta ($p \geqslant 18mc \cdot 10^{-3}$) is close to unity, compared to the much larger enhancement required to fit the annihilation of the conduction electrons. The enhancement seems reasonably momentum independent, at least at these large momenta. CARBOTTE and SALVADORI [142] predict theoretically this momentum independence, but their total core enhancement is much larger (~ 2.7); on the

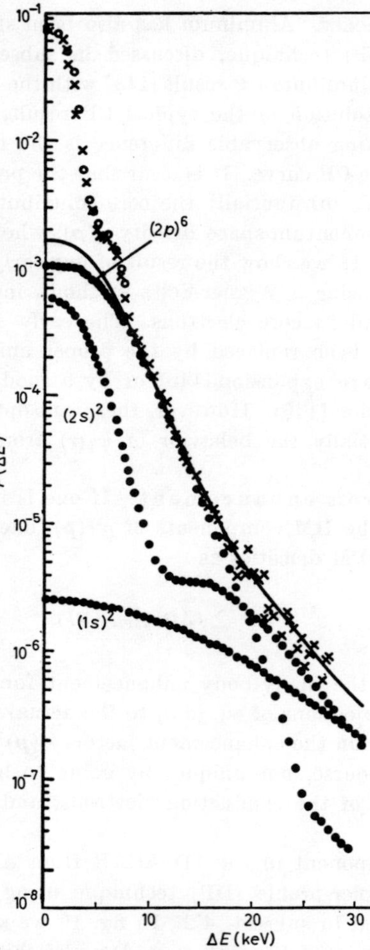

Fig. 19. – **Doppler-profile measurements in Al using two Ge detectors in coincidence [71]; experimental points ×××, theoretical core annihilation profiles (••• for different core electrons, —— for total core). Open circles are 1D ACAR data in Al [144].**

other hand, a theoretical calculation by BONDERUP *et al.* [147] yields a low enhancement ($\varepsilon \sim 1$) for tightly bound core electrons. At lower momenta the core annihilation has to be separated from the appearance of Umklapp annihilations from the conduction electrons. These effects are much easier to study with 2D ACAR than with the more usual long-slit systems.

The existence of e^+-e^- many-body enhancement effects renders the ACAR experiments much less valuable than the Compton-profile experiments for the study of the details of the electronic wave functions. On the other hand, the large decrease in the core component due to positron overlap projects out the momentum density of the conduction electrons, and makes the study of the Fermi surface for high-Z materials much easier by ACAR than by CP; in Pb, for example, the core function in ACAR is similar to that in Al, while in CP it reflects the large total number of core electrons to that of conduction electrons. Thus for high Z the conduction electrons in the CP measurement appear superimposed on a large core « background », that renders the FS measurement hard to achieve, even after resolution deconvolution.

5·6.3. $B^{2\gamma}(r)$ for Al. Because the Fermi surface of Al is nearly spherical in the extended-zone scheme, the CP and 1D ACAR are expected to be nearly parabolic. Al thus provides a good test of studying the theorem relating the zeros of the $B(r)$ function to the value of the Fermi momentum—see subsect. 3·5 for a theoretical discussion. PATTISON and WILLIAMS [41] used CP data of Al (as well as Li and Na) to study the behavior of $B(r)$, *i.e.* of the 1D Fourier transform of $J(p_z)$. From the zeros of $B(r)$ they obtain the correct p_F for Al to within 0.01 a.u. of the accepted value, even though the CP resolution function had a width of 0.37 a.u. In fig. 20 we plot $B^{2\gamma}(r)$, *i.e.* the 1D Fourier transform of 1D ACAR data for Al obtained with two angular resolutions: 2 mrad

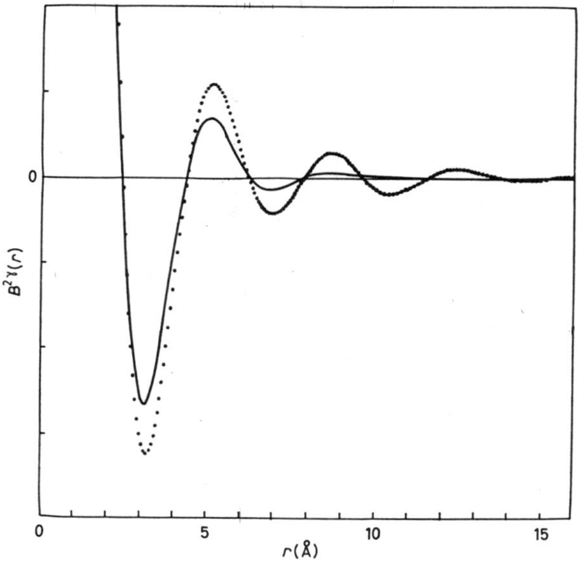

Fig. 20. – The Fourier transform $B^{2\gamma}(r)$ of the 1D ACAR data in Al (110) for two different momentum resolutions [148]: ··· 0.5 mrad, —— 2.0 mrad; p_z along the (110) crystallographic direction.

and 0.5 mrad. We note that the zeros of the two curves are the same within the width of the lines even though the amplitudes of the oscillations are very different due to the damping of the Fourier transform by their respective resolution functions. A study [148] of these zeros indicates that all but the first one are within 1 % of their theoretical values for a free-electron parabola (see subsect. **3˙**5). The first zero, however, is much closer to that predicted from a « Kahana enhanced » $\varrho^{2\gamma}(p)$. Thus we find that (for nearly free metals) the $B^{2\gamma}(r)$ is sensitive to the e^+-e^- many-body interactions.

An interesting use of the Fourier transform of $N(p_z)$ was introduced by GUPTA and SIEGEL [149], who seemed unaware of the $B(r)$ literature; they point out that the FT of the core annihilation component in $N(p_z)$ for Al has a narrow distribution in $\boldsymbol{r}$ space, reflecting the localized nature of core electrons. A separation into core and conduction electrons is thus feasible in FT space, and they use this to obtain the core *vs.* conduction electron enhancement factors, obtaining $\varepsilon_{\text{core}} = 2.7$, $\varepsilon_{\text{cond}} = 4.2$, *i.e.* a core enhancement in good agreement with CARBOTTE and SALVADORI, and in disagreement with Lynn's result [71].

5˙6.4. 2D ACAR Al results. We now turn to the recent 2D ACAR measurements. In 1976 we published the results of a high-precision Al experiment using our prototype 22-detector 2D apparatus [150]. The results showed clear evidence for a momentum-dependent many-body enhancement superimposed on anisotropies due to Brillouin-zone (BZ) interactions. Figure 21 shows $N(p_y, p_z)$ for Al, with the crystal orientation indicated in the inset [82]. By analyzing the 2D derivatives of this surface we conclude that the FS is

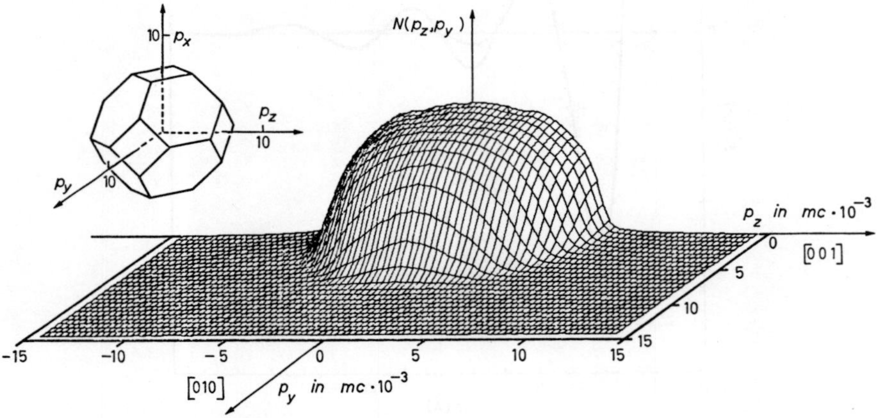

Fig. 21. – 2D ACAR surface of Al by BERKO *et al.* [82]. The orientation of the crystal is illustrated in terms of the Brillouin zone in the inset. Each crossing of lines is an independent measurement. Sample at 100 K.

circular within 1% in the extended-zone scheme. Clear deviations appear in the region where the BZ intersects the momentum distribution, particularly about the (200) face. We obtain [4, 150, 151] a good fit between the experiment and an OPW computation (provided the OPW result is multiplied by a Kahana momentum-dependent enhancement). An even better fit seems to result with the APW computation by WAKOH [152].

When we scale up the 2D surface of fig. 21, we observe at high momenta a marked anisotropic structure due to the high-momentum components (HMC) of the conduction electrons. The OPW computation that includes the effect of an anisotropic $\psi^+(r)$ predicts well these HMC, which are but a few percent

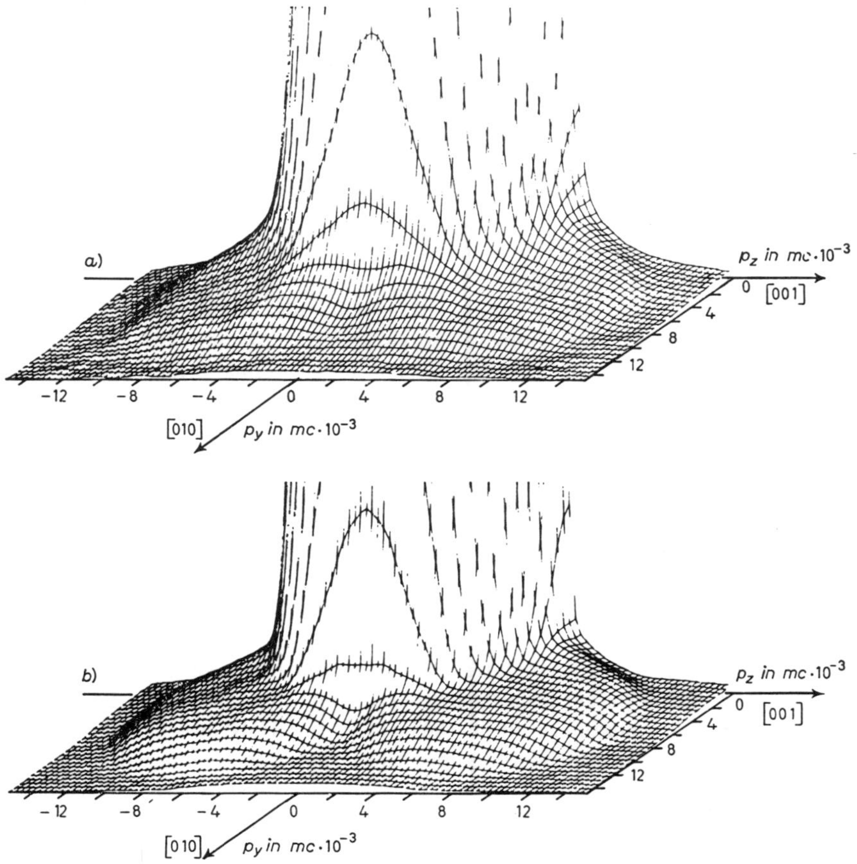

Fig. 22. – High-momentum region of *a*) the experimental 2D ACAR aluminum surface compared with *b*) an OPW computation [150]. The experimental data [4] correspond to a vertical enlargement of the scale of fig. 21.

in amplitude of the $N(p_y = 0, p_z = 0)$ peak intensity. It was found that the electronic wave functions had to be true OPW's rather than pseudo wave functions in order to fit the HMC correctly [150]. We plot the re-scaled experimental as well as theoretical (OPW) HMC region in fig. 22 for the (100) plane [4, 151]. It was found that in order to fit the very-high-momentum region ($p_z > 10$ mrad), upon normalization to the experimental value at $p_y = p_z = 0$, the theoretical core intensity had to be reduced by a factor of two [150]. This was interpreted as due to a core enhancement $\varepsilon_{\text{core}}$ that is less than the conduction band enhancement $\varepsilon_{\text{cond}}$; using the full 2D ACAR curves we obtained $\varepsilon_{\text{cond}}/\varepsilon_{\text{core}} = 2.15$, larger than the theoretical value of ~ 1.5 predicted by CARBOTTE and SALVADORI [142] (see also ref. [149]), but smaller than the value of Lynn *et al.* [71, 72]. These numbers depend sensitively on the IPM positron wave function used; the IPM annihilation probability with core electrons can vary as much as 30 %, depending on the computed positron wave function. The deduced $\varepsilon_{\text{core}}$ is also sensitive to the accuracy of the experimental data at high momenta, where accidental rate problems can become important, as well as to the quality of single crystals used in the experiments. A crystal with some defects will tend to decrease the core contribution due to defect-trapped positrons. It is also possible that the core enhancement is slowly varying with momentum, decreasing as p increases. More detailed work on this problem will be required.

We note that in the alkali metals, divalent metals, as well as in Al, the 1D and 2D ACAR results were compared to various IPM theories that include the e^+-e^- many-body interaction only via an added « enhancement » borrowed from jellium computations [111] for the conduction electrons and from tight-binding theory [142, 147] for core electrons. HEDE and CARBOTTE [153] have studied theoretically the positron annihilation in a nearly-free-electron band, using a modified ladder approximation that includes in lowest order the electron-lattice interaction. They conclude that the Umklapp components of the conduction band are enhanced similarly to the $\boldsymbol{G} = 0$ component, *i.e.* a large constant enhancement plus a momentum dependence that increases as $\boldsymbol{p} \to \boldsymbol{p}_{\text{F}}$. More recently CHAKRABORTY [154] developed a two-component density functional scheme for Al using a self-consistent pseudopotential method which incorporates directly the e^+-e^- interaction into the band computation, in the spirit of the Kohn-Sham theory. The predicted « bulge » in the 2D ACAR fits well our experimental result [4, 150] albeit the HMC are underestimated due to the use of pseudo wave functions. No doubt this scheme will set a new trend in ACAR band computations. It would be most desirable to obtain experimentally a reconstructed $\varrho^{2\gamma}(\boldsymbol{p})$ for Al, in order to check the behavior of this function close to the BZ boundaries and to the FS cut-offs; this detailed behavior would test the theoretical predictions by FUJIWARA *et al.* [155], who considered the effect of the e^+-e^- interaction as a perturbation in a two-band nearly-free-electron model.

5.7. *Copper.*

5.7.1. 1D and 2D ACAR results. Another extensively used metal to test the various ACAR techniques has been copper. Copper exhibits the typical band structure of metals having d electrons, it is very well understood theoretically, and possesses an experimentally well-known, yet sufficiently complex Fermi surface [156]. Copper is also important since it forms the basis of many classical alloys of interest to fundamental alloy theories. The history of the many 1D ACAR experiments has been amply reviewed [29-31, 69].

FUJIWARA and SUEOKA [85] were the first to observe the main anisotropic feature of the Cu FS, the [111] FS « necks », by using the « crossed-slit » technique (rough 2D, (0.85×5.5) mrad^2); similar experiments were reported by WILLIAMS and coworkers [86]. The [111] « necks » were also studied by the « rotating-specimen » method [87, 88]. Since then many high-precision 1D ACAR measurements obtained details of the Cu FS; in particular MIJNARENDS reconstructed the full 3D $\varrho^{2\gamma}(\boldsymbol{p})$ from eight 1D ACAR curves [50], BERKO *et al.* [157] tested the anisotropy of the 1D ACAR curves with a modified Wigner-Seitz model, using the Halse Cu FS model [156] and ROZENFELD *et al.* [158] also investigated finer details of the Cu FS. HOWELLS and OSMON [159] used a spark chamber method to produce the first 2D ACAR surface in Cu, albeit with poor statistics in view of the low efficiency of their setup.

Using the Brandeis 2D machine we obtained full 2D ACAR in Cu along various integration directions p_x. In fig. 23 we plot the 2D ACAR surface with p_x along [111], so as to exhibit the FS « necks » around $p_y = p_z = 0$ [4, 82]. The 2D ACAR surfaces of Cu were compared to the APW calculations of Wakoh [152]. In order to exhibit these anisotropies, but without using other ACAR orientations to produce the unphysical difference curves required in long-slit measurements [157], we construct a smooth rotationally symmetric surface $R(p_y, p_z)$ with a flat region at low momenta and a nearly Gaussian radial fall-off for $p > 5$ mrad. This surface is constructed to be below all experimental data, joining smoothly past $(12 \div 13)$ mrad. We emphasize that R does not represent an attempt to account for the contribution of the filled bands; it was used simply to exhibit clearly the anisotropic components of $N(p_y, p_z)$. Figure 24 shows the surface obtained from experiment as a contour map of $N(p_y, p_z) - R(p_y, p_z)$ and the corresponding APW computation of Wakoh. Inside the first zone the contours are plotted in steps of 5.74 % of the value at $p_y = p_z = 0$; past the sharp slope due to the FS in the first zone the contour steps have been decreased to 0.53 % of the (0, 0) peak. The same technique was used in a paper where we compare $N(p_y, p_z)$ with a simple modified Wigner-Seitz band model [160]. The APW theory fits the anisotropies clearly better, as expected. We see that the anisotropies, which stem from the Umklapp annihilation of the partially filled « $4s$-$3d$ » band as well as from anisotropies of the filled « $3d$ » bands due to hybridization with the $4s$ band,

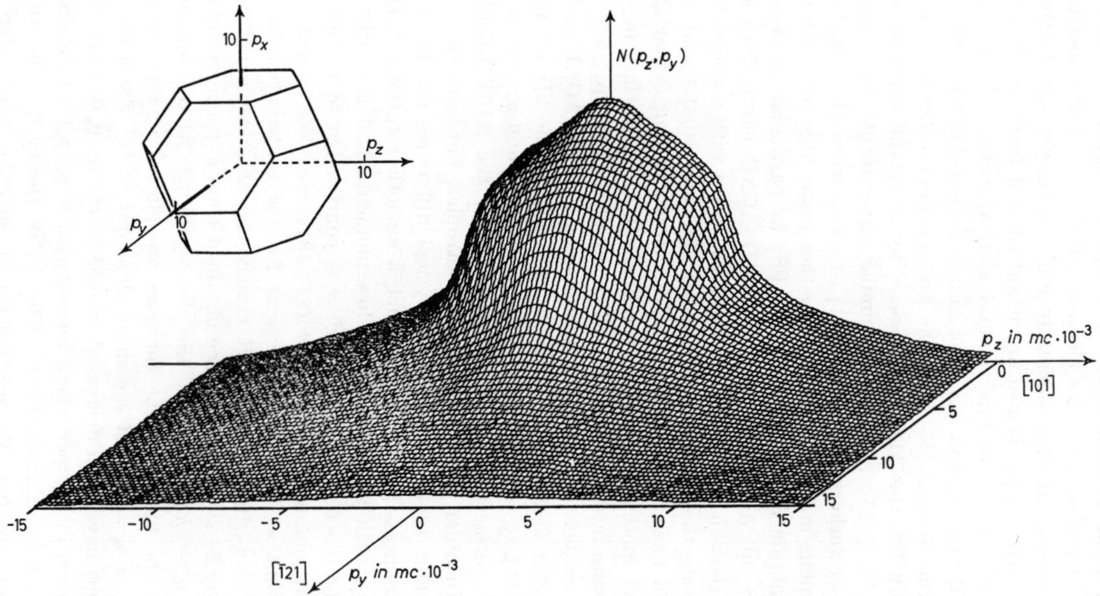

Fig. 23. – 2D ACAR surface of Cu by BERKO *et al.* [82]. The orientation of the crystal is illustrated in terms of the Brillouin zone in the inset. Sample at 100 K.

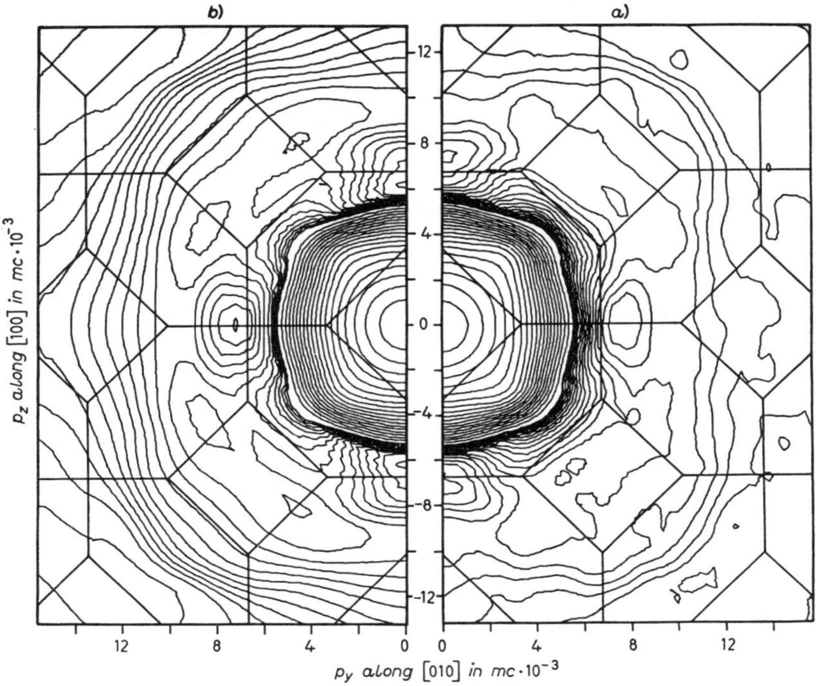

Fig. 24. – Contour map [4] of the experimental (*a*)) *vs.* theoretical (APW) (*b*)) $N(p_y, p_z) - R(p_y, p_z)$ for Cu, with p_x along the [100] direction—see text for details.

are very well reproduced by the APW theory. A more careful comparison of the full shape shows, however, a considerable deviation between the APW theory and experiment [161], deviations attributed to many-body effects.

Prior to the full 2D measurements, we measured [80] several slices of $N(p_y, p_z)$ with $p_y = \text{const}$ using the earlier 22-counter system [80]. Mijnarends and Singru [162] have obtained excellent theoretical fits to these data by using a Hubbard approximation scheme for the IPM momentum density sampled by an IPM positron wave function, but combining this result with a modified Kahana-type many-body enhancement factor. Instead of using the jellium $\varepsilon(p)$ (eq. (5.2)), they propose an enhancement factor that depends on the energy of the conduction band electrons counted from the bottom of the band, $\varepsilon(E) = a + b(E/E_F) + c(E/E_F)^2$, where E_F is the Fermi energy, but with b/a and c/a originally computed for an electron gas by Kahana [111]. (A similar energy-dependent enhancement was proposed by Sob [163].) The energy-dependent $\varepsilon(E)$ is used in the first zone: the Umklapp terms are enhanced in a momentum-independent way using the enhancement factor of the Γ_1 state at $\boldsymbol{k} = 0$. Finally, the IPM core annihilation contribution is reduced by a factor of 0.2 (empirically). The obtained fits to the data as well as to the derivative of the data are excellent, as can be seen in fig. 25 for $N(p_y = 0, p_z)$ with p_x along [100] and p_z along [001]. It will be most important

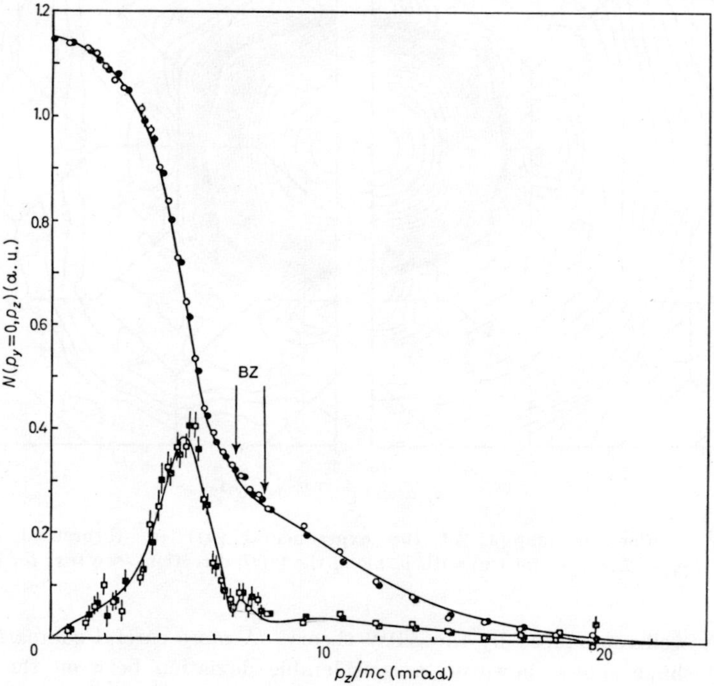

Fig. 25. – Experimental $N(p_y = 0, p_z)$ and its derivative with respect to p_z [80], compared to the theoretical result of Mijnarends and Singru [162], that includes an energy-dependent enhancement factor as discussed in the text; • $p_z > 0$, ∘ $p_z < 0$.

to see whether the empirical enhancement formula used for the (3*d*-4*s*) conduction band in Cu in this work can be derived from fundamental many-body physics (perhaps via the local-density formalism [154]).

5·7.2. 3D reconstruction of $\varrho^{2\gamma}(\boldsymbol{p})$ from 2D ACAR data for Cu. In order to test in more detail the effect of e⁺-e⁻ interactions on the ACAR result, we reconstructed the 3D momentum-space density $\varrho^{2\gamma}(\boldsymbol{p})$ for Cu [4, 126]. Using the Brandeis multicounter system with circular collimator slits 0.75 mrad in diameter, we obtained $N(p_y, p_z)$ on Cu crystals using an integration direction p_x in successive 9° steps around the p_y direction, which was held fixed, first along [100] and then along [110]; we reconstructed $\varrho^{2\gamma}(\boldsymbol{p})$ in several (p_x, p_z)-planes. We tested two reconstruction methods, described in subsect. **3.4**: the Fourier-transform method based on eq. (3.14) and the « filtered back-projection » method (eq. (3.16)). We use the symmetry of $\varrho^{2\gamma}(\boldsymbol{p})$ in both methods to increase the statistical precision of our result; the two methods gave nearly identical results. In fig. 26 we show a contour map of the reconstructed $\varrho^{2\gamma}(\boldsymbol{p})$

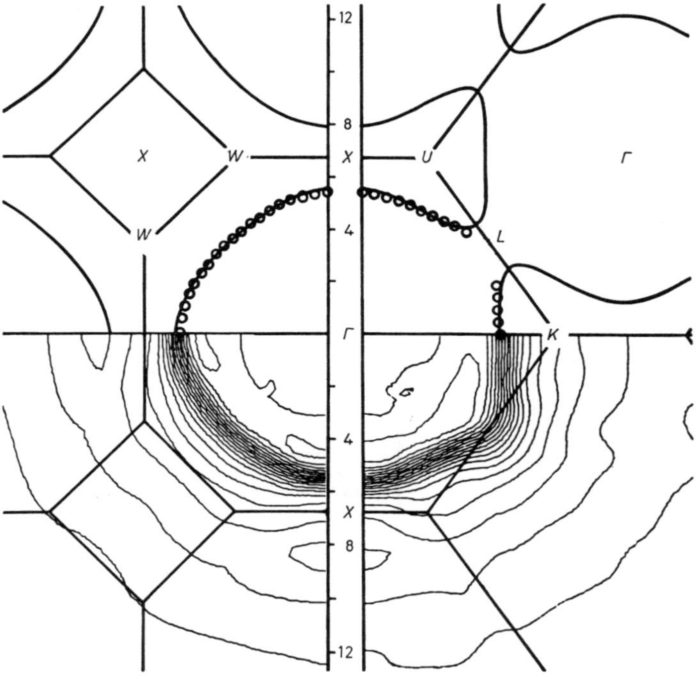

Fig. 26. – Contour map of the reconstructed $\varrho^{2\gamma}(\boldsymbol{p})$ in Cu (bottom half) for the (100) (left side) and (110) (right side) planes. Top half: the FS of Halse [156] (solid line) and the measured FS from $\varrho^{2\gamma}(\boldsymbol{p})$ (open circles) [126].

in two planes, and the 7-parameter analytical expression of the corresponding Cu FS of Halse [156]. The effective resolution in the reconstructed $\varrho^{2\gamma}(\boldsymbol{p})$ along the radial direction is ~ 0.8 mrad, determined primarily by the p_z resolution of the 2D ACAR data; in the circunferential direction the resolution is $\sim 9°$, the angle between adjacent projections, *i.e.* it increases with $|\boldsymbol{p}|$ being ~ 0.8 mrad at the FS. We measure the FS radii p_F by finding the position of the maximum radial slope of $\varrho^{2\gamma}(\boldsymbol{p})$. In the [111] neck region we find an experimental slope less than one-half of that in the [001] direction, indicating no FS (see fig. 5, 6); these points are thus omitted. We find the FS radii to be within 0.15 mrad, *i.e.* $\sim 3\%$ of the Halse values. The Umklapp annihilations in the second zone are clearly visible in both planes.

In fig. 27 we plot the radial sections through $\varrho^{2\gamma}(\boldsymbol{p})$ for the three symmetry directions, together with the Mijnarends-Hubbard IPM computation [48]

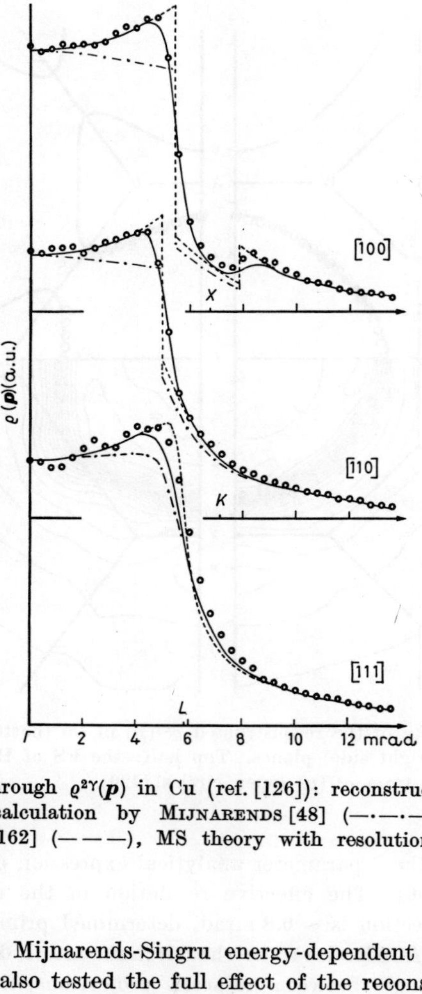

Fig. 27. – Sections through $\varrho^{2\gamma}(\boldsymbol{p})$ in Cu (ref. [126]): reconstruction from 2D ACAR curves (ooo), IPM calculation by MIJNARENDS [48] (—·—·—), Mijnarends-Singru energy enhancement [162] (– – –), MS theory with resolution and reconstruction effects (———).

(see fig. 6) and the Mijnarends-Singru energy-dependent enhancement computation [162]. We also tested the full effect of the reconstruction: the theoretical MS model was integrated in each direction, convoluted with the instrumental resolution and then processed by the data reconstruction programs; the full curve of fig. 27 shows the result, *i.e.* the excellent fit to the data, except at the [100] Umklapp region close to the FS break. In this region there is a hint that the Umklapp component also requires a momentum (or energy)-dependent enhancement, as predicted by HEDE and CARBOTTE [153] and FUJIWARA [155].

By studying the radial dependence of the experimental $\varrho^{2\gamma}(\boldsymbol{p})$ from 0 to 4 mrad, we obtain momentum-dependent enhancement parameters defined by $\varepsilon(\boldsymbol{p}) = \varrho^{2\gamma}(\boldsymbol{p})/\varrho^{2\gamma}_{\mathrm{IPM}}(\boldsymbol{p})$, perhaps useful for further theoretical work. Using the Mijnarends IPM model based on the Hubbard scheme [48], we obtain $0.19\pm$ ± 0.020 for b/a and 0.110 ± 0.036 for c/a (see eq. (5.2)), close to the Kahana values of 0.195 and 0.139, respectively (p_{F} chosen to have its free-electron value of 5.26 mrad). On the other hand, using the APW computation by WAKOH [152] as well as a recent KKR computation by MIJNARENDS [164], we obtain 0.131 ± 0.020 for b/a and 0.110 ± 0.036 for c/a, values closer to those given by ARPONEN and PAJANNE [116] (*i.e.* 0.126 and 0.102). We see that

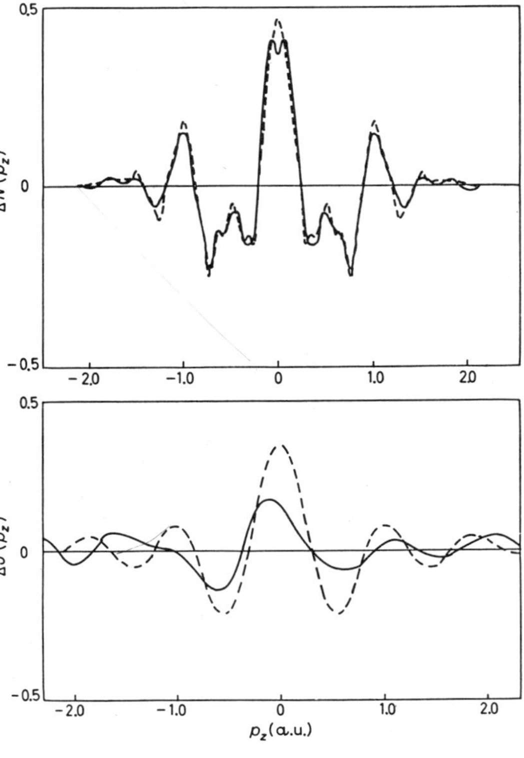

Fig. 28. – Difference curves (« anisotropy curves ») $\Delta N(p_z)$ between 1D ACAR along [110] and [100] directions of Nb (top figure) and the corresponding $\Delta J(p_z)$ between Compton profiles [165]; experimental (——) and theoretical (— — —) results.

great care has to be taken with the IPM computations before one can assign a « true » many-body enhancement to the data. All these enhancement methods are, of course, based on « empty-lattice » many-body theories; a more complete fundamental theory of $\varrho^{2\gamma}(\boldsymbol{p})$ for Cu is still in the future.

In 1979 PECORA and EHRLICH [60] tested their reconstruction method with data obtained by their somewhat unusual 2D ACAR apparatus using « Soller slit »-like Ta collimators and only two γ detectors. They obtain slices in $\varrho^{2\gamma}(\boldsymbol{p})$ for Cu, with less statistics and larger oscillations than those of fig. 27.

5·8. *Transition metals.* – Fundamentally, the electronic structure of transition metals is similar to that of the noble metals Cu, Ag and Au; however, the Fermi level, instead of being well above the « *d* »-like bands, intersects the « *d* » states, leading to complex Fermi-surface topologies [166]. Detailed discussion of the many 1D ACAR results can be found in the reviews by BERKO [30] and MIJNARENDS [31]. KANHERE and SINGRU [167] discussed the systematic behavior of the electronic momentum-space density in the transition metals V, Cr, Fe, Ni (and in Cu), using the Mijnarends-Hubbard approximation.

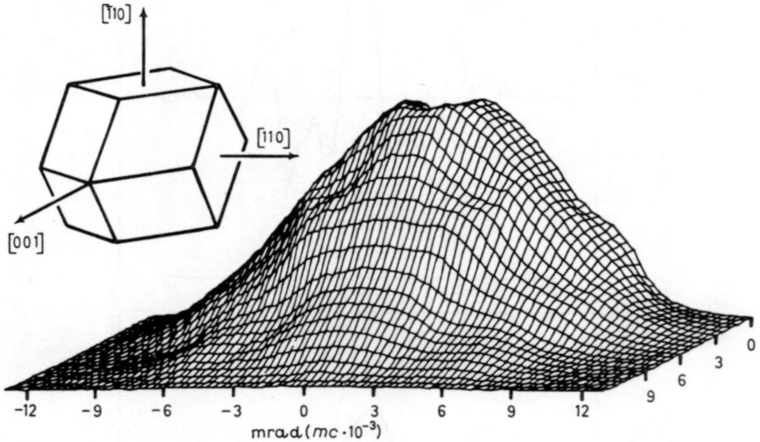

Fig. 29. – 2D ACAR for Ta by SINCLAIR *et al.* [168] with integration direction along (110). The inset shows the Brillouin zone.

As for Cu, the existing IPM band computations that include the e^+ wave function can predict well the anisotropies found by 1D ACAR transition metal experiments, but fail to explain the overall shape of the individual curves. As an example, we reproduce in fig. 28 the experimental and theoretical results for Nb from the work by WAKOH *et al.* [165] in form of the difference $\Delta N(p_z)$ between 1D ACAR along [110] and [100] directions, and the difference $\Delta J(p_z)$

between the corresponding Compton profiles. We note the excellent fit for the ACAR result as well as the much greater detail obtainable by ACAR than by CP, in view of the resolution problem mentioned before.

Recently we measured the 2D ACAR in Ta [168], in order to start a series of 2D experiments on transition metal hydrides, in particular Ta hydrides. Figure 29 shows the 2D ACAR surface obtained with the p_x direction along [110], exhibiting the large anisotropies due to the complex FS of Ta [169]. Since we did not have a band computation for Ta incorporating $\psi_+(\boldsymbol{r})$, we form the « reduced momentum density » $\tilde{N}(p_y, p_z) = \int \tilde{\varrho}^{2\gamma}(\boldsymbol{p}) = \sum_{\boldsymbol{H}} N(p_y + H_y, p_z + H_z)$,

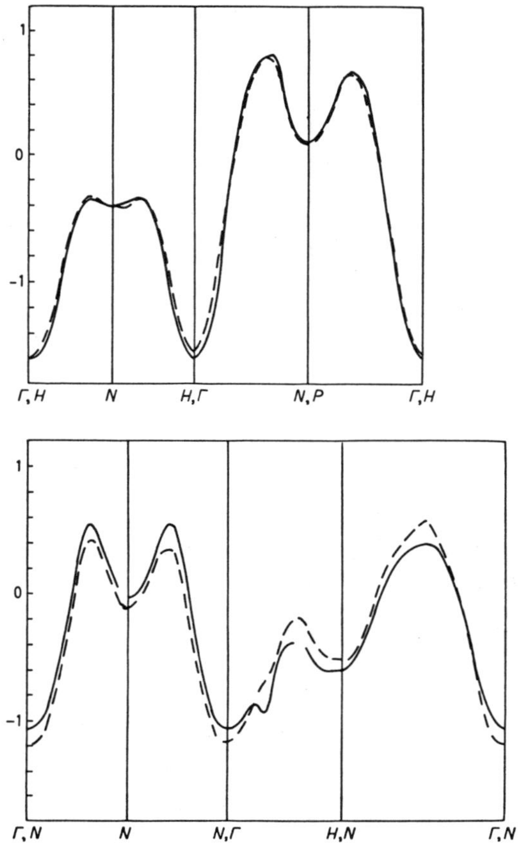

Fig. 30. – Sections through the reduced $\tilde{N}(p_y, p_z)$ distributions (see text). Top: (100) integration; bottom: (110) integration. The solid line is the model, the dashed line the data.

where $\boldsymbol{H}$ is the two-dimensional lattice formed by the perpendicular projection of the reciprocal lattice onto the (p_y, p_z)-plane. We thus test the extension of the LCW theorem to 2D ACAR data—see subsect. 3˙6. We find that the theorem works remarkably well: The experimental $\tilde{N}(p_y, p_z)$ can be reproduced well with a model that involves only the shape and size of the Ta Fermi surface (and includes the resolution of the 2D ACAR measurement). In fig. 30 we exhibit this fit by showing sections through $\tilde{N}(p_y, p_z)$ corresponding to some symmetry directions, and compare them with the FS model predictions. More fundamental theoretical work will be necessary to explain the underlying theoretical reason for the surprisingly good fit obtained by the use of the extended LCW theorem, perhaps along the lines of ref. [65].

So far we have discussed ACAR experiments in metals with well-understood electronic band structures and Fermi surfaces (with the exception of Li). We described experiments in alkali metals, in aluminum and in copper in some detail, since these experiments form the basis of e^+-e^- many-body studies and can be used to test the various ACAR analysis techniques. We now turn briefly our attention to the application of the ACAR technique to less well understood metallic systems.

5˙9. *Rare earths.* – The rare earths have been the subject of many experimental and theoretical investigations; they form mostly h.c.p. structures, have a «transition-metal-like» hybridized «*s-d*» conduction band, and exhibit complex magnetic properties [170]. One of the earliest applications of the 1D ACAR technique was to rare-earth studies by Mackintosh and collaborators [171]. Their experiments indicated a large anisotropy in the conduction band of Gd, Tb, Dy, Ho and Er; a similar large anisotropy was found in the h.c.p. 4*d* transition metals Y and Zr. These results were most important even in a qualitative way, because they demonstrated conclusively that the three electrons forming the conduction band in these metals produce a complex, highly anisotropic FS, as predicted by theory [172]. The low purity and the various phase transitions at low temperature prohibit the FS measurement by standard techniques in many rare earths. Important in the analysis of these experiments were the APW computations of the momentum density of Loucks *et al.* [173], based on a paramagnetic band structure.

Recently we performed at Brandeis a 2D ACAR study of Gd [174]. The measured 2D ACAR surface with the integration direction p_x along the *c*-axis reveals a minimum around $p_y = p_z = 0$, reflecting the theoretically predicted cylindrical hole FS around Γ, with the axis along the *c*-direction. A more detailed comparison with theory requires the use of ferromagnetic-band computations [175]. The 3D reconstructed and $\boldsymbol{k}$-space reduced density agrees qualitatively with the FS proposed by Mattocks and Young [176], based on their recent de Haas-van Alphen measurement. We are presently in the process of applying the spin-polarized positron annihilation technique, developed some time ago [28] for transition metal ferromagnets, to 2D ACAR studies of the spin density in Gd. A theoretical momentum density computation for rare earths that includes the positron wave function would be most desirable.

5·10. *Ferromagnetic studies using polarized positrons.* – Due to parity nonconservation in weak interactions, positrons are emitted from radioactive sources partially polarized with their spin along their direction of motion (positive chirality). This fact led early on to the development of the use of polarized positrons to study the spin-aligned electrons in ferromagnetic materials. The theory has been developed by us in a paper that also reviews the early history of the spin-dependent ACAR technique [28]. Briefly, if a polarized positron enters a ferromagnet and if, upon thermalization, the positron is left in a state with partial spin polarization P_p, the ACAR result will change when the spin of the ferromagnetic electrons is flipped by the reversal of an external saturating magnetic field. This effect is due to the annihilation selection rules discussed in subsect. 1·2: only spin singlet e^+-e^- overlap can lead to 2γ annihilation. The typical experiments measure 1D ACAR with a magnetic field parallel $(N_\uparrow(p_z))$ and antiparallel $(N_\downarrow(p_z))$ to the positron spin. Defining $\varrho^{2\gamma}_{+(-)}(\boldsymbol{p})$ as the positron-sampled momentum density of the majority (+) and minority (—) electrons in a ferromagnet, one can derive [28, 177] the expression for the measured 1D ACAR curves:

$$N_{\uparrow(\downarrow)}(p_z) = \int [A^{\uparrow(\downarrow)}\varrho^{2\gamma}_{+}(\boldsymbol{p}) + B^{\uparrow(\downarrow)}\varrho^{2\gamma}_{-}(\boldsymbol{p})]\,\mathrm{d}p_x\,\mathrm{d}p_y\,, \tag{5.4}$$

where A and B are constants depending on P_p and the spin-dependent total 3γ yield, that has to be measured in a separate experiment. It is then possible to obtain from the experimental data, using eq. (5.4), the physically interesting quantities $\int [\varrho^{2\gamma}_{+}(\boldsymbol{p}) \pm \varrho^{2\gamma}_{-}(\boldsymbol{p})]\,\mathrm{d}p_x\,\mathrm{d}p_y$ for direct comparison with theoretical predictions.

The direct interpretation of such results in terms of majority and minority FS effects is complicated by the presence of the positron; the positron may sample the majority and minority electrons differently. The many-body enhancement factors might also be spin dependent [178].

The early experiments in Fe and Ni [28] have indeed shown that the positrons preserve some of their spin direction. In a detailed experimental and theoretical study of iron [179], Mijnarends has reconstructed from 1D ACAR the 3D $\Delta\varrho^{2\gamma}(\boldsymbol{p}) \equiv \varrho^{2\gamma}_{+}(\boldsymbol{p}) - \varrho^{2\gamma}_{-}(\boldsymbol{p})$ and analyzed it in terms of a ferromagnetic band structure of Fe. Polarization experiments were also performed in Gd by Hohenemser *et al.* [180] and Berko and Mills [177]; these experiments measure mainly the spin density of the conduction electrons in Gd, since the positron barely samples the deep-lying f electrons, known to be mainly responsible for the magnetism of the rare earths.

In analogous manner to $\Delta\varrho^{2\gamma}(\boldsymbol{p})$ one can introduce a « positron undisturbed » « spin density » in momentum space $\Delta\varrho(\boldsymbol{p}) = \varrho_+(\boldsymbol{p}) - \varrho_-(\boldsymbol{p})$, paralleling $\varrho_+(\boldsymbol{r}) - \varrho_-(\boldsymbol{r})$, the spin density in real space—as measured by the polarized-neutron diffraction technique [181]. It is interesting to note that one can actually measure $\Delta\varrho(\boldsymbol{p})$ (albeit with very low counting rates), by studying the spin-dependent Compton profiles obtained from Compton scattering of circularly polarized γ-rays—as was done by SAKAI and ÔNO [182].

The extension of the polarized-positron technique to 2D ACAR experiments is straightforward, and it is hoped that in the near future we shall have spin-dependent 2D ACAR results for ferromagnetic transition metals as well as for rare earths.

For a more comprehensive list of references on ferromagnetic ACAR experiments see the reviews by BERKO [30] and MIJNARENDS [31].

5˙11. *Disordered alloys.* – During the last decade the central interest in metal physics has shifted from the study of pure ordered metals to the investigation of disordered systems, such as random substitutional alloys and amorphous metals. The theory of the electronic structure of concentrated random substitutional alloys has undergone a major development, with the coherent potential approximation (CPA) playing a central role [183, 184]. Realistic calculations of energy bands, density of states, as well as of electron lifetimes have been performed for transition and noble metal alloys, using the average t-matrix approximation ATA [183] and, more recently, the full CPA treatment [184, 185]. One of the essential conclusions of these computations is that, in the disordered alloys treated, the Fermi surface is still well defined, since the smearing of the FS, although appreciably larger than the thermal smearing, is still a small fraction of the FS dimension.

It is in the area of high-concentration disordered substitutional alloys that the ACAR technique provides an almost unique tool for measuring the FS. It is well known that standard solid-state FS techniques, such as the use of the de Haas-van Alphen effect, cannot be applied to high-concentration random alloys, because the electron scattering leads to a drastic reduction in signal [186]. In fact, most standard FS measurements cannot provide data for impurity concentrations of higher than one percent. Positron annihilation, not requiring coherent electron wave packets, does not suffer from such limitations; it has, of course, its own characteristic problems, involving the presence of the positron. The theoretical computation of the momentum density $\varrho(\boldsymbol{p})$ within the CPA or ATA is far more complex than for pure metals. An important step has been taken by MIJNARENDS and BANSIL [187], who developed the formalism within the ATA. The correct theoretical treatment of the positron is an additional complicating factor; HONG and CARBOTTE [188] conclude, for example, that the positron's presence has the effect of additional FS smearing, over and above the inherent smearing due to electron scattering. Such computations are further complicated by the question of properly taking the configuration average over the positron-electron system. By taking the average to be a product of positron and electron averages one neglects the so-called « vertex » corrections; the possible tendency of the positron to sample preferentially one of the alloy constituents is not as yet properly taken into account by the theoretical models.

The many positron experiments on various random alloys have been discussed in the general positron reviews [30, 31] and, in some more detail, in specific disordered-alloy reviews [189, 190]. Although some interesting work on light-Z alloys such as Li-Mg has been performed [191], most recent experiments have studied mainly the copper-based alloys. Long-slit as well as crossed-slit ACAR experiments have been performed on various random alloys, and the results have been analyzed in terms of deviations from the shape of the FS of copper. Usually the neck radii k_N and the FS dimensions k_{100} and k_{110} have been obtained by various fitting techniques or by differentiation of the ACAR data, in order to establish the FS discontinuities in $\varrho^{2\gamma}(\boldsymbol{p})$. The FS changes only slightly with concentration, and the analysis of the experiments

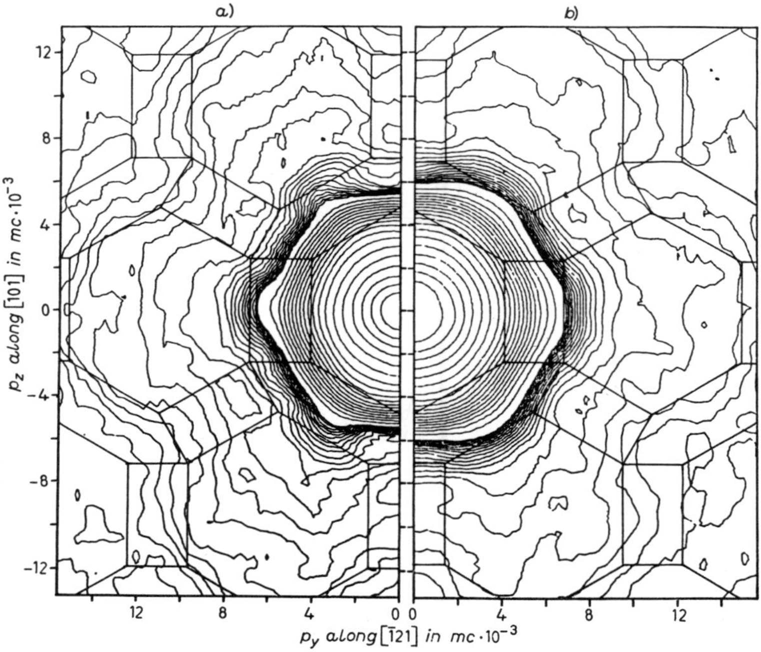

Fig. 31. – The contour map of $N(p_y, p_z) - R(p_y, p_z)$ for *a*) Cu and *b*) Cu-30 at.% Zn, with p_x along [111]. $R(p_y, p_z)$ is a smooth, rotationally symmetric surface, as discussed in the text.

is quite sensitive to the exact effects of the momentum resolution; early FS results contain most probably systematic errors that make the comparison between different sets of data somewhat questionable. The classic rigid-band model predicts a universal curve for the change of the FS parameters of various copper-polyvalent metal alloys, when these parameters are plotted against e/a, the electron-per-atom ratio. Such plots of the present data indicate a deviation from rigid scaling; much more precise data will be required to clearly demonstrate such trends. Since most of the existing data have been carefully compiled and plotted in the reviews [189, 190], we forego such an analysis, and describe only some recent experiments.

Using the Brandeis multicounter system, we performed full 2D ACAR measurements in several oriented α-brass single crystals [192]. The 2D ACAR results from Cu-11.6 at.% Zn and Cu-30 at.% Zn have been compared with those of pure Cu; we find the α-brass momentum density to be quite similar to that of copper, except for the growth of the FS. We demonstrate this similarity in fig. 31, where the contour maps of $N(p_y, p_z) - R(p_y, p_z)$ have been plotted for Cu and Cu-30 at.% Zn for p_x along the [111] direction (corresponding to the orientation of fig. 23). The isotropic R for Cu was the same as used in

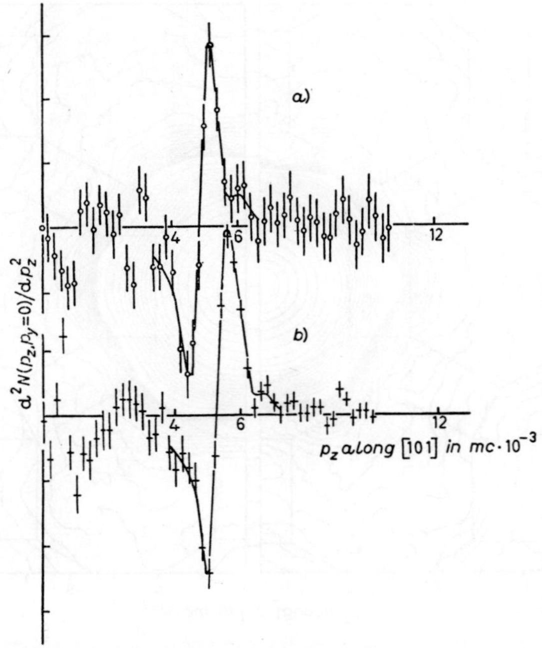

Fig. 32. – Experimental second derivative of the 2D ACAR data for *a*) pure Cu and *b*) Cu-30 at.% Zn.

fig. 24; the same shape R but with a 12 % reduced amplitude was used for Cu-30 at. % Zn. We note the clear growth of the FS features in the first zone. The anisotropy in higher zones is quite similar between Cu and the Cu-30 at. % Zn sample. In order to analyze quantitatively these 2D ACAR surfaces, we constructed from the raw $N(p_y, p_z)$ a set of radial distributions $N(p, \theta)$ in the (p_y, p_z)-plane by linear interpolation, and differentiated these distributions to obtain the breaks due to the FS. Figure 32 exhibits the second derivatives, indicating the precision one can obtain with our present 2D setup. We estimate from such data that the FS smearing due to disorder is less than 0.15 mrad, *i.e.* less than 3 % of k_{110}. After correcting these derivatives for the effects of finite resolution, we obtain the FS dimensions; we show in fig. 33 the [111]

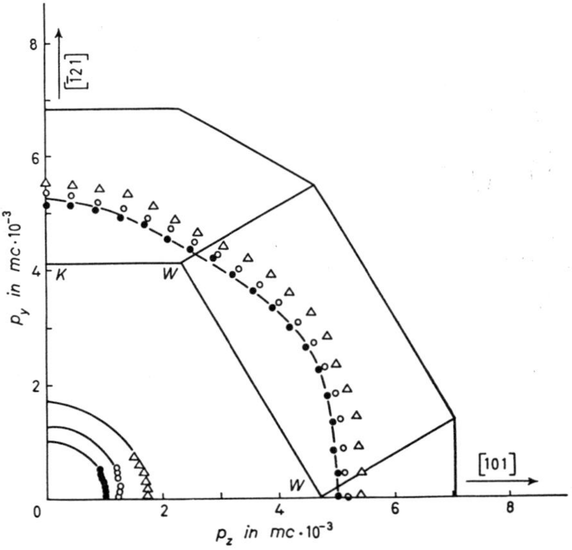

Fig. 33. – FS dimensions of the [111] neck and the (111) « belly »: • Cu, ○ Cu-11.6 at. % Zn, △ Cu-30 at. % Zn. The projection of the BZ onto the (p_z, p_y)-plane is also shown, indicating the 2D ACAR orientation used to obtain the FS dimensions.

neck and the FS « belly » radii. We also plot the pure Cu FS of Halse [156]. The FS is found to grow essentially uniformly with Zn concentration. The [111] necks are found to be circular to within ±5 %. The observed rate of growth of the « belly » Fermi radii is somewhat slower than predicted by the rigid-band model, as well as those of the more recent theories, including CPA [184] (but the actual values are, within their estimated statistics, still in agreement with these theoretical calculations). The [111] neck radii fit the rigid-band model: (1.24 ± 0.07) mrad and (1.71 ± 0.07) mrad for the 11.6 at. % and 30 at. % samples, respectively, compared to the 1.3 mrad and 1.7 mrad rigid-band values. We compared the shape of the first and second derivative of the radial distributions around the FS and conclude that the smearing of the FS, if present, is not observable with the present resolution. Clearly higher-resolution 2D ACAR data will be needed to study the question of disorder smearing at the FS. Future random-alloy work will no doubt include high-precision 3D reconstruction from such high-resolution 2D data.

From the theoretical point of view the noble metal-transition metal alloys such as Cu-Ni are more challenging, since they contain the unfilled *d* bands of the transition metal solutes (leading to the well-known *d* resonances for the dilute alloys). Fewer ACAR experiments have been performed to date on these important alloys than on the Cu-polyvalent systems. The reason is twofold: *a*) it is not easy to grow good homogeneous single crystals of these materials (Cu-Ni is a prime example), and *b*) because the [111] neck shrinks, rather than expands upon alloying, it is harder to measure its value with reliable accuracy. Among the more reliable results are perhaps those of Hasegawa *et al.* [193], who find a nearly linear dependence of the [111] neck radius on e/a, in good agreement with a CPA prediction by GORDON *et al.* [194], and deviating from the standard rigid-band model. Substantial deviations from rigid-band behavior were also found in Cu-Pd alloys [195]. We can look forward to much more work in these alloys, using the new 2D ACAR techniques.

A most interesting study on the Fermi surface of the so-called « Hume-Rothery » phase in Ag-Al and Cu-Ge alloys has been performed by KOIKE *et al.* [196]—see also the discussion in ref. [190]. In order to interpret the lattice parameter dependence on e/a for the hexagonal closed-packed phases of these alloys, JONES formulated a long time ago a FS-BZ interaction model along the lines of Hume-Rothery. The change of the c/a lattice ratio and of the electronic specific heat with e/a [186] support this old model, but no FS measurements were performed prior to the ACAR experiments. Using 1D ACAR data, KOIKE *et al.* reconstruct by the Mijnarends method the 3D $\varrho^{2\gamma}(\boldsymbol{p})$ of Cu-Ge and Ag-Al, for several e/a ratios between 1.5 and 1.8, corresponding to the h.c.p. « ζ phases » of these alloys. They find a highly anisotropic FS, having several dominant features that change with e/a; the FS of the two systems show very similar topology for equal e/a; their result supports Jones' model regarding the attractive interaction between the FS and the zone faces.

Finally we should briefly mention that several experiments have been performed on various metal hydrides. The experiments on polycrystalline samples are hard to interpret; the few experiments on oriented crystals such as VD and TaH [197] seem to support the band-theoretic model that predicts that a certain percentage of the hydrogen electrons goes into the conduction band to fill hole pockets in the FS. The question of the interaction between the hydrogen and the positron has not been studied in detail [198], and many

unanswered questions remain in this very interesting application of the ACAR technique.

5·12. *Amorphous alloys.* – A great deal of research has been devoted recently to the study of various properties of amorphous metallic alloys [199]. Most of the experiments designed to study the nature of positron behavior in amorphous alloys use lifetime and Doppler-profile techniques. Although the earlier studies were interpreted as indicating no positron trapping (the observation of a more-or-less single annihilation lifetime for example [200]), almost all subsequent experiments indicate that the positrons are trapped into the « quenched-in » defect regions of the alloy [201]. Most informative for the recent studies has been the coincident momentum (Doppler profile) and lifetime measurement technique [202]. The problem of positron localization in these amorphous systems is a most interesting one, and will be discussed in other lectures at this Fermi School. Fewer experiments have been designed to obtain information on the Fermi momentum of amorphous metals. Perhaps the most thorough study along such lines is that of Shiotani *et al.* [203], who investigated the electronic structure of $Mg_{70}Zn_{30}$ alloys by the Compton-profile as well as the 1D ACAR technique. From the Compton-profile data they conclude that the conduction electrons are nearly-free-electronlike, with a Fermi momentum ((0.77 ± 0.02) a.u.) in agreement with the simple electron gas calculation. In the 1D ACAR data they find evidence for smearing at k_F, which is interpreted as evidence for positron localization. They assume a localized isotropic Gaussian positron wave function $\psi_+(r) = A \exp[-r^2/2r_0^2]$ annihilating with a free-electron Fermi gas, and show that the FT of the 1D ACAR data, *i.e.* $B^{2\gamma}(z)$ (eqs. (3.4), (3.11) and (3.12)), becomes

$$B^{2\gamma}(z) \propto \frac{1}{p_F^2 z^2}\left[\frac{\sin p_F z}{p_F z} - \cos p_F z\right] \exp\left[-\frac{1}{4}\frac{z^2}{r_0^2}\right]. \tag{5.5}$$

This expression is to be compared with that of $B^{2\gamma}$ for a free positron annihilating with a free-electron gas, eq. (3.17); we notice that the effect of positron localization is the same as that of an increased momentum resolution, as already noted for positron trapping in defects [204]. SHIOTANI *et al.* fit eq. (5.4) to the FT of the experimental 1D ACAR curve, to obtain the extent of the positron localization r_0, as well as the « local Fermi momentum » p_F. They find that the positron is more localized at 80 K ($r_0 = 8.6$ a.u.) than at 300 K ($r_0 = 11.6$ a.u.) and that the effective Fermi momentum is quite close to that of the bulk obtained by the Compton-profile measurement. The localization of the positron is, however, much more diffuse even at 80 K than it is in a vacancy ($(2 \div 3)$ a.u.), indicating a shallower and wider trap in amorphous metals. The temperature dependence of this interesting effect needs further study.

5'13. 2D ACAR *study of* V_3Si. – We conclude our discussion of the experimental data in metals and alloys with the example of V_3Si, studied recently with 2D ACAR by the Brandeis [205] as well as the University of Geneva [206] groups. Many of the techniques introduced in the earlier sections of these lectures have been used in these studies.

V_3Si, a high-temperature superconductor, belongs to the family of *A*-15 intermetallic compounds that has been intensively studied during the last decade [207]. Recent experimental indications of the shape of the FS in some *A*-15 compounds have been provided by magneto-thermal data and de Haas-van Alphen (dHvA) experiments [208], suggesting a FS consisting of several nested cylindrical or ellipsoidal hole surfaces aligned along the [100] directions; however, no detailed dHvA experiments exist for V_3Si.

Long-slit measurements on oriented crystals of V_3Si were reported by BERKO and WEGER [209] for several orientations. Fine structure in the derivative dN/dp_z was attributed to planar sheets in the FS normal to [100] as predicted by the linear-chain model [210]. Recently, SAMOILOV and WEGER [211] have performed a similar experiment yielding a derivative structure again attributed to planar FS sheets.

Early 2D data by BERKO and MADER [80], corresponding to a few slices in the 2D (p_y, p_z)-plane, revealed a much more complex FS than the simple planar one used to explain the 1D ACAR results. The full 2D ACAR data were obtained more recently, using the multicounter setup [205]. The raw 2D ACAR data $N(p_y, p_z)$ look like a smooth Gaussian, without revealing any

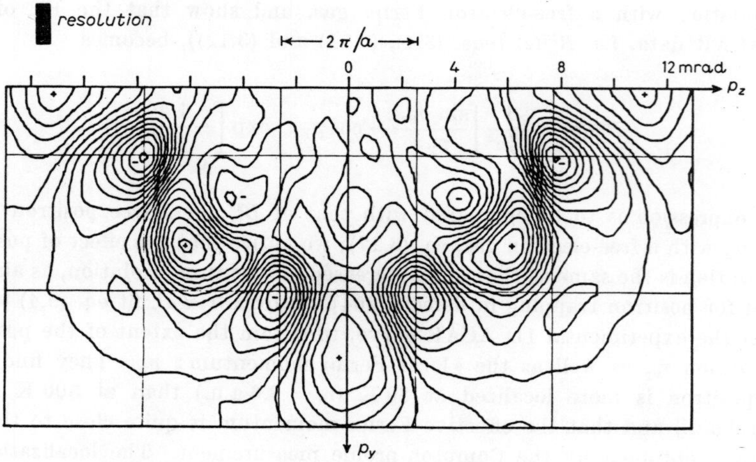

Fig. 34. – Contour map of $N(p_y, p_z) - R(p_y, p_z)$ in V_3Si for the (100) plane, with p_z along the [100] direction, contour interval 0.25% $N(0, 0)$. The resolution of the apparatus is indicated in the black inset (FWHM).

clear anisotropies. The low-intensity (a few percent) FS signal is not surprising in view of the many filled bands. To reveal the structure, we construct the « anisotropic distribution » $N'(p_y, p_z) = N(p_y, p_z) - R(p_y, p_z)$, where R is the isotropic function obtained by a rotation of the data: $R(|\boldsymbol{p}|) = (1/2\pi)\int N(|\boldsymbol{p}|, \theta)\,d\theta$. The contour plot of $N'(p_y, p_z)$ for the (100) plane is shown in fig. 34, exhibiting small but clear deviations from isotropy. The minima at the $(\pm\pi/a, \pm 3\pi/a)$ points are clearly commensurate with the reciprocal lattice, indicating a concentration of hole densities around the corners of the (100) projection of the repeated zone in $\boldsymbol{p}$ space. The direct analysis of these anisotropies requires care because the rotational averaging may cause aliasing of a sharp feature around a circle of the same radius. In fig. 34, the sharp minima can generate a ring of compensatory maxima. For these reasons we carried out a transformation from $\boldsymbol{p}$ space to $\boldsymbol{k}$ space of the full 2D ACAR surface, by forming the reduced density $\tilde{N}(p_y, p_z)$, as discussed in subsect. **5**'8 for Ta. The analysis of $\tilde{N}(p_y, p_z)$ led us to formulate a « cubic jungle gym » FS, consisting of a set of empty cylinders running along the cube edges of the single cubic BZ. This FS model did not fit all the details of $\tilde{N}(p_y, p_z)$, and we decided to produce a full 3D reconstruction of $\varrho^{2\gamma}(\boldsymbol{p})$. Using six orientations rotated about a fixed p_y direction (along [100]) we reconstructed $\varrho^{2\gamma}(\boldsymbol{p})$ in (p_x, p_z)-planes, using the filtered back-projection technique discussed in subsect **3**'4. Finally, we produce the 3D reduced density

$$\tilde{\varrho}^{2\gamma}(\boldsymbol{p}) = \text{const}\cdot \sum_{\boldsymbol{k},j,\boldsymbol{G}} n_j(\boldsymbol{k})\,\delta(\boldsymbol{p}-\boldsymbol{k}-\boldsymbol{G})\Big\{\sum_{\boldsymbol{G}'} |A_j^{+-}(\boldsymbol{k},\boldsymbol{G}')|^2\Big\} \tag{5.6}$$

as explained in subsect. **3**'6 (eq. (3.22)). We assume that the 3D version of the LCW holds, *i.e.* that the sum in the parentheses is constant, or at least varies slowly with respect to $\boldsymbol{k}$ compared to $n_j(\boldsymbol{k})$, the occupation density due to the various FS sheets belonging to the various partially occupied bands j. In fig. 35 we show the experimentally obtained reduced density $\tilde{\varrho}^{2\gamma}(\boldsymbol{p})$ for a section through the center of the zone ($p_y = 0$) and through the face ($p_y = 2.57$ mrad). Once again we see the low electron density along the zone edges confirming the overall topology of the jungle gym model, but more details appear. We have tried to model a more complex FS that reproduces these details. In fig. 35 we also show the contour plots of this FS model, after folding in the effective resolution. Also shown is a theoretical FS computed by Klein *et al.* [212], by a self-consistent APW technique, and again folded into the resolution.

Our FS model consists of a hole structure along the zone edges, made up of nested cylindrical surfaces, that meet at the corners of the zone. An additional electron density placed around the center of the zone faces is also included. (For a sketch of the model FS see ref. [205].) We note in fig. 35 that the density $\tilde{\varrho}^{2\gamma}(\boldsymbol{p})$, based on the model FS, differs somewhat from the experimental result, particularly around the corners of the zone, where the model

Fig. 35. – Comparison between two sections through $\boldsymbol{k}$ space for the reconstructed and « reduced » experimental data, the FS model described in the text and the APW calculation for V_3Si. For all three the lowest electron density in the $k_y = 0$ section ((a)) is at the M point and in the $k_y = \pi/a$ section ((b)) at the R point.

has too much hole density. It is possible that the inner jungle gym does not continue all the way to the corner, but terminates, forming closed ellipsoids along the cube edges. The APW FS also reproduces the overall topology of the observed $\tilde{\varrho}^{2\gamma}(\boldsymbol{p})$; this time the theoretical model indicates more hole surfaces at the R point than observed. Clearly much better-resolution ACAR data will be required to justify a further refinement of the FS model; it is, however, encouraging to see such a good overall topological fit between theory and experiment. A theoretical band computation incorporating the positron wave function would be most desirable, in order to test the extent to which one can trust the approximations inherent in the use of the LCW theorem. The Brandeis 2D ACAR data agree quite well with the data from the University of Geneva [206]; the FS model proposed by that group, based directly on the 2D ACAR results without a full 3D reconstruction, differs, however, in some details from the Brandeis model. Further detailed work will be required to clarify these models. The FS results to date are sufficient, however, to indicate a substantial deviation from the early simple linear-chain model prediction in V_3Si, and favor the APW computations.

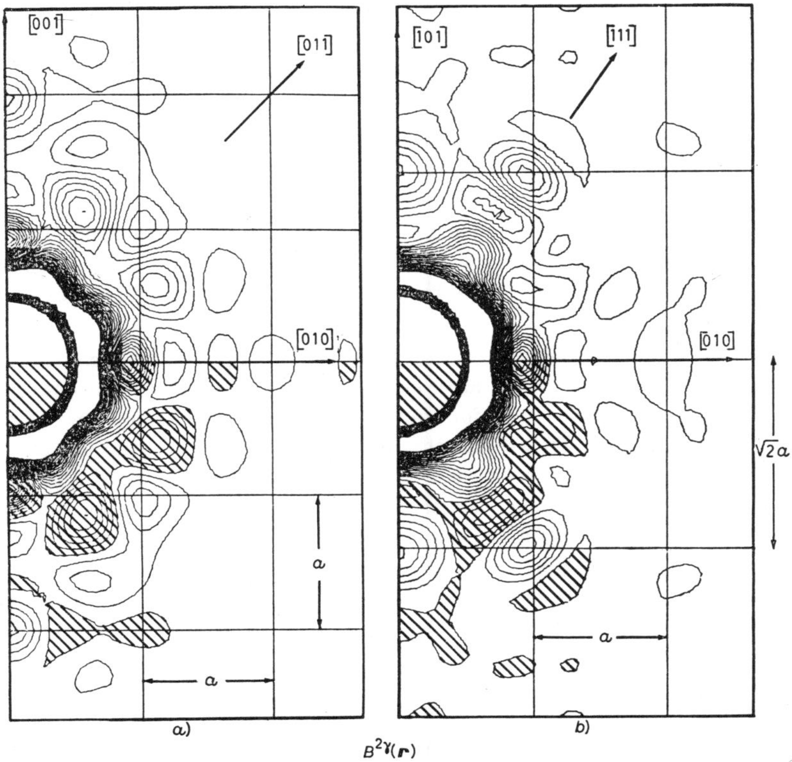

Fig. 36. – Contour map of $B^{2\gamma}(\boldsymbol{r})$ in V_3Si in the *a*) (100) and *b*) (10$\bar{1}$) planes. The regions shaded in the bottom half of the figures have positive $B^{2\gamma}(\boldsymbol{r})$ values. The two white regions around the center correspond to large positive and negative excursions beyond the limit of the maximum or minimum contours used.

Finally, we show in fig. 36 the computed $B^{2\gamma}(\boldsymbol{r})$, the FT of the 2D ACAR surfaces, for the (100) and (10$\bar{1}$) planes of V_3Si [213]. It is clear that the small anisotropies in the 2D ACAR curves are reflected in the strongly anisotropic features of the $B^{2\gamma}(\boldsymbol{r})$ function. As was discussed in subsect. **3**.2, **3**.4 and **3**.5, the $B^{2\gamma}(\boldsymbol{r})$ can serve different purposes: it can be used to reconstruct mathematically the 3D $\varrho^{2\gamma}(\boldsymbol{p})$, and its values at the lattice points $\boldsymbol{R}$ can be used to obtain the reduced density $\tilde{\varrho}^{2\gamma}(\boldsymbol{p})$ via eqs. (3.20) and (3.24); in itself it contains important physical information, since it is the autocorrelation of the positron-sampled electron wave functions (subsect. **3**.2). We note from fig. 36 that several features of $B^{2\gamma}(\boldsymbol{r})$ are indeed centered close to the various interatomic vectors in V_3Si. We also note that the locus of the first zero crossing coincides surprisingly well with the closest V-V interatomic distance ($a/2$); we fail to understand at this stage, however, why these zeros should lie on a nearly spherical pattern.

It is clear that the behavior of the $B^{2\gamma}(\boldsymbol{r})$ function will become an important clue when comparing the experimental data with various band-theoretic computations, as has been done in Compton-profile analysis. Comparison between different data sets can also be facilitated by the use of $B^{2\gamma}(\boldsymbol{r})$, because of the simplicity with which different resolutions influence this function—*i.e.* by a smooth multiplicative function. We can, therefore, look forward to an increased use of the $B^{2\gamma}$ function in the analysis of ACAR data.

With this section on V_3Si we finish our review of ACAR experiments in metals and alloys. We shall now discuss very briefly the status of momentum density measurements by ACAR techniques in some nonmetallic solids.

5.14. ACAR *experiments in semiconductors and in other solids.* – Several 1D and 2D ACAR experiments have been performed in various semiconductors, particularly in Si and Ge—for details see the general reviews [29-31]. Since the early 1D ACAR experiments on Si [214], the substantial anisotropies observed in semiconductors have been correlated with the cross-sections of the « Jones zone », assuming that in a semiconductor with reasonably small energy gaps $\varrho^{2\gamma}(\boldsymbol{p})$ is determined mainly by the occupation of states in $\boldsymbol{k}$ space. Stroud and Ehrenreich [215] improved greatly on this simple model, by performing a pseudopotential computation that included the positron wave function. High-precision long-slit ACAR data in Si by Shulman *et al.* [216] revealed a minimum at the origin of $N(p_z)$, thus showing a large deviation from the simple Jones-zone picture. In addition, the Umklapp components at high momenta were also observed. The small differences obtained between 5 K and 300 K were attributed to the thermal motion of the positron, and the results were analyzed with a derivative technique to obtain a positron effective mass in Ge of $m^* = (1.23 \pm 0.17)m$; one expects a smaller mass in semiconductors than in metals (subsect. **5**.4.3), because of the less efficient electron screening due to the presence of the energy gap. The first detailed 2D

ACAR experiments in Ge [80] revealed an even larger anisotropy than that obtained with the long-slit geometry; this anisotropy was confirmed in the recent full 2D ACAR measurement by WEST and collaborators [217]. Theoretical work is needed to analyze these newer 2D results.

The applicability of the LCW theorem to semiconductors was the topic of another positron study [218]. The theorem was found not to be obeyed in detail: the reduced density $\tilde{N}(p_z)$ did not stay a constant for all directions of p_z, a deviation presumably due to the presence of the positron. Recently, the $B^{2\gamma}(\boldsymbol{r})$ function has been studied for Ge [213], indicating some deviations from the predicted zero crossings for insulators (see subsect. **3**˙5). Much experimental and theoretical work on the electron momentum density of semiconductors has been performed by investigators using the Compton-profile technique [219].

In some insulators, such as quartz, as well as in some alkali halides, a certain fraction of positrons was found to form Ps in a delocalized Bloch state. The physics of such quasi-Ps states is most interesting, and is the topic of the lectures by DUPASQUIER at this Fermi School. Annihilation from the Ps state leads to a nearly delta-function-like narrow distribution, which can be used effectively to align the ACAR machines and measure their angular resolution [4].

A most interesting set of experiments has been recently performed on graphite intercalation compounds [220]. The remarkable anisotropic properties of these « quasi-two-dimensional » layered structures have been the subject of intense solid-state research during the last several years [221], and it is pleasing to see ACAR techniques applied to such an up-to-date topic. It is amusing to note that these experiments make contact with the very first ACAR measurement on an oriented single crystal, the 1D ACAR study of graphite performed twenty-five years ago [11]. The large anisotropic results found in that early study—which remain the strongest anisotropies observed in any crystal—were interpreted as being due to preferential annihilation with the π electrons of graphite, since the positrons tend to reside in the regions between the carbon layers of graphite. Because of this tendency, the positron is very sensitive to the charge distribution of the « intercalates »; the large changes observed upon intercalation were successfully explained as being due to « positron-halide states », such as $[Br^-; e^+]$ and $[Cl^-; e^+]$. Experiments with alkali metal intercalates prove most interesting; questions of the possible formation of quasi-2D Fermi distributions can be studied.

5˙15. *Future prospects*. – With the advent of high-efficiency, high-resolution 2D ACAR machines, the prospects for interesting new applications look most promising. Clearly, high resolution is of paramount importance in the study of more and more complex Fermi surfaces. The high-precision study of the small changes in Fermi-surface dimensions upon alloying calls for increased counting efficiency. One hopes that it will become possible to build high-density proportional wire chambers of increased size, with multiple lead converters for increased efficiency. Besides future high-precision studies on the more standard metallic systems discussed in the bulk of our lectures, much 2D ACAR work remains to be done on various more unusual systems of current interest to solid-state physics, such as the intercalation compounds, the layered dichalcogenides, metallic transition metal oxide compounds, etc. Recent 1D ACAR work on metal-oxide compounds, such as the tungsten bronzes Na_xWO_3 [222], indicates a complex FS topology, clearly requiring high-resolution 2D ACAR studies.

Fundamental studies on the momentum-dependent e^+-e^- many-body interaction, the positron effective mass, etc. will no doubt be continued, with improved statistics and better 3D reconstruction algorithms. In the near future we shall also see the 2D ACAR technique applied to various defect work in search for crystal orientation effects, as well as to temperature-dependent phenomena such as anisotropic phonon effects, various phase transitions and so on.

With the impressive developments of the last few years in producing variable-energy slow-positron beams of increasing intensity, one hopes to be able to apply ACAR techniques to the study of the various e^+-surface interaction phenomena. Given sufficiently high-intensity slow-e^+ beams, 2D ACAR measurements to study the nature of the recently observed positron surface state could be performed; the study of atoms and molecules on solid surfaces can also be anticipated.

Let me, therefore, end my lectures by quoting the closing sentence of A. T. Stewart's invited talk at the First International Conference on Positron Annihilation in 1965: « There seems to be endless possibilities for imaginative, interesting and useful work in the subject of positron annihilation in metals ».

* * *

I thank K. R. Hoffmann and R. R. Lee for proofreading the manuscript of these lectures, and M. Fricks and J. Jordan for their hard work and endless patience in typing it. I am grateful to F. Sinclair for many illuminating discussions regarding various aspects of positron physics and for a critical reading of the manuscript.

I wish to express my gratitude to W. Brandt and A. Dupasquier for inviting me to present these lectures. The work was supported by the National Science Foundation under Grant DMR-79-26035.

REFERENCES

[1] For brief reviews of the early history of positron physics, see papers by S. DeBenedetti and A. T. Stewart: in *Positron Annihilation*, edited by A. T. Stewart and L. O. Roellig (New York, N. Y., 1967), p. 3 and 17.

[2] For latest review articles, see *Positron Annihilation—Proceedings of the V International Conference on Positron Annihilation*, edited by R. R. HASIGUTI and K. FUJIWARA (Sendai, 1979).
[3] R. BERINGER and C. G. MONTGOMERY: *Phys. Rev.*, **61**, 222 (1942).
[4] S. BERKO: in ref. [2], p. 65.
[5] S. BERKO and F. L. HEREFORD: *Rev. Mod. Phys.*, **28**, 299 (1956).
[6] R. A. FERRELL: *Rev. Mod. Phys.*, **28**, 308 (1956).
[7] S. DEBENEDETTI, C. E. COWAN, W. R. KONNEKER and H. PRIMAKOFF: *Phys. Rev.*, **77**, 205 (1950).
[8] G. LANG and S. DEBENEDETTI: *Phys. Rev.*, **108**, 914 (1957).
[9] A. T. STEWART: *Can. J. Phys.*, **35**, 168 (1957).
[10] S. BERKO and J. S. PLASKETT: *Phys. Rev.*, **112**, 1877 (1958).
[11] S. BERKO, R. E. KELLEY and J. S. PLASKETT: *Phys. Rev.*, **106**, 824 (1957).
[12] G. E. LEE-WHITING: *Phys. Rev.*, **97**, 1557 (1955).
[13] E. DANIEL and J. FRIEDEL: *Phys. Chem. Solids*, **4**, 111 (1958).
[14] W. BRANDT: in ref. [1], p. 155.
[15] A. T. STEWART: in ref. [1], p. 17.
[16] I. YA. DEKHTYAR, D. A. LEVINA and V. S. MIKHALENKOV: *Sov. Phys. Dokl.*, **9**, 492 (1964).
[17] S. BERKO and J. C. ERSKINE: *Phys. Rev. Lett.*, **19**, 307 (1967).
[18] A. T. STEWART, J. H. KUSMISS and R. H. MARCH: *Phys. Rev.*, **132**, 495 (1963).
[19] I. K. MACKENZIE, G. F. O. LANGSTROTH, B. T. A. MCKEE and C. G. WHITE: *Can. J. Phys.*, **42**, 1837 (1964).
[20] I. K. MACKENZIE, T. L. KHOO, A. B. MCDONALD and B. T. A. MCKEE: *Phys. Rev. Lett.*, **19**, 946 (1967).
[21] D. C. CONNORS and R. N. WEST: *Phys. Lett. A*, **30**, 24 (1969).
[22] B. BERGERSEN and M. J. STOTT: *Solid State Commun.*, **7**, 1203 (1969).
[23] See, for example, M. DOYAMA: in ref. [2], p. 13, and references therein.
[24] For a review of positron thermalization see B. BERGERSEN and E. PAJANNE: *Appl. Phys.*, **4**, 25 (1974); for a general review of positron dynamics in solids see W. BRANDT: *Appl. Phys.*, **5**, 1 (1974).
[25] W. BRANDT and R. PAULIN: *Phys. Rev. B*, **15**, 2511 (1977).
[26] For a brief review of slow-positron beam production see the positronium review of S. BERKO and H. N. PENDLETON: *Annu. Rev. Nucl. Part. Sci.*, **30**, 543 (1980).
[27] For an account of QED selection rules, as well as fundamental positronium physics, see ref. [26].
[28] For an early account and the outline of the theory of spin-polarized experiments see S. BERKO: in ref. [1], p. 61.
[29] R. N. WEST: *Adv. Phys.*, **22**, 263 (1973).
[30] S. BERKO: in *Compton Scattering*, edited by B. WILLIAMS (London, 1977), p. 273.
[31] P. E. MIJNARENDS: in *Positrons in Solids*, edited by P. HAUTOJARVI (Berlin, 1979), p. 25.
[32] See review papers in *Positron Annihilation*, edited by P. C. JAIN and R. M. SINGRU (New Delhi, 1979); see also reviews in ref. [2].
[33] R. W. SIEGEL: *Annu. Rev. Mater. Sci.*, **10**, 393 (1980).
[34] For a brief « mini-review », see S. BERKO: *Scr. Metall.*, **14**, 23 (1980).
[35] CHANG LEE: *Žurn Ėxp. Teor. Fiz*, **33**, 365 (1957).
[36] P. PLATZMAN and N. TZOAR: in *Compton Scattering*, edited by B. WILLIAMS (London, 1977), p. 28.
[37] J. W. M. DUMOND: *Rev. Mod. Phys.*, **5**, 1 (1933).
[38] R. H. STUEWER and M. J. COOPER: in ref. [30].

[39] V. H. Smith jr.: in *Electron and Magnetization Densities in Molecules and Crystals*, edited by P. Becker (New York, N. Y., 1980), p. 3.
[40] R. Benesch, S. R. Singh and V. H. Smith jr.: *Chem. Phys. Lett.*, **10**, 151 (1971).
[41] P. Pattison and B. Williams: *Solid State Commun.*, **20**, 585 (1976).
[42] P. Thulstrup: *J. Chem. Phys.*, **65**, 3386 (1976).
[43] W. Schülke: *Phys. Status Solidi B*, **82**, 229 (1977); **80**, K67 (1977).
[44] W. Schülke: *Jpn. J. Appl. Phys.*, **17**, *Supp.* 17-2, 332 (1978).
[45] W. Weyrich, P. Pattison and B. G. Williams: *Chem. Phys.*, **41**, 271 (1979).
[46] M. J. Stott and R. N. West: *J. Phys. F*, **8**, 635 (1978).
[47] W. Brandt, L. Eder and S. Lundqvist: *Phys. Rev.*, **142**, 165 (1966).
[48] P. E. Mijnarends: *Physica (Utrecht)*, **63**, 235 (1973).
[49] R. Harthoorn and P. E. Mijnarends: *J. Phys. F*, **8**, 1147 (1978), and references therein.
[50] P. E. Mijnarends: *Phys. Rev.*, **160**, 512 (1967); **178**, 622 (1969).
[51] P. E. Mijnarends: in ref. [30], p. 323.
[52] See, for example, review by J. A. Correia, C. A. Burnham, D. A. Chesler, D. R. Elmaleh, N. M. Alpert and G. L. Brownell: in ref. [2], p. 391, and references therein.
[53] See the review by R. A. Brooks and G. Di Chiro: *Phys. Med. Biol.*, **21**, 698 (1976).
[54] *Image Reconstruction from Projections*, edited by G. T. Herman (Berlin, 1979).
[55] E. L. Hall: *Computer Image Processing and Recognition* (New York, N. Y., 1979).
[56] J. Radon: *Ber. Verh. Sächs. Akad. Wiss. Leipzig, Math. Phys. Kl.*, **69**, 262 (1917).
[57] See, for example, R. N. Bracewell: *Proc. IRE*, **46**, 106 (1958); see also R. N. Bracewell: in ref. [54].
[58] F. M. Mueller: *Phys. Rev. B*, **15**, 3039 (1977).
[59] C. K. Majumdar: *Phys. Rev. B*, **4**, 2111 (1971).
[60] L. Pecora and A. C. Ehrlich: *Phys. Rev. B*, **19**, 719 (1979).
[61] N. K. Hansen: Report No. B342, Hahn-Meitner-Institute (Berlin, 1980).
[62] D. A. Chesler and S. J. Riederer: *Phys. Med. Biol.*, **20**, 632 (1975).
[63] W. Farmer: Ph. D. Thesis, Brandeis University (1981).
[64] D. G. Lock, V. H. C. Crisp and R. N. West: *J. Phys. F*, **3**, 561 (1973).
[65] D. G. Lock and R. N. West: *Appl. Phys.*, **6**, 249 (1975).
[66] G. M. Beardsley, S. Berko, J. J. Mader and M. A. Shulman: *Appl. Phys.*, **5**, 375 (1975).
[67] P. E. Mijnarends: *J. Appl. Phys.*, **40**, 3027 (1969); see also ref. [31] and other references therein.
[68] E. A. Lorch: in ref. [2], p. 403.
[69] R. M. Singru: *Phys. Status Solidi A*, **30**, 11 (1975).
[70] R. M. Singru: *Phys. Lett. A*, **46**, 61 (1973).
[71] K. G. Lynn, J. R. MacDonald, R. A. Boie, L. C. Feldman, J. D. Gabbe, M. F. Robbins, E. Bonderup and J. Golovchenko: *Phys. Rev. Lett.*, **38**, 241 (1977).
[72] J. R. MacDonald, K. G. Lynn, R. A. Boie and M. F. Robbins: *Nucl. Instrum. Methods*, **153**, 189 (1978).
[73] I. K. MacKenzie and P. Sen: *Phys. Rev. Lett.*, **37**, 1296 (1976).
[74] K. Hinode, S. Tanigawa, Y. Suzaki, K. Ohno and M. Doyama: in ref. [2], p. 703.
[75] *Compton Scattering*, edited by B. Williams (London, 1977).
[76] R. J. Weiss, W. A. Reed and P. Pattison: in ref. [75], p. 43.
[77] M. J. Cooper: *Nucl. Instrum. Methods*, **166**, 21 (1979).
[78] J. R. Schneider, P. Pattison and H. A. Graf: *Nucl. Instrum. Methods*, **166**, 1 (1979), and references therein; P. Pattison and J. R. Schneider: *Nucl. Instrum. Methods*, **158**, 145 (1979).

[79] G. Loupias, J. Petiau, A. Issolah and M. Schneider: *Phys. Status Solidi B*, **102**, 79 (1980).
[80] S. Berko and J. Mader: *Appl. Phys.*, **5**, 287 (1975).
[81] S. Berko, G. M. Beardsley, M. Haghgooie, I. Tal and J. Mader: Paper H13 of the *IV International Conference on Positron Annihilation* (Helsingor, 1976).
[82] S. Berko, M. Haghgooie and J. J. Mader: *Phys. Lett. A*, **63**, 335 (1977).
[83] W. S. Farmer, F. Sinclair, S. Berko and G. M. Beardsley: *Solid State Commun.*, **31**, 481 (1979).
[84] P. Colombino, B. Fiscella and L. Trossi: *Nuovo Cimento*, **27**, 589 (1963); **31**, 950 (1964).
[85] K. Fujiwara and O. Sueoka: *J. Phys. Soc. Jpn.*, **21**, 1947 (1966).
[86] D. Ll. Williams, P. Petijevich and G. Jones: *Bull. Am. Phys. Soc.*, **10**, 1181 (1965).
[87] O. Sueoka: *J. Phys. Soc. Jpn.*, **23**, 1246 (1967).
[88] D. Ll. Williams, E. H. Becker, P. Petijevich and G. Jones: *Phys. Rev. Lett.*, **20**, 448 (1968).
[89] A. A. Manuel, S. Samoilov, O. Fischer and M. Peter: in ref. [2], p. 685; A. Jeavons: in ref. [2], p. 355.
[90] J. Mayers, J. D. McGervey, P. A. Walters and R. N. West: in ref. [2], p. 417; R. N. West: *Inst. Phys. Conf. Ser.*, **55**, 35 (1980).
[91] G. Riepe, D. Protic, R. Kurz, W. Triftshäuser, Zs. Kajcsos and J. Winter: in ref. [2], p. 371.
[92] R. J. Douglas and A. T. Stewart: Paper H14 of the *IV International Conference on Positron Annihilation* (Helsingor, 1976).
[93] H. O. Anger: *Rev. Sci. Instrum.*, **29**, 27 (1958).
[94] H. O. Anger: in *Instrumentation in Nuclear Medicine*, Vol. **2**, edited by G. J. Hine and J. A. Sorenson (New York, N. Y., 1974), p. 61.
[95] W. Triftshäuser: in *Proceedings of the II International Conference on Positron Annihilation* (Kingston, Ont., 1971), 4.77.
[96] A. P. Jeavons, D. W. Townsend, N. L. Ford, K. Kull, A. Manuel, O. Fischer and M. Peter: *IEEE Trans. Nucl. Sci.*, NS-**25**, 164 (1978).
[97] A. A. Manuel, O. Fischer, M. Peter and A. P. Jeavons: *Nucl. Instrum. Methods*, **156**, 67 (1978).
[98] A. A. Manuel, S. Samoilov, O. Fischer, M. Peter and A. P. Jeavons: *Helv. Phys. Acta*, **52**, 255 (1979).
[99] C. K. Majumdar: *Phys. Rev. Sect. A*, **140**, 227 (1965).
[100] See, for example, P. Paatero, S. Manninen and T. Paakkari: *Philos. Mag.*, **30**, 1281 (1974).
[101] G. E. Lee-Whiting: *Phys. Rev.*, **97**, 1557 (1955).
[102] J. P. Carbotte and H. L. Arora: *Can. J. Phys.*, **45**, 387 (1967).
[103] E. J. Woll jr. and J. P. Carbotte: *Phys. Rev.*, **164**, 985 (1967).
[104] H. J. Mikeska: *Z. Phys.*, **232**, 159 (1970).
[105] B. B. J. Hede and J. P. Carbotte: *Can. J. Phys.*, **48**, 2661 (1970).
[106] A. Perkins and J. P. Carbotte: *Phys. Rev. B*, **1**, 101 (1970).
[107] B. Bergersen and E. Pajanne: *Appl. Phys.*, **4**, 25 (1974).
[108] S. M. Kim, A. T. Stewart and J. P. Carbotte: *Phys. Rev. Lett.*, **18**, 385 (1967); S. M. Kim and A. T. Stewart: *Phys. Rev. B*, **11**, 2490 (1975).
[109] P. Kubica and A. T. Stewart: *Phys. Rev. Lett.*, **34**, 852 (1975).
[110] M. J. G. Lee: *Crit. Rev. Solid State Sci.*, **2**, 85 (1971).
[111] S. Kahana: *Phys. Rev.*, **129**, 1622 (1963); see also the review by S. Kahana in ref. [1], p. 51; J. P. Carbotte: *Phys. Rev.*, **155**, 197 (1967).
[112] H. Weisberg and S. Berko: *Phys. Rev.*, **154**, 249 (1967).

[113] P. BHATTACHARYYA and K. S. SINGWI: *Phys. Rev. Lett.*, **29**, 22 (1972); *Phys. Lett. A*, **41**, 457 (1972).
[114] A. SJÖLANDER and M. J. STOTT: *Phys. Rev. B*, **5**, 2109 (1972).
[115] J. CROWELL, V. E. ANDERSON and R. H. RITCHIE: *Phys. Rev.*, **150**, 243 (1966).
[116] J. ARPONEN and E. PAJANNE: *J. Phys. F*, **9**, 2359 (1979); *J. Phys. C*, **12**, L161 (1979); *Ann. Phys. (N. Y.)*, **121**, 343 (1979).
[117] P. F. MALDAGUE: *Phys. Rev. B*, **20**, 21 (1979).
[118] E. BORONSKI, Z. SZOTEK and H. STACHOWIAK: *Phys. Rev. B*, **23**, 1785 (1981).
[119] W. BRANDT, D. ISAACSON and R. PAULIN: *Phys. Rev. Lett.*, **35**, 1180 (1975).
[120] I. K. MACKENZIE, T. E. JACKMAN and N. THRANE: *Phys. Rev. Lett.*, **34**, 512 (1975).
[121] W. BRANDT and J. REINHEIMER: *Phys. Lett. A*, **35**, 109 (1971).
[122] I. K. MACKENZIE, R. LEBLANC and B. T. A. MCKEE: *Phys. Rev. Lett.*, **27**, 580 (1971).
[123] P. KUBICA and M. J. STOTT: *Can. J. Phys.*, **53**, 450 (1975).
[124] J. J. DONAGHY and A. T. STEWART: *Phys. Rev.*, **164**, 396 (1967).
[125] D. R. GUSTAFSON and J. D. WILLENBERG: *Phys. Rev. B*, **13**, 5193 (1976).
[126] F. SINCLAIR, W. FARMER and S. BERKO: to be published.
[127] J. A. ARIAS-LIMONTA and P. G. VARLASHKIN: *Phys. Rev. B*, **1**, 142 (1970).
[128] For the earliest m^* measurement see A. T. STEWART, J. B. SHAND and S. M. KIM: *Proc. Phys. Soc.*, **88**, 1001 (1966).
[129] C. K. MAJUMDAR: *Phys. Rev.*, **149**, 406 (1966).
[130] D. R. HAMANN: *Phys. Rev.*, **146**, 277 (1966); B. BERGERSEN and E. PAJANNE: *Phys. Rev.*, **186**, 375 (1969).
[131] M. BALDO and R. PUCCI: unpublished result referred to in P. KUBICA: Ph. D. Thesis, Queen's University, Canada (1975) (unpublished).
[132] H. J. MIKESKA: *Z. Phys.*, **232**, 159 (1970); B. BERGERSEN and E. PAJANNE: *Phys. Rev. B*, **3**, 1588 (1971).
[133] S. L. BASINSKI, R. J. DOUGLAS and A. T. STEWART: in ref. [2], p. 665.
[134] J. J. PACIGA and D. L. WILLIAMS: *Can. J. Phys.*, **49**, 3227 (1971).
[135] S. B. NICKERSON and S. H. VOSKO: *Phys. Rev. B*, **14**, 4399 (1976).
[136] A. T. STEWART, J. B. SHAND, J. J. DONAGHY and J. H. KUSMISS: *Phys. Rev.*, **128**, 118 (1962).
[137] S. BERKO: *Phys. Rev.*, **128**, 2166 (1962).
[138] O. MOGENSEN and K. PETERSEN: *Phys. Lett. A*, **30**, 542 (1969).
[139] G. KONTRYM-SZNAJD and H. STACHOWIAK: *Appl. Phys.*, **5**, 361 (1975).
[140] Z. PAWLOWSKA: *Phys. Status Solidi B*, **97**, 197 (1980).
[141] N. SHIOTANI, T. OKADA, H. SEKIZAWA and S. WAKOH: *J. Phys. Soc. Jpn.*, **50**, 498 (1981).
[142] J. P. CARBOTTE and A. SALVADORI: *Phys. Rev.*, **162**, 290 (1967).
[143] P. PATTISON, S. MANNINEN, J. FELSTEINER and M. COOPER: *Philos. Mag.*, **30**, 973 (1974).
[144] P. HAUTOJÄRVI: *Solid State Commun.*, **11**, 1049 (1972).
[145] D. STROUD and H. EHRENREICH: *Phys. Rev.*, **171**, 399 (1968); A. G. GOULD, R. N. WEST and B. G. HOGG: *Can. J. Phys.*, **50**, 2294 (1972).
[146] P. KUBICA and M. J. STOTT: *J. Phys. F*, **4**, 1969 (1974).
[147] E. BONDERUP, J. U. ANDERSEN and D. N. LOWY: *Phys. Rev. B*, **20**, 883 (1979).
[148] S. BERKO, F. SINCLAIR and W. FARMER: to be published.
[149] R. P. GUPTA and R. W. SIEGEL: *J. Phys. F*, **9**, 2353 (1979).
[150] J. MADER, S. BERKO, H. KRAKAUER and A. BANSIL: *Phys. Rev. Lett.*, **37**, 1232 (1976).
[151] J. MADER, M. HAGHGOOIE and S. BERKO: to be published.

[152] S. WAKOH: in ref. [2], p. 49.
[153] B. B. J. HEDE and J. P. CARBOTTE: *J. Phys. Chem. Solids*, **33**, 727 (1972).
[154] B. CHAKRABORTY: *Phys. Rev. B*, **24**, 7423 (1981).
[155] K. FUJIWARA: *J. Phys. Soc. Jpn.*, **29**, 1479 (1970); K. FUJIWARA, T. HYODO and J. OHYAMA: *J. Phys. Soc. Jpn.*, **33**, 1047 (1972).
[156] M. R. HALSE: *Philos. Trans. R. Soc. London Ser. A*, **265**, 507 (1969).
[157] S. BERKO, S. CUSHNER and J. C. ERSKINE: *Phys. Lett. A*, **27**, 668 (1968); S. CUSHNER, J. C. ERSKINE and S. BERKO: *Phys. Rev. B*, **1**, 2852 (1970).
[158] B. ROZENFELD, S. CHABIK, J. PAJAK, K. JERIE and W. WIERZCHOWSKI: *Acta Phys. Pol. A*, **44**, 21 (1973).
[159] M. R. HOWELLS and P. E. OSMON: *J. Phys. F*, **2**, 277 (1972).
[160] M. HAGHGOOIE, J. J. MADER and S. BERKO: *Phys. Lett. A*, **69**, 293 (1978).
[161] S. WAKOH, S. BERKO, M. HAGHGOOIE and J. J. MADER: *J. Phys. F*, **9**, L231 (1979).
[162] P. E. MIJNARENDS and R. M. SINGRU: *Phys. Rev. B*, **19**, 6038 (1979).
[163] M. SOB: in *Proceedings of the VIII Annual International Symposium on the Electronic Structure of Metals and Alloys*, Gaussig, GDR, edited by P. ZIESCHE (1978), p. 170.
[164] P. E. MIJNARENDS: private communication.
[165] S. WAKOH, T. FUKAMACHI, S. HOSOYA and J. YAMASHITA: *J. Phys. Soc. Jpn.*, **38**, 1601 (1975).
[166] For a recent review of transition metals, see A. R. MACKINTOSH and O. K. ANDERSEN: in *Electrons at the Fermi Surface*, edited by M. SPRINGFORD (Cambridge, 1980), p. 149.
[167] D. G. KANHERE and R. M. SINGRU: *J. Phys. F*, **7**, 2603 (1977).
[168] F. SINCLAIR, K. R. HOFFMANN and S. BERKO: to be published.
[169] M. H. HALLORAN, J. H. CONDON, J. E. GRAEBNER, J. E. KUNZLER and F. S. L. HSU: *Phys. Rev. B*, **1**, 366 (1970).
[170] See, for example, the review by A. J. FREEMAN: in *Magnetic Properties of Rare Earth Metals*, edited by R. J. ELLIOT (New York, N. Y., 1972), p. 245; also B. COGBLIN: *The Electronic Structure of Rare-Earth Metals and Alloys: the Magnetic Heavy Rare Earths* (New York, N. Y., 1977); also S. H. LIU: in *Handbook on the Physics and Chemistry of Rare Earth Metals*, Vol. **1**, *Metals*, edited by K. A. GSCHNEIDER jr. and L. EYRING (Amsterdam, 1978), p. 233.
[171] D. R. GUSTAFSON and A. R. MACKINTOSH: *J. Phys. Chem. Solids*, **25**, 389 (1964); R. W. WILLIAMS and A. R. MACKINTOSH: *Phys. Rev.*, **168**, 679 (1968), and references therein.
[172] S. C. KEETON and T. L. LOUCKS: *Phys. Rev.*, **168**, 672 (1968).
[173] R. P. GUPTA and T. L. LOUCKS: *Phys. Rev.*, **176**, 848 (1968), and references therein.
[174] K. R. HOFFMANN, S. BERKO and B. J. BEAUDRY: to be published.
[175] B. N. HARMON and A. J. FREEMAN: *Phys. Rev. B*, **10**, 1979 (1974).
[176] P. G. MATTOCKS and R. C. YOUNG: *J. Phys. F*, **7**, 1219 (1977).
[177] S. BERKO and A. P. MILLS: *J. Phys. (Paris)*, **32**, C1-287 (1971).
[178] Y. NAKAO: *Appl. Phys.*, **7**, 81 (1975).
[179] P. E. MIJNARENDS: *Physica (Utrecht)*, **63**, 248 (1973), and references therein.
[180] C. HOHENEMSER, J. M. WEINGART and S. BERKO: *Phys. Lett. A*, **28**, 41 (1968).
[181] See *Electron and Magnetization Densities in Molecules and Crystals*, edited by P. BECKER (New York, N. Y., 1980).
[182] N. SAKAI and K. ÔNO: *J. Phys. Soc. Jpn.*, **42**, 770 (1977).
[183] For reviews of the subject see *Electrons in Disordered Metals and at Metallic Surfaces*, edited by P. PHARISEAU, B. L. GYÖRFFY and L. SCHEIRE, NATO Adv. Study Inst. Ser., Ser. B, **42** (New York, N. Y., 1979).

[184] R. Prasad, S. C. Papadopoulos and A. Bansil: *Phys. Rev. B*, **23**, 2607 (1981), and references therein.
[185] G. M. Stocks, W. M. Temmerman and B. L. Györffy: *Phys. Rev. Lett.*, **41**, 339 (1978); W. M. Temmerman, B. L. Györffy and G. M. Stocks: *J. Phys. F*, **8**, 2461 (1978).
[186] For a recent review of the experimental literature, see D. J. Sellmeyer: in *Solid State Physics*, Vol. **33**, edited by H. Ehrenreich, F. Seitz and D. Turnbull (New York, N. Y., 1978), p. 83; see also T. B. Massalski and U. Mizutani: *Prog. Mater. Sci.*, **22**, 151 (1978).
[187] P. E. Mijnarends and A. Bansil: *Phys. Rev. B*, **19**, 2912 (1979); see also the CPA treatment by A. Bansil, R. S. Rao, P. E. Mijnarends and L. Schwartz: *Phys. Rev. B*, **23**, 3608 (1981).
[188] K. M. Hong and J. P. Carbotte: *Can. J. Phys.*, **55**, 1335 (1977); see also J. P. Carbotte: in ref. [2], p. 629.
[189] S. Berko: in ref. [183], p. 239.
[190] M. Hasegawa, M. Hirabayashi and S. Koike: in ref. [2], p. 673.
[191] A. T. Stewart: *Phys. Rev. Sect. A*, **133**, 1651 (1964); P. Kubica, B. T. A. McKee, A. T. Stewart and M. J. Stott: *Phys. Rev. B*, **11**, 11 (1975).
[192] M. Haghgooie, S. Berko and U. Mizutani: in ref. [2], p. 291.
[193] M. Hasegawa, To. Suzuki and M. Hirabayashi: *J. Phys. Soc. Jpn.*, **37**, 85 (1974).
[194] B. Gordon, W. M. Temmerman, B. L. Györffy and G. M. Stocks: *Inst. Phys. Conf. Ser.*, **39**, 402 (1978).
[195] M. Hasegawa, To. Suzuki and M. Hirabayashi: *J. Phys. Soc. Jpn.*, **43**, 89 (1977).
[196] S. Koike, M. Hasegawa, To. Suzuki and M. Hirabayashi: in ref. [2], p. 275; also, to be published in *Philos. Mag. B.*
[197] M. Hasegawa, S. Koike, M. Hirabayashi, H. Asano and To. Suzuki: *J. Phys. Soc. Jpn.*, **46**, 481 (1979); see also ref. [190].
[198] See, however, the discussion by I. Ya Dekhtyar and V. I. Shevchenko: *Phys. Status Solidi B*, **83**, 323 (1977).
[199] See, for example, the review by G. S. Cargill III: in *Solid State Physics*, Vol. **30**, edited by H. Ehrenreich, F. Seitz and D. Turnbull (New York, N. Y., 1975); *Metallic Glasses*, edited by H. J. Güntherodt and H. Beck (Berlin, 1980).
[200] S. Y. Chuang, S. J. Tao and H. S. Chen: *J. Phys. F*, **5**, 1681 (1975).
[201] See the review by P. Hautojarvi and J. Yli-Kaupila: in *Proceedings of the International Conference on Amorphous Systems Investigated by Nuclear Methods*, edited by Zs. Kajcsos, I. Dezsi, D. Horvath, T. Kernéuy, L. Marczis and D. L. Nagy, to be published in *Nucl. Instrum. Methods.*
[202] K. Suzuki, F. Itoh, M. Hasegawa, T. Fukunaga and T. Honda: in ref. [2], p. 861.
[203] N. Shiotani, N. Sakai, H. Sekizawa and T. Mizoguchi: *J. Phys. Soc. Jpn.*, **50**, 828 (1981).
[204] M. A. Shulman and S. Berko: in *Abstracts of the XIV International Conference on Positron Annihilation* (Helsingor, 1976), E31.
[205] S. Berko, W. S. Farmer and F. Sinclair: in *Superconductivity in d- and f-Band Metals*, edited by H. Suhl and M. G. Maple (New York, N. Y., 1980), p. 281; W. S. Farmer, F. Sinclair, S. Berko and G. M. Beardsley: *Solid State Commun.*, **31**, 481 (1979); see also ref. [4].
[206] A. A. Manuel, S. Samoilov, R. Sachot, P. Descouts and M. Peter: in *Superconductivity in d- and f-Band Metals*, edited by H. Suhl and M. G. Maple (New York, N. Y., 1980), p. 273; A. A. Manuel, S. Samoilov, R. Sachot, P. Descouts and M. Peter: *Solid State Commun.*, **31**, 955 (1979).
[207] See, for example, the review by M. Weger and I. B. Goldberg: in *Solid State*

Physics, Vol. **28**, edited by H. EHRENREICH, F. SEITZ and D. TURNBULL (New York, N. Y., 1973), p. 2.
[208] A. J. ARKO, Z. FISK and F. M. MUELLER: *Phys. Rev. B*, **16**, 1387 (1977); A. J. ARKO, D. H. LOWNDES, F. A. MULLER, L. W. ROELAND, J. WOLFRAT, A. T. VAN KESSEL, H. W. MYRON, F. M. MUELLER and G. W. WEBB: *Phys. Rev. Lett.*, **40**, 1590 (1978).
[209] S. BERKO and M. WEGER: in *Computational Solid State Physics*, edited by F. HERMAN, N. W. DALTON and T. KOEHLER (New York, N. Y., 1972), p. 59.
[210] G. BARAK, I. B. GOLDBERG and M. WEGER: *J. Phys. Chem. Solids*, **36**, 847 (1975).
[211] S. SAMOILOV, M. WEGER, I. NOWIK, I. B. GOLDBERG and T. ASHKENAZI: *J. Phys. F*, **11**, 1281 (1981), and references therein.
[212] B. M. KLEIN, L. L. BOYER, D. A. PAPACONSTANTOPOULOS and L. F. MATTHEISS: *Phys. Rev. B*, **18**, 6411 (1978).
[213] S. BERKO, W. S. FARMER and F. SINCLAIR: to be published.
[214] J. C. ERSKINE and J. D. MCGERVEY: *Phys. Rev.*, **151**, 615 (1966).
[215] D. STROUD and H. EHRENREICH: *Phys. Rev.*, **171**, 399 (1968).
[216] M. A. SHULMAN, G. M. BEARDSLEY and S. BERKO: *Appl. Phys.*, **5**, 367 (1975).
[217] J. MAYERS, J. D. MCGERVEY, P. A. WALTERS and R. N. WEST: in ref. [2], p. 417.
[218] G. M. BEARDSLEY, S. BERKO, J. J. MADER and M. A. SHULMAN: *Appl. Phys.*, **5**, 375 (1975).
[219] A. SETH and D. E. ELLIS: *J. Phys. C*, **10**, 181 (1977).
[220] E. CARTIER, F. HEINRICH, P. PFLUGER and H. J. GÜNTHERODT: *Phys. Rev. Lett.*, **46**, 272 (1981).
[221] See, for example, *Intercalated Layered Materials*, edited by F. A. LEVY, *Physics and Chemistry of Materials with Layered Structures*, Vol. **6** (Dordrecht, 1979).
[222] T. AKAHANE, K. UCHIDA, T. CHIBA and N. TSUDA: in ref. [2], p. 649.

INDEX